THE ROLE OF VIRUSES IN HUMAN CANCER
Volume II

THE ROLE OF VIRUSES IN HUMAN CANCER
Volume II

Proceedings of the Second International Congress sponsored by the T. and L. de Beaumont Bonelli Foundation for Cancer Research held in Naples, Italy, September 22-24, 1983

Editors:

GAETANO GIRALDO and ELKE BETH

Istituto Nazionale Tumori "Fondazione G. Pascale"
Naples, Italy

1984
ELSEVIER SCIENCE PUBLISHERS
AMSTERDAM • NEW YORK • OXFORD

ISBN 0 444 80584 2
ISSN 0270-2118

Published by:
Elsevier Science Publishers B.V.
P.O. Box 211
1000 AE Amsterdam
The Netherlands

Sole Distributors for the USA and Canada:
Elsevier Science Publishing Co. Inc.
52, Vanderbilt Avenue
New York, NY 10017
USA

Library of Congress Cataloging in Publication Data
(Revised for volume 2)

International Congress of Viral Oncology (1st : 1979 : Naples, Italy)
The Role of viruses in human cancer.

Proceedings of the first- International Congress of Viral Oncology sponsored by the T. and L. de Beaumont Bonelli Foundation for Cancer Research.
Includes bibliographical references and indexes.
1. Viral carcinogenesis--Congresses. I. Giraldo, G. (Gaetano) II. Beth, E. (Elke) III. T. and L. de Beaumont Bonelli Foundation for Cancer Research. IV. International Congress of Viral Oncology (2nd : 1983 : Naples, Italy) V. Title.
RC268.57.I574 1979 616.99'40194 81-113457
ISBN 0-444-00440-8

Printed in The Netherlands

CONTENTS

ORGANIZING COMMITTEE

Elke Beth
Errico di Lorenzo
Giovan Giacomo Giordano
Gaetano Giraldo
Ariel C. Hollinshed
Donato Zarrilli

PREFACE

The content of the book is based on the Second International Congress entitled THE ROLE OF VIRUSES IN HUMAN CANCER, held in Naples, Italy, September 22-24, 1983, and sponsored by the T. and L. de Beaumont Bonelli Foundation for Cancer Research. The book is dedicated to Dr. Denis P. Burkitt, the recipient of the first "De Beaumont Bonelli Award for Cancer Research' for outstanding achievements in human cancer research.

The major objective of the book is to present the actual status on the role of viruses in human cancer taking into account achievements, challenges and perspectives. It comprises comprehensive reviews and recent experimental data on oncogenic human viruses, including herpesviruses, hepatitis B virus, papovaviruses and retroviruses, and their potential association with various human malignancies.

The content of several chapters identifies productive interface studies in the furtherance of knowledge in cancer biology and biochemistry by multimodality approaches, current perspectives on multifactorial and multistep carcinogenesis hypotheses proposed for carcinomas in general and their specific fit for certain human cancers. Between the themes reported are: Studies concerning molecular aspects of oncogenic viruses with identification and cloning of viral and cellular oncogenes, their interrelationships, and their gene products *in vitro* and *in vivo* for better understanding of their functions; data evaluation of prospective studies in selected human populations to establish epidemiologic evidence for viral involvement in certain human cancers; current state of malignancy outcome in immunodeficient individuals, particularly in AIDS developing Kaposi's sarcoma; possible therapeutic approaches of cancer by vaccines, interferon and monoclonal antibodies.

Based on informations obtained on viral and cellular oncogenes and thier products through DNA recombinant techniques, DNA sequence analyses, transfection experiments and monoclonal antibodies, we are now beginning to realise that in the process of cell transformation viral agents, as chemical and physical ones, are acting through similar steps ending in the activation of cellular oncogenes. At that point, considering oncogenic viruses, the presence of viral markers (products of their oncogenes) could be no more necessary for the maintenance of cell transformation.

We are indebted to Dr. Errico di Lorenzo who, in his function as Secretary of the Congress, has contributed significantly to the organization of this meeting. We thank also Dr. Emilia Bracco of his department for valuable assistence before and during the meeting.

Gaetano Giraldo
Elke Beth

FOREWORD

In Honour of Denis Burkitt

Modern biomedical science can be said to have had its origins in the Renaissance. During the 16th century it was the great Italian Schools with their studies in anatomy which laid the foundations, and as a result we still today speak of the Fallopian tubes and bursa of Fabricius. As with much else in the Renaissance it took about 100 years for such work to spread to Northern Europe and it was in the 17th century that Niels Stensen in Denmark gave his name to a salivary duct and William Cowper in England to a gland. With the perfection of the achromatic microscope in the 1830s, anatomical studies were extended to tissue components and as a result of pioneer work in Germany we have the loop of Henle in Kidney tubules and the giant cells of Langhans in the inflammatory response.

As accurate anatomy, physiology, and histology developed, the definition of clinical conditions became more certain in the 19th century and such giants as Bright, Hodgkin and von Recklinghausen were able to achieve immortality along with many others by delineating and giving their names to diseases. By the 20th century, most of the great descriptions had been made and it has become increasingly unusual for a new and outstanding condition in medicine to be recognised by a single worker, and for that worker to be honoured by the attachment of his name to the syndrome.

Denis Parsons Burkitt is one of the rare individuals in recent years to have made such a contribution and thus to have been accepted throughout the world as an eponym; it is because of his work with the lymphoma which bears his name that he is receiving the Beaumont-Bonelli Award today.

Before going further and considering the importance of Burkitt's lymphoma I think it might be interesting to see how its discovery came about. Denis Burkitt was born in February 1911 in Enniskillen, Ulster. The family had a strong tradition both of public service in the British Colonies and India, and of deep Christian faith. Denis Burkitt graduated from Trinity College, Dublin, with the MB Degree in 1935, and three years later took the Fellowship of the Royal College of Surgeons of Edinburgh. From 1941 until 1946 Denis Burkitt was a Surgeon in Britain's Royal Army Medical Corps serving mostly in East Africa with African troops, and then in Ceylon and in liberated Singapore. During his army service he visited Uganda and felt greatly drawn to that country because it was *African* rather than a *settler* country, a difference which even in those days was extremely important. There was also much successful Christian activity in Uganda and when Denis Burkitt was demobilised he decided that his vocation lay there since he felt that he could help the people both medically and spiritually. He joined the Colonial Medical Service and became in Uganda what he has often described as "a simple bush surgeon". There can be no doubt that Denis Burkitt made an important and valuable contribution to the welfare of the Ugandans in this rather humble and undistinguished post, but being the man he is, that was by no means the end of the story. By dint of the true scientific spirit of enquiry and keen observation, Denis Burkitt came to realize in the course of his surgical work that the many seemingly different lymphoid cancers of children in Africa were not disparate entities, but all facets of a single, hitherto unrecognised tumour syndrome, commoner in endemic regions than all other children's cancers added together. It had long been known that various childhood lymphoid tumours were especially frequent in tropical Africa, but although observations on these go back to the very first arrival of missionary doctors, it was not until Denis Burkitt's work that they were recognised as a single disease.

Such a discovery would have been sufficient achievement for most research workers and many have indeed been honoured for comparable findings. However, Denis Burkitt showed himself unique not only in bringing his work to fruition in the most primitive conditions and without any help or back-up, but also by further inspired discoveries. Rightly unsatisfied with descriptive study alone he proceeded to investigate the epidemiology of the tumour. With his first cancer research grant, which added up in total to the amazing sum of

£15 (sterling), Denis Burkitt organized a postal questionnaire throughout sub-Saharan Africa from which, during three years intensive work, he was able to piece together the outlines of the tumour's distribution. Then, with a further research grant - this time the British Medical Research Council actually gave him £250 - and an ancient Ford stationwagon, Denis Burkitt proceeded to drive through East Africa on a 10,000 mile Safari to check the data in person. During this incredible journey he interviewed doctors throughout this portion of the Continent visiting tiny medical missions, village dispensaries, and palm-thatched district hospitals, so that at the end he was able accurately to map the extent of the Burkitt lymphoma belt. From this it became clear that the distribution of the tumour was determined by geographical factors governing temperature and rainfall, and this second contribution has proved of even greater moment. For, a human tumour dependent on climate for its distribution must have some biological agent involved in its cause. By analogy with some animal tumours, a cancer-causing virus was judged from the outset to be the most likely such biological agent, and it is for this reason that Burkitt's lymphoma has proved of such enormous significance as a kind of Rosetta Stone for human tumour virology.

All this was achieved while at the same time ministering as a skilled surgeon to the needs of a population largely lacking even rudimentary medical care. Denis Burkitt progressed from Government Surgeon in the Colonial Medical Service to be a lecturer in Surgery at Makarere University Medical School in Kampala, and was later promoted to Senior Consultant Surgeon to the Uganda Ministry of Health. He left Uganda shortly before independence and continued his geographical pathology studies in London as a member of the External Scientific staff of the Medical Research Council, and has in recent years become eminent again for his widely-known advocacy of high residue diet as an alleviator of many of Western man's ills.

I have told you about Dr Burkitt's progress through his career. Now I must add a word about the reactions of the world of science to his contributions. In 1964 the Harrison Prize from the Royal Society of Medicine; in 1966 the Stuart Prize from the British Medical Association; in 1968 the Arnott Gold Medal of the Irish Hospitals and Medical Schools Association; in 1969 the Katherine Berkan Judd Award of the Sloan

Kettering Institute, New York; in 1970 the Robert de Villiers Award of the American Leukaemia Society; in 1971 the Walker Prize from the Royal College of Surgeons of England; in 1972 the Paul Ehrlich Ludwig-Darmsdaeter Prize and Medal, the London Society of Apothecaries Gold Medal, and the Albert Lasker Clinical Chemotherapy Award; in 1973 the Gairdner Foundation Award; in 1978 the Gold Medal of the British Medical Association; in 1982 the Bristol Myers Award and the General Motors Mott Prize; and in 1983 the Gold Medal of the French Académie de Médicine - each of these a major scientific accolade, and to match this catalogue let me tell you of the honours he has received from Academic bodies. An Honorary DSc Degree of the University of East Africa in 1970; Fellowship of the Royal Society in 1972; Honorary Fellowship of the Royal College of Surgeons of Ireland in 1973; Honorary Fellowship of the Royal College of Physicians of Ireland in 1976; an Honorary MD Degree of the University of Bristol in 1979; and an Honorary DSc Degree of the University of Leeds in 1982. He has been made an Honorary Fellow of the East African Association of Surgeons, of the Brazilian Society of Surgery, of the Sudan Association of Surgeons and of the International Medical Club of Washington. In 1970 Her Majesty the Queen made him a Companion of the Order of St Michael and St George.

Each contribution from Denis Burkitt has revealed his astonishing powers of observation and the originality with which he has recognised simple things and developed them to make concepts of outstanding significance. Each innovatory idea has provided a step forward of great importance in medical and biological science. It is a priviledge for me to welcome Denis Burkitt before you here, as we join other distinguished institutions and groups in honouring him. I have much pleasure in presenting to you Denis Parsons Burkitt as eminently worthy of the Beaumont-Bonelli Award for Cancer Research.

M. A. Epstein

PARTICIPATING CONTRIBUTORS AND CHAIRMEN

Laure Aurelian
Division of Comparative Medicine
The Johns Hopkins University
BALTIMORE (USA)

Giuseppe Barbanti-Brodano
Institute of Microbiology
University of Ferrara
FERRARA (ITALY)

Francis Barin
Institut de Virologie
Facultés de Médecine et de Pharmacie
TOURS (FRANCE)

Yechiel Becker
Department of Molecular Virology
The Hebrew University
Hadassah Medical School
JERUSALEM (ISRAEL)

Elke Beth
Division of Viral Oncology
Istituto Nazionale Tumori
"Fondazione Pascale"
NAPLES (ITALY)

Beda Břicháček
Department of Experimental Virology
Institute of Sera and Vaccines
PRAGUE (CZECHOSLOVAKIA)

Elena Buimovici-Klein
Laboratory of Virology
St. Luke's Hospital
NEW YORK (USA)

Denis P. Burkitt
Unit of Geographical Pathology
St. Thomas's Hospital Medical School
LONDON (UK)

Enzo Cassai
Institute of Microbiology
School of Medicine
University of Ferrara
FERRARA (ITALY)

Ysolina M. Centifanto-Fitzgerald
Department of Ophthalmology
LSU Eye Center
Louisiana State University
Medical Center
NEW ORLEANS (USA)

Max A. Chernesky
Regional Virology Laboratory
St. Joseph's Hospital
McMaster University
HAMILTON (CANADA)

Gabriel Ernödi
Molecular Biology and Immunology
Virogen A.G.
BASEL (SWITZERLAND)

M.A. Epstein
Department of Pathology
School of Medicine
University of Bristol
BRISTOL (UK)

Bernhard Fleckenstein
Institute of Clinical Virology
University of Erlangen-Nürnberg
ERLANGEN (FED. REP. GERMANY)

Erhard Geissler
Zentralinstitut für Molekularbiologie
Akademie der Wissenschaften der DDR
Forschungszentrum für Molekularbiologie
und Medizin
BERLIN-BUCH (DEM. REP. GERMANY)

Giovan G. Giordano
Division of Pathology
Istituto Nazionale Tumori
"Fondazione Pascale"
NAPLES (ITALY)

Gaetano Giraldo
Divison of Viral Oncology
Istituto Nazionale Tumori
"Fondazione Pascale"
NAPLES (ITALY)

Alain M. Goudeau
Laboratoire de Virologie
Centre Hospitalier Régionale de Tours
TOURS (FRANCE)

William D. Hardy, Jr.
Laboratory of Veterinary Oncology
Memorial Sloan-Kettering Cancer Center
NEW YORK (USA)

Yorio Hinuma
Institute for Virus Research
Kyoto University
KYOTO (JAPAN)

Ariel Hollinshead
Department of Medicine
The George Washington University
Medical Center
WASHINGTON (USA)

Eng-Shang Huang
Department of Medicine
and Cancer Research Center
The University of North Carolina
at Chapel Hill
CHAPEL HILL (USA)

Abraham Karpas
Department of Haematological Medicine
University of Cambridge
Clinical School
CAMBRIDGE (UK)

Herbert E. Kaufman
Department of Ophthalmology
LSU Eye Center
Louisiana State University Medical Center
NEW ORLEANS (USA)

Elliott Kieff
The M.B. Kovler Viral Oncology Labs.
University of Chicago
CHICAGO (USA)

Cesare Maltoni
Institute of Oncology
BOLOGNA (ITALY)

James K. McDougall
Department of Tumor Biology
Fred Hutchinson Cancer
Research Center
University of Washington
SEATTLE (USA)

David T. Purtilo
Department of Pathology
and Laboratory Medicine
University of Nebraska
Medical Center
OMAHA (USA)

Giancarlo F. Rabotti
Laboratoire de Médecine Expérimentale
Collège de France
PARIS (FRANCE)

Bernard Roizman
The M.B. Kovler Viral
Oncology Laboratories
University of Chigaco
CHICAGO (USA)

Nurul H. Sarkar
Laboratory of Molecular Virology
Memorial Sloan-Kettering Cancer Center
NEW YORK (USA)

Rupert Schmidt-Ullrich
Department Therapeutic Radiology
Tufts-New England Medical Center
BOSTON (USA)

Carlo Sirtori
Fondazione Carlo Erba
MILAN (ITALY)

Francesco Squartini
Institute of Morbid Anatomy
School of Medicine
University of Pisa
PISA (ITALY)

Jesse Summers
The Institute for Cancer Research
The Fox Chase Cancer Center
PHILADELPHIA (USA)

David A. Thorley-Lawson
Department of Pathology
Tufts University School of Medicine
BOSTON (USA)

Pierre Tiollais
Unité de Recombination et Expression
Génétique
Institut Pasteur
PARIS (FRANCE)

L. Váczi
Department of Microbiology
Medical University
DEBRECEN (HUNGARY)

D. Zarrilli
Department of Medicine
Istituto Nazionale Tumori
"Fondazione Pascale"
NAPLES (ITALY)

The Role of Viruses in Human Cancer. Volume II
G. Giraldo and E. Beth editors

POSSIBILITIES AND PRIORITIES OF CANCER PREVENTION

DENIS BURKITT

Unit of Geographical Pathology, St. Thomas's Hospital,
London, S.E.1., England.

INTRODUCTION

Modern scientific medicine is too prone to consider man as a machine. When some component of the machine functions badly the approach to the problem is focused almost exclusively on identification of the fault, representing medical diagnosis, and on efforts to repair, as far as possible, the defective part, the counterpart of treatment. Hardly a thought is given to the possibility that malfunction of the machine may be caused by the environment in which it operates, and consequently no serious effort is made to rectify this all-important factor. The machine is left out in the rain or snow, surrounded by skilled mechanics who are elaborately equipped, so intent on their job that the thought never crosses their mind that the machine might run better if protected from the elements.

I do not believe this picture to be an exaggeration of the overwhelming preponderance of effort and expenditure currently devoted to the diagnosis and treatment of disease, including cancer, relative to the low priority given to preventive measures. This in spite of the fact that it is now generally accepted that environmental factors are of paramount importance in the causation of most disease, and not least in the case of cancer. Early detection is often of greater prognostic significance than is the choice of therapeutic procedures, but prevention is of incomparably greater value than is early diagnosis or skilful therapy.

In that highly authorative document "Nutrition and Cancer"(1) it was stated "It is highly likely that the United States will eventually have the option of adopting a diet that reduces the incidence of cancer by approximately one-third, and it is absolutely certain that another third could be prevented by abolishing smoking."

Knowledge now available makes it clear that several of the commonest cancers in the world, together making up a considerable proportion of the whole, are likely soon to be potentially preventable, the "potentially" referring to the fact that scientific knowledge alone is impotent, being dependant for success on the will of individuals to change their life-style and accept recommendations calculated to reduce their risk of developing disease.

The emerging possibilities of preventing a significant proportion of tumours which are major cancer killers in large areas of the world will now be considered.

TUMOURS RELATED TO VIRUSES

The conference being held in Naples this week is predominantly devoted to the role played by viruses in the induction of cancer. It is therefore appropriate that I begin by considering two of the worlds most lethal cancers which are also among the commonest tumours encountered over extensive areas of the earth. These are nasopharyngeal carcinoma (NPC) and primary hepatocellular carcinoma (PHC). Not only is there convincing evidence pointing to a viral aetiology in the cause of both of these tumours, but in addition the use of a protective vaccine has become a feasible proposition in the case of each of these tumours. I will also discuss briefly some aspects of another tumour which was not only the first to offer prospects of being causally related to a virus, the now famous and ubiquitous Epstein-Barr Virus (E.B.V.) but was the tumour in which this virus was first described by Epstein, Achong and Barr (2) and it is this same virus that is the prime suspect as an aetiological factor in the development of NPC (3).

Cancers in which E.B.V. has been implicated

Naso-pharyngeal cancer (NPC).

This tumour develops in the mucous membrane of the nasopharynx. It is the commonest cancer in men and the second commonest in women in some of the most densely populated areas of the world including much of China and many countries in South-East Asia including Taiwan, Hong Kong and Singapore. Immigrant Chinese populations in America have higher rates than other ethnic groups in that country, and there would appear to be a genetic susceptibility to the development of this tumour in Chinese people. When using NPC cases as controls while investigating possible relationships between B.L. and E.B. Virus it was discovered that this virus was related to both these tumours. This observation led to studies indicating that throughout the world E.B.V. was related to nasopharyngeal cancer. Patients with this tumour have significantly higher mean antibody titres against E.B.V. than have controls. Antibody titres have also been shown to rise as tumour growth progresses. Of greater significance is the observation that E.B.V. specific nuclear antigen can be detected in the malignant epithelial cells of the tumour. Infectious virus has been isolated from the cells of some tumours.

Burkitt's Lymphoma.

This tumour, which has become in Sir Harold Himmsworth's words "May well prove to be something of a Rosetta Stone" has been shown to be common only in, and yet always in regions where malaria is holo or hyper-endemic (4,5). In these situations B.L. is much the commonest tumour of childhood. It was in this tumour that the ubiquitous and now famous E.B.virus was first identified by Epstein and his colleagues (2).

Possibility of Prevention.

The two factors which appear to be necessary for the occurrence of high incidence rates of this tumour are infection with EBV early in life and the presence of intense and persistent malarial infection which produces severe immuno-depression (6). If either of these co-factors could be reduced or eliminated the disease should be reducible or preventable.

Malaria eradication or suppression.

This disease has been successfully eliminated from Europe, in parts of which it was common before the 20th Century, and reduced in intensity and extent in most of Asia. It has however, not yet proved possible to conquer its ravages, other than in urban and other controlled situations, in tropical Africa, Papua New Guinea, parts of Malaysia and the Amazon basin. If the mosquito vectors responsible for its transmission could be exterminated, or the population protected from them, which is virtually impossible in native dwellings, malaria might be reduced to less lethal levels.

Alternatively maintained chemotherapy prophylaxis could have the same effect. Either of these measures should reduce or eliminate B.L.

Provision of E.B.V. Vaccine

Another approach could be the immunisation of young children with an EBV vaccine when present efforts to prepare such a vaccine are successfully completed. It is however unlikely that in the foreseeable future such a programme could be economically justifiable in regions in which the tumour is most common, as in Africa and Papua, New Guinea. In these poor countries other expenditure of available resources could save infinitely more lives.

Cancer in which Hepatitis B Virus is Implicated

Primary hepatocellular carcinoma (PHC) is one of the most frequently encountered cancers in sub-saharan Africa and in much of South East Asia, and particularly in China and in the Chinese population in Singapore. The highest rates in the world have been reported in Mozambique and parts of Zimbabwe.

The role of hepatitis B virus in the causation of hepatic cirrhosis is now well established, and this cirrhosis is believed to predispose to the development of liver cancer. There is in addition evidence that the virus is directly involved in neoplastic transformation of liver cells.

Persistent hepatitis B infection has been shown to be associated with increased risk of PHC development.

In regions where the incidence of this tumour is high HBV infection occurs before, during and soon after birth from carrier mothers. A vaccine against HBC has already been developed and this opens the way for potential immunisation of high risk populations (7). Such immunisation together with the use of HB immunoglobulin to prevent mother to child infection should prevent the development of hepatitis B infection and consequently prevent the development of HPC.

TUMOURS RELATED TO CULTURAL HABITS

Pan Chewing

In parts of India (8) and South-East Asia, and in Papua, New

Guinea, one of the commonest forms of neoplasm encountered is cancer of the buccal cavity resultant on chewing a "pan". This consists of various substances often including pan, tobacco and lime concocted together and usually wrapped in a leaf. Not only is it chewed during the day, but it is often retained in the cheek pouch during sleep. This practice may be considered to correspond to cigarette smoking, and both are responsible for cancers which are obviously preventable if these practices could be discontinued.

Tobacco Smoking

There is no longer any doubt that the major cause of the predominent killing cancer in Western countries, that is lung tumours, is smoking cigarettes. And this neoplasm is among the cancers for which therapy has little to offer and yet prevention is clearly possible.

DIET AND CANCER

Benign and malignant diseases associated with modern Western life-style

Some of the cancers most frequently observed in Western countries have been shown to be causally related to modern Western life-style. In addition to lung tumours, related to smoking, these cancers include tumours of the large bowel, the breast, the endometrium and the prostate. The last four are closely associated epidemiologically with a group of diseases which are now recognised to be common only in the context of modern Western culture (9,10). These vary from gall-stones to coronary heart disease, from diabetes (Type II) to diverticulosis and from appendicitis to hiatus hernia. Not only do these benign and malignant conditions share the same geographical distribution, but they all increase in prevalence following emigration from low to high risk situations. As a consequence black Americans and the descendants of Japanese immigrants to Hawaii now have rates of these diseases comparable to members of other ethnic groups in these communities. The demonstrated associations between these conditions suggests that they share some causative factors common to each, and there is a considerable body of evidence to incriminate diet in the pathogenesis of these disorders.

The main contrasts between diets of populations at minimal and at maximal risk of developing these diseases are that the former consume diets with an abundance of starch and fibre and with a low content of fat, salt and sugar while the reverse is true of communities at minimal risk. The proportion of energy provided by protein is similar in both situations though its largely of vegetable origin in the Third World and animal derived in the affluent West.

Large Bowel Cancer

The characteristics of Western diets which are currently held responsible for the high rates of colo-rectal cancer in more affluent societies are the high fat and low fibre content of the diet. Excessive fat has been postulated to play a causative role, (11) partly by increasing the bile-acids and other natural bowel constituents which bacteria can degradate into potential carcinogens and partly by fostering the growth of bacteria endowed with this property. Dietary fibre on the other hand is generally held to confer protection against the development of colo-rectal cancer (12,13,14). This is believed to be mediated through such diverse mechanisms as:-

1. Fibre lowers faecal pH and this reduces bacterial degradation of bile acids and cholesterol into potentially carcinogenic metabolites (15).
2. Fibre increases stool volume (16) and thus dilutes faecal constituents including any carcinogens and thus presumably diminishes their cancer producing potential.
3. Fibre shortens intestinal transit time and thus reduces contact time between faecal carcinogens and bowel mucosa(16)
4. The increase in bacterial population resultant in increased fibre intake requires additional nitrogen and thereby reduces the nitrogen that could be used to produce ammonia, and ammonia is known to enhance malignant transformation in cells (18).

   In addition fibre increases production of butyrate which inhibits malignant transformation (19).

Breast Cancer

Breast cancer rates are significantly lower in Seventh-Day Adventists and Mormons than they are in other Americans (Walker 20)

When comparing breast cancer rates in vegetarians and omnivores the former were shown to have significantly lower breast-cancer rates. They also had higher fibre intakes which resulted in increased stool weights. Faecal oestrogen excretion was shown to be directly related to stool volume, and was inversely related to serum oestrogen levels (21). This observed relationship between serum oestrogen levels and fibre intake may help to explain why breast cancer levels are low where fibre consumption is high. Epidemiologically breast cancer incidence is directly related to fat intake (22) and a change in energy intake from vegetable to animal sources is characteristic of populations at high risk of developing both colon and breast cancer (23). In experimental studies increased fat consumption and the deposition of fat resultant on excessive calorie intake have been shown to predispose to the development of breast tumours.

There are of course many other factors involved in the pathogenesis of breast cancer.

Endometrial Cancer

This tumour is also related to economic development, being much more frequently observed in affluent than in poorer populations. This tumour is hormone dependant and epidemiological and experimental evidence suggests that dietary fat may play a significant role in its development (24).

CONCLUSIONS

The observations outlined above clearly indicate that all these common cancers are to a large extent potentially preventable. Moreover many of them are extremely resistant to therapy of any kind.

Nasopharyngeal, liver and lung tumours, which are among the commonest neoplasms in large areas of the world are benefitted hardly at all by treatment, but prevention of each is now an obvious possibility. Treatment of bowel and breast cancer offers reasonably good prospects when the tumours are detected early, but the potential of prevention must be viewed as a far superior achievement than early detection.

Burkitt's lymphoma and endometrial cancer are very treatable when detected early and good results can follow therapy for prostatic cancer. Even so, prevention is a preferable goal. So what in practical terms might be the changes to recommend in Western life-style?

Reduction in cigarette smoking would drastically lower rates of lung cancer and other smoking related tumours, and in addition would reduce ischaemic heart disease, bronchitis and emphysema.

The dietary changes recommended would consist of the following:-

Consumption of starch and fibre should be approximately doubled. That of salt and sugar halved, and fat reduced by a third. A greater reduction of fat would not be acceptable. This in practical terms implies eating more fibre-rich cereals as brown or whole-meal bread, more fibre-rich breakfast cereals, all legumes and tubers, and potatoes in particular, provided they are neither cooked nor eaten with fat. Much less meat should be consumed, and fish and fowl should be eaten in preference to red meats. Fried foods should be avoided as far as possible. Such changes would not only offer prospects of a considerable reduction in cancer risk but in addition would almost certainly provide protection against many of the commonest diseases in all Western countries. Recommendations along these lines have already been made in many countries including the U.S.A., Canada, the United Kingdom, Eire, Norway, Sweden, Denmark, France and Australia.

If water is running from a fawcet, the basin over-flowing and the floor flooded, there can be no question that it is more important to turn off the fawcet than to concentrate exclusively on mopping the floor. Every effort must be made to reverse the flood on the floor representing developed disease, for any lasting results priority must be given to efforts to quench the flow at the source and turn off the running tap.

REFERENCES

1. Diet, Nutrition and Cancer. (1975) National Academy Press Sect. 2-9, Washington DC.

2. Epstein, M.A., Achong, B.G. and Barr, Y.M. (1964). Lancet, 1, 702-703.

3. Epstein, M.A. (1978) in: de The, G. and Ito, Y. (Ed), Naso-pharyngeal Carcinoma: Etiology & Control. International Agency for Research on Cancer, Lyon.

4. Kafuko, G.W. and Burkitt, D.P. (1970) Int.J.Cancer, 6, 1-9.

5. Burkitt, D.P. (1969) J.Nat.Cancer Inst, 42, 19-28.

6. Ziegler, J.L. (1982) Cancer J. for Clinicians, 32, 144-157.

7. Blumberg, B.S. and London, W.T. (1982) Cancer, 50, 2657-2665.

8. Paymaster, J.C. (1979) J.Ind.Med.Ass, 57, 37-44.

9. Burkitt, D.P. and Trowell, H.C. (1975) Refined Carbohydrate Foods and Disease. Academic Press, London, New York.

10. Trowell, H.C. and Burkitt, D.P. (1981) Western Diseases, their emergence and prevention. Harvard University Press, Boston, Mass

11. Wynder, E.L. and Reddy, B.S. (1975) J.Nat.Cancer Inst, 54, 7-10.

12. Burkitt, D.P. (1971) Cancer, 28, 3-13.

13. Walker, A.R.P. (1977) Am.J.Clin.Nut, 29, 1417-1426.

14. Cummings, J.H. and Branch, W.J. (1982) in: Vahounev, G. and Kritchevsky, D. (Ed), Dietary Fibre in Health and Disease. Plenum Press, New York and London.

15. Macdonald, I.A. and Rao, B.G. (1979) Medical Progress, 34, 136-140.

16. Burkitt, D.P., Walker, A.R.P. and Painter, N.S. (1972) Lancet, 2, 1408-1412.

17. Cummings, J.H. (1981) Proc: Nutr.Soc, 40, 7-14.

18. Visek, W.J. (1972) Fed.Proc, 31, 1178-1193.

19. Hagopian, H.K., Riggs, M.G., Swartz, L.A. and Ingram, V.M. (1977) Cell, 12, 855-860.

20. Walker, A.R.P. (1981) S.Afr.Med.J, 60, 405-406.

21. Goldin, B.R., Adlercreutz, H., Gorbach, S.L. et al, (1982) New Eng.J.Med, 307, 1542-1547.

22. Armstrong, B. and Doll, R. (1975) Int.J.Cancer, 15, 612-631.

23. Correa, P. (1981) Cancer, 41, 3685-3689.

24. Elwood, J.M., Cole, P., Rothman, J. and Kaplan, S.D. (1977) J.Nat.Cancer Inst, 59, 1055-1060.

© 1984 Elsevier Science Publishers B.V.
The Role of Viruses in Human Cancer. Volume II
G. Giraldo and E. Beth editors

THE FUNCTIONAL ORGANIZATION OF THE HERPES SIMPLEX VIRUS GENOMES

BERNARD ROIZMAN The Marjorie B. Kovler Viral Oncology Laboratories, The University of Chicago, 910 East 58th Street, Chicago Il 60637 USA

INTRODUCTION

A quarter of a century ago when these studies were first begun, there was not yet a plaque assay for herpes simplex viruses (HSV), and evidence that there existed a serotype other than that known today as HSV-1 was not yet available. The composition of the genome was uncertain, and the structure of the virus particle itself was not known. We have learned a lot in the interim. Foremost, the studies done in the past dozen years in my laboratory and in that of many others have laid a sound foundation for our understanding of the structure and function of the HSV genome and it is both possible and of heristic value to begin to delineate not only its structure but its function. The purpose of this paper is to summarize very briefly what we know about the informational content of the HSV-1 genome.

The structure of herpes simplex virus DNA. The HSV genome is linear, double stranded and approximately 96 million in molecular weight[1,2,3]. It consists of two covalently linked components, L (large) and S (small) representing 82 and 18 percent of total DNA, respectively[3,4]. Each component consists of unique sequences flanked by inverted repeats[4]. The inverted repeats of the L component were designated as ab and b'a' whereas those of the S component were designated as a'c' and ca[3]. An unusual characteristic of wild-type HSV genomes is that the two components can invert during viral replication. Consequently, viral DNA extracted from virions consists of four isomers differing solely in their orientation of L and S components relative to each other[5]. The focus of this paper is on the informational DNA sequences contained in the HSV-1 genome. It is convenient to classify the informational sequences contained in HSV genomes into two classes. The first class comprises trans-acting genes. These genes code the genetic information expressed in the form of RNA and proteins. The gene products act in trans; they combine with the DNA of the virus to form viral particles, they ensure an orderly expression of the trans-acting genes and contribute to the replication of the genome.

The second class of informational sequences consists of cis-acting sites which might best be described as homing signals. Their function is to provide recognition sites for specific proteins, which bind either to DNA or to the

RNA transcript, and carry specific instruction for events which involve specific sites or even the entire molecule. There are many different cis-acting signals. One subset defines the expression of individual genes. These are the determinants of the rate of transcription, transcription initiation and termination, translation initiation and termination, polyadenylation signals, etc. A second subset of cis-acting signals is involved in the replication and packaging of the viral DNA into virions.

HSV-1 TRANS-ACTING GENES

To date approximately 50 gene products - proteins specified by the virus - have been identified[6,7,8]. Painstaking analyses of the transcription of the HSV genome [reviewed by E.K. Wagner[9]] suggest that as many as 75 genes may be encoded in the HSV-1 genome. The HSV-1 genes can be classified into five groups on the basis of the kinetics and requirements for their expression.

α genes. The viral genes expressed first after entry of the virus into cells were designated as α genes[10,11]. There are five α genes, 0, 4, 22, 27 and 47. These genes are coordinately regulated. Except for α 4, a high molecular weight protein with pleiotropic functions essential for the optimal expression of later genes[12-14], the function of α gene is not known. α 22 gene was deleted without effecting the ability of the recombinant virus to multiply in continuous lines of primate cells[15], but the ability of the virus to multiply in primary human cell cultures and in rodent cells is impaired (A. Sears and B. Roizman, manuscript in preparation). The five α genes examplify to some extent the structural organization of herpes simplex virus genes. Two of the genes, α0 and α4, are each present in two copies per genome, because they map inside the inverted repeat sequences b and c, respectively, of the viral genome. However, one copy of these genes appears to be sufficient inasmuch as the deletion of one copy of each did not profoundly affect the ability of the virus to multiply[16]. The five genes map at or near the termini of both L and S components on the same and opposite strands. Of the 5, two (#22 and #47) appear to contain introns and may overlap at least in part with other genes. Neither introns nor overlapping genes have been detected within the domains of α 4 and 0 genes[17]. All α genes are expressed from independent promoters[17].

β genes. The second and third groups are both designated as β. Both groups require for their expression functional α gene products and are currently differentiated solely on the basis of the kinetics of their appearance. Subgroup β_1 is the earliest expressed and we generally recognize at least two proteins, #6 - a component of ribonucleotide reductase[18,19], and #8 - a DNA-binding protein - as belonging to this subgroup. β_2 proteins appear later; the major examples of this subgroup are the DNA polymerase and the thymidine kinase (TK). The function of β genes studied to date appears to be that of DNA binding proteins and enzymes involved in replication of viral DNA. Unlike papova and adenoviruses, HSV specify a number of enzymes with functions similar to those of the host and which are involved in replication of viral DNA. Except for the genes which specify the major DNA binding protein and DNA polymerase and which map near the L component origin of DNA synthesis, β genes do not appear to cluster.

γ genes. The two groups which follow, γ_1 and γ_2, require viral DNA synthesis for maximum expression. Whereas γ_1 genes can be expressed in the absence of DNA synthesis without however reaching the maximum level of expression, γ_2 stringently require the onset of viral DNA synthesis. Most γ proteins appear to be structural components of the HSV virion. With the exception of the genes specifying glycoproteins and mapping in the S component, γ genes do not appear to cluster.

HSV-1 and HSV-2 are generally colinear with respect to all of the genes essential for virus multiplication. Non-colinear sequences that appear to be biologically active were reported at this symposium by McDougall and Galloway. An apparent instance of a non colinear gene that specifies a glycoprotein not essential for replication in cell culture has also been reported [20,21].

CIS-ACTING SITES INVOLVED IN REPLICATION AND STRUCTURAL REARRANGEMENTS OF HSV DNA

Available data indicate that HSV DNA circularizes after infection and that at least late in infection it is replicated by a "rolling circle" mechanism[22]. Circularization undoubtedly involves the a sequences located at the termini and at the junction of the L and S components[3] whereas origins of DNA synthesis are very likely involved in the replication of the viral DNA. This section concerns primarily the a sequences and the origins of DNA synthesis in HSV DNA.

The structure and function of a sequences. The a sequences are the terminal portions of the reiterated sequences flanking L and S components and shared by both components. The terminal a sequences are in the same orientation and the initial definition was operational. Thus, digestion of the ends of HSV-1 DNA by a processive exonuclease exposed complementary sequences which annealed to circularize the DNA[23]. The complementary sequences exposed by the exonuclease digestion and which permitted the DNA to circularize were defined as the a sequence[3]. The structural and functional aspects of the a sequences may be summarized as follows:

(i) The complete a sequence in HSV-1(F) DNA (Table 1) is expressed by the formula:

$$DR1 - Ub - (DR2)_{19-22} - DR4_{2-3} - Uc - DR1$$

where DR1, DR2, and DR4 are sequences containing 20, 12, and 37 bp repeated once, 18-22, and 2-3 times, respectively. U denotes unique sequences. The sequence nearest the b repeat (Ub) contains 64 bp whereas the sequence nearest the c repeat (Uc) contains 58 bp[24]. Although significant variation has been noted in DNAs from different HSV-1 isolates, the available data on a sequences derived from HSV-1(F) and two other HSV-1 strains[25] indicate considerable conservation of nucleotide sequences (Table 1). The single published HSV-2 a sequence[25] differs considerably from HSV-1 a nucleotide sequences.

(ii) The number of a sequences at the junction between the L and S components and at the L component terminus vary from one to more than ten[24,26,27]. Adjacent a sequences share the DR1 sequence[24,28]. DNA molecules derived from a plaque-purified stock show considerable heterogeneity in the number of a sequences at the L component terminus and at the junction between the L and S components. Studies on concatemeric structures derived from infected cells transfected with intact viral (helper) DNA and fragments containing one or more sequences and an origin of HSV DNA synthesis are particularly revealing. These structures consist of head to tail concatemers of fragments amplified by a rolling circle mechanism[29]. The significant finding is that the packaged concatemers terminate with at least one a sequence and that the number of a sequences contained in the repeat unit may increase or decrease relative to the number of a sequences in the fragment with which the cells were transfected (L.P. Deiss and N. Frenkel, personal communication). The mechanism by which the a sequences are amplified to form tandem repeats or are reduced in number is not known. Of the various possible mechanisms - inter- and intra- molecular recombination, slippage replication, and the use of internal inverted repeat a sequences as templates[24,28] - only the latter was shown to be untenable since (a) the terminal a sequence was amplified in a recombinant from which most of the internal inverted repeats including the a sequences had been deleted[16] and (b) internal inverted repeats were not present in concatemers resulting from rolling circle replication of a repeat containing an a sequence and an origin of DNA replication[29].

```
                                  L                                    S
|a b|---------------------------------------------|b'a'c'|-------|ca|

                         b'
1  GGTTACCTGGG G CTGTGCGGTTGGGACGGCGCCCGTGGGCCCGGGCGGCCGGGGGCGGCGGG
2  GGTTACCTGGG A CTGTGCGGTTGGGACGGCGCCCGTGGGCCCGGGCGGCCGGGGGCGGCGGG
3  GGTTACCTGGG A CTGTGCGGTTGGGACGGCGCCCGTGGGCCCGGGCGGCCGGGGGCGGCGGG

                         b'
1  GGCCGGCATGGCGGCG   CGGCGGGCCATGGAGACAGAGAGCGTGCCGGGGTGGTAGAGTTTG
2  GGCCGCGATGGCGGCG G CGGCGGGCCATGGAGACAGAGAGCGTGCCGGGGTGGTAGAGTTTG
3  GGCCGCGATGGCGGCG G CGGCGGGCCATGGAGACAGAGAGCGTGCCGGGGTGGTAGAGTTTG

                         b'
1  ACAGGCAAGCATGTGCGTGCAGAGGCGAGTAGTGCTTGCCTGTCTAACTCGCTCGTCTCGG
2  ACAGGCAAGCATGTGCGTGCAGAGGCGAGTAGTGCTTGCCTGTCTAACTAGCTCGTCTCGG
3  ACAGGCAAGCATGTGCGTGCAGAGGCGAGTAGTGCTTGCCTGTCTAACTCGCTCGTCTCGG

   |      DR1'           |                        Ub'
1  CCGCGGGGGGCCCGGGC T  GC G CCGCCGC GAC     TTTAAAGGGCCGC G CGCGACCC
2  CCGCGGGGGGCCCGGGC T  GC   CCGCCGC CACCGC TTTAAAGGGCCGC G CGCGACCC
3  CCGCGGGGGGCCCGGCC GC GC   CCGCCGC  CGCGC TTTAAAGGGCCGC C CGCGACCC

                      Ub'                                | DR2'       |
1  CCGGGGGGTGTGTTT C GGGGGGGGCCCGTTTT                   [CGCTCCTCCCCC]19
2  CCGGGGGGTGTGTTT T GGGGGGGGCCCGTTTT CGGGGTCTGGC [CGCTCCTCCCCC]18
3  CCGGGGGGTGTGTTT C GGGGGGGGCCCGTTTT TGGGGTCTGGC [CGCTCCTCCCCC]8

   |                                    DR4'                   |  Uc'
1               [C GCTCCTCCCC     GCTCCCGCGGCCCCGCCCCC A ACGCC]3 CGCCGC
2                                 GCTCCCGCGGCCCCGCCCCC C ACGCC   CGCCGC
3  [GTCTGTGGGTGGGGCTCCTCCCCC]6 GCTCCCGCGGCCCCGCCCCC C ACGCC   CGCCGC

                              Uc'                                   |
1  GCGCGCGC A CGCCGCCCGGACCGCCGCCCGCCTTTTTGCGCGC CGC  C C CG C
2  GCGCGCGC A CGCCGCCCGGACCGCCGCCCGCCTTTTTGCGCGC GCG  C G CG C
3  GCGCGCGC G CGCCGCCCGGACCGCCGCCCGCCTTTTTGCGCGC CGC  C C CG C

   |      DR1'           |                        c'
1  CCGCGGGGGGCCCGGGC T GC CACAGG T G T AA CAACA CCAACAG AACACCAAC
2  CCGCGGGGGGCCCGGGC T GC CACAGG T G A AA       CCAACAG
3  CCGCGGGGGGCCCGGGC C GC CACAGG C G T AA CAACA CCAACAG AACACCAAC

          c'
1  AGCACGGCGCAC C GG C G AC T C
2  AGCACGGCGCAC T CC G C AC G T
3  AGCACGGCGCAC C GG C G AC T C
```

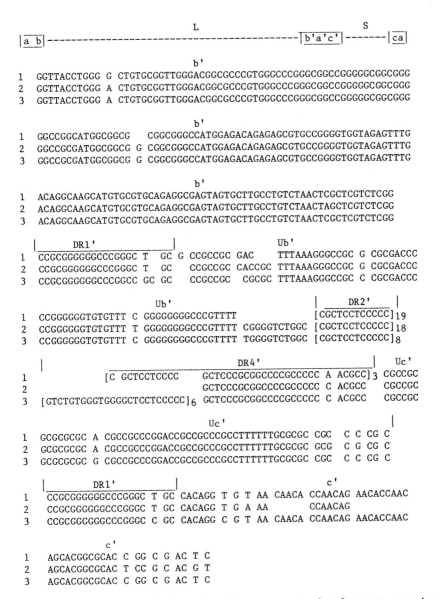

Table 1. Comparison of the nucleotide sequences of a fragment spanning the HSV-1 L-S Junction. The top line shows the sequence arrangement in HSV-1 DNA. The lines below show the nucleotide sequences 5' to 3' of a portion of b', a complete a' sequence, and the adjacent portion of c'. The numbers 1, 2, and 3 refer to HSV-1 strain F[24], 17[25], and USA-8[25], respectively. Blank spaces outline nucleotide columns showing sequence divergence. The arrangement of reiterated sequences in lines 2 and 3 is slightly different from that appearing in the original paper.

(iii) The terminal a sequences of the L and S components if juxtaposed, would form two a sequences sharing an intervening DR1[28]. The L component terminal a sequence contributes 18bp and a single 3' nucleotide extension whereas the S component contributes 1bp and a single 3' nucleotide extension to the shared DR1[28]. This finding suggested that (a) circularization could occur by the mere juxtaposition and ligation of the termini, (b) that cleavage of concatemers into unit length molecules occurs within a DR1 shared by adjacent a sequences, and (c) that the termini generated by the cleavage are not further repaired before packaging.

At least one of these conclusions presents a particular problem inasmuch as packaged concatemers made up of repeats with a single a sequence contain at least one a sequence at both termini (L.P. Deiss and N. Frenkel, personal communication). These data suggest that at the site of cleavage the single a sequence had to be amplified either before or during cleavage in order to generate at least one a sequence on both sides of the cleavages sites.

(iv) The a sequence appears to be a cis-acting site required for and through which the inversion of L and S components occurs. This conclusion is based on the observation that insertion into the unique sequences of the L component of a sequence-containing DNA derived from the junction between the L and S components, or the termini of L or S component causes additional inversions to occur. The segments that invert are flanked by inverted repeats of the a sequence[24,28,30,31]. A key question is whether a specific gene product or the mere presence of inverted repeats within non-coding regions of the genome is sufficient to cause inversion of segments flanked by such repeats. Although the latter cannot be excluded and in fact has been observed (P.G. Spear, personal communication), it would be expected that inversion occurring through a specific sequence at high frequency would involve specific viral gene products interacting with a cis-acting site. Initial experiments designed to test this hypothesis have shown that cells converted to TK+ phenotype with a plasmid containing an origin of DNA synthesis and a TK gene flanked by inverted repeats of the a sequence retain the transfected sequences in circular or head-to-tail concatemeric form but that the sequences flanked by the a sequence invert on infection of the cells with HSV[32]. The gene products involved in the inversion have not been identified nor is the function satisfied by the inversions known; the problem in part is that inversions are not essential for virus multiplication and hence mutations in genes causing inversions may not be lethal. Thus, deletion of the internal inverted repeats yielded a virus which does not invert[16].

The origins of DNA replication in the HSV genome. The origins of HSV DNA replication were defined operationally as the sequence which must be present in a fragment of HSV DNA in order to be amplified in permissive cells containing the fragment and intact, helper viral DNA. The available data indicate that HSV-1 and, presumably, HSV-2 DNAs contain 3 origins of DNA replication. Two of the origins map in the c reiterated sequence of the S component[32, 33,34,35] whereas a third origin maps in the middle of the L component near the genes specifying the major DNA binding protein (#8) and DNA polymerase[36,37].

The S component origin of HSV-1 has been sequenced[35,38,39] and it has been reported that all the functions of the origin are contained in a 90bp stretch of DNA containing a near perfect AT rich palindromic sequence 45bp long. Because the S component origins are contained within reiterated repeats, they would be expected to be identical. Little is known concerning the nucleotide sequence of the L component origin. It is perhaps noteworthy that no homologous sequences have been reported among the fragment containing the S and L component origins.

Very little is known about the function of the origins. The major questions are whether all or any of these sequences are actually involved in the replication of viral DNA and whether they are equivalent or subordinate to each other. It is noteworthy that the S component origins are situated upstream from the sites of initiation of transcription of α genes 4, 22 and 27[17,40,41], but are distinct from the major sequences regulating the expression of these genes[32,42,43,44].

CIS-ACTING SITES INVOLVED IN REGULATION OF VIRAL GENE EXPRESSION

Among the most interesting as yet undefined cis-acting sites are those involved in the regulation of HSV gene expression at transcriptional and translational levels. As noted earlier in the text, HSV genes form several major groups whose expression is coordinately regulated in a cascade fashion. Thus, α gene products are required to induce β genes; the latter shut off the expression of the α genes and turn on the synthesis of γ genes which, in turn, shut of β gene expression[10,11]. Identification of the structure and function of cis-acting sites involved in regulation of gene expression is fragmentary, but remains the focus of much of the work being currently done.

Regulation of α genes. Until recently, it has been thought that α gene expression required the presentation of competent HSV DNA to the transcriptional machinery of the cell. Moreover, the observation that ts mutants in the α 4 gene resume the synthesis of α 4 gene product on shift-up from per-

permissive to non- permissive temperature[14,45] suggested that at least α 4 was autoregulatory and responded to negative signals. Recent studies indicate however that α genes are inducible[42,43,45,46,47,48]. The fundamental design in all of these studies was to construct chimeric genes consisting of promoter and regulatory sequences of α genes fused to the structural sequences and other selected regions of the HSV TK gene. The TK gene served as the indicator of activity; because the wild-type gene is regulated as a β gene, it was readi- differentiated from α genes. An added advantage is that available drugs and and procedures allow the selection of viruses or of cell lines which either express or fail to express the natural or chimeric TK.

The chimeric α-TK recombined into the virus[46] was regulated as an α gene in that the transcript of the gene accumulated in the presence of inhibitory concentrations of cycloheximide and was expressed after the drug was withdrawn in the absence of further transcription. Analyses of the regulation of TK expression in cells converted to TK phenotype with the chimeric α TK genes supported this conclusion in that α-TK expression was induced by α 4 ts mutants at non-permissive temperature. These mutants do not induce the expression of the enzyme in cells converted to TK+ phenotype by the wild-type β-TK. These studies showed that α-TK was induced in the absence of de novo protein synthesis after infection[46] and that viruses inactivated by UV light and a ts mutant not uncoated at the non-permisive temperature were efficient in inducing α-TK but not the wild type TK gene[49]. While the identity of the structural protein(s) responsible for the induction of α genes is not known, considerable progress has been made in the attempts to identify the cis-acting sequences which impart to genes the regulation characteristic of α genes. Mackem and Roizman[42] reported that the promoter and regulatory sequences of α 4 gene could be seperated and independently transferred to appropriate domains of the TK gene to convert these to α-regulated genes (Figure 1). Furthermore, they[42,43,48] also reported that the regulatory domains of all α genes share two sets of sequences. The first set comprised G+C rich inverted repeats whereas the second comprised an A+T rich sequence (Table 2). More recent analyses of the function of these sequences[50] (Kristie and Roizman, manuscript in preparation) indicate that the nucleotide sequences within or near the domains of the G+C rich inverted repeats determine the promoter function and especially the base level of α gene expression whereas the sequences within the domain of the A+T rich sequences may be at least a part of the cis-acting sequences responsible for the induction of the α genes.

Figure 1. Nucleotide sequence (5' to 3') of the +33 to -332 domain of the α 4 gene. The presentation emphasizes the presence of inverted repeats capable of forming stems, but there is no evidence that they actually form. The nucleotides are numbered by an asterisk (every ten) or arabic numeral (every hundred) from +1 - the site of initiation of transcription. The underlined nucleotides indicate A+T rich sequences shared by all α genes. The dashes indicate altered modes of base pairing with complementary sequences unavailable for annealing in the presentation shown above. Fusion of the DNA segment -110 to +33 to the +50 nucleotide of the TK gene imparts on the chimeric construct the capacity to convert cells to TK+ phenotype, but the chimeric gene is not induced by superinfection with TK- virus. The construct made by fusing the -332 to -110 segment to the -80 nucleotide of the TK gene is regulated as an α gene in cells converted to TK+ phenotype. Data from Mackem and Roizman[42,43].

Table 2: Directly repeated A+T-rich sequences present in the promoter-regulatory regions of α genes[a]

Consensus:	5'	G$^{C}_{T}$AT GN TAAT GA$^{G}_{A}$ TTC $^{C}_{T}$TT GN GGG	3'
α Gene 0:	-310	GTAT GG TAAT GAGT TTC TTC GG GAA	-286
	-228	GCAT GC TAAT GGGG TTC TTT GG GGG	-204
	-166	GCAT GC TAAT GATA TTC TTT GG GGG	-142
α Gene 4:	-268	GGGC GG TAAT GAGA TGC CAT GC GGG	-244
	-119	CGTG CA TAAT GGAA TTC CGT TC GGG	-95
α Gene 27:	-158	ATAT GC TAAT TAAA TAC ATG CC ACG	-134

	+158	ATAT GC TAAT TGAC CTC GGC CT GGA	+182
α Gene 47:	-397	GCAT GC TAAC GAGG AAC GGG CA GGG	-375
	-343	CGGC GG TAAT GAGA TAC GAG CC CCG	-321
	-178	GCCG GG TAAA AGAA GTG AGA AC GCG	-154

[a] The possible consensus sequence shown at the top of the table was derived by comparison of the six sequences above the horizontal dashed line. Note that a central, 11-nucleotide region is most highly conserved. The homologs shown for α gene 47 are taken from sequence data by Watson and Vande Woude[39]. Data from Mackem and Roizman[43].

Regulation of β genes. Induction of β genes requires the presence of functional α gene products[11] and stringent *ts* mutants in α 4gene express only α gene products at the non-permissive temperature. In cells converted to TK+ phenotype with the wild type viral TK gene the expression of TK is enhanced by infection with wild type virus[51], but not with *ts* mutants in the α 4 gene[52].

Detailed analyses of the promoter and regulator sequences are available only for the wild type TK gene. Specifically, McKnight and colleagues[53,54,55] and others[57,58] have carefuly delineated the *cis*-acting sites required for the expression of the wild type TK gene in oocytes and in cell culture. In addition to sequences responsible for proper punctuation of the transcript, the data indicate that the domain of the TK gene contains at least two *cis*-acting sites to -120 bp from the capping site and possible additional upstream sites necessary for optimal promoter function in the absence of *trans*-acting signals. There seems to be a general agreement that the sites responsible for efficient expression coincide with those required for regulation. At least in the β TK gene, the promoter and regulator sequences are not readily separable.

γ genes. As indicated earlier in the text, the expression of γ genes is amplified by the replication of viral DNA and is adversely affected in the presence of inhibitors of DNA synthesis or in cells infected with ts mutants in β genes at non-permissive temperature[10,59]. Experiments designed to determine which cis-acting sequences of γ genes are capable of imparting upon other genes (eg TK) the capacity to be regulated as γ genes were generally confined to a few γ_1 genes (E.K. Wagner, personal communication; J.R.Smiley, personal communication; S. Silver and B. Roizman, manuscript in preparation). To date, these experiments have not yielded unambiguous results largely because the γ-TK chimeras cannot be differentiated from wild-type TK genes with respect to requirements for optimal expression in cells transformed to TK+ phenotype.

CONCLUSION

Complete sequencing of the HSV genomes is likely to pave the way for a thorough analysis and recognition of all informational sequences specifying viral gene products. Specifically, once the predicted amonoacid sequence of a putative protein becomes apparent, it is readily possible to determine with the help of monoclonal antibodies to synthetic peptides whether the putative proteins are actually made in the infected cells.

Recognition of cis-acting sites other than those specifying transcription initiation, termination or initiation of translation is more complex simply because viral genes may differ from each other with respect to the nature of regulation and the regulatory trans-acting proteins which affect the regulation. Another problem is that not enough is known to set operational definitions for the various components of cis-acting sites involved in regulation of viral genes. Further studies on the function of these components by construction of deletions or by reconstruction of promoter-regulatory regions may lead to their enumeration and operational definition. Once operational definitions become available, it should be relatively simple to identify the structure and function of cis-acting sites involved in HSV gene expression and regulation. We have as yet little or no data on the mechanism by which trans-acting gene products interact with cis-acting sites to regulate gene expression or alter the structure and arrangement of viral DNA sequences.

ACKNOWLEDGEMENTS

The studies done at the University of Chicago were aided by grants from the National Cancer Institute, United States Public Health Service (CA08494 and CA19264), the American Cancer Society (ACS MV-2R), and in part by an unrestricted grant from the Bristol-Myers Company.

REFERENCES

1. Becker, Y., Dym, H. and Sarov, I: Virol. 36:184, 1968.
2. Kieff, E.D., Bachenheimer, S.L. and Roizman, B.: Virol. 8:125, 1971.
3. Wadsworth, S., Jacob, R.J. and Roizman, B.: Proc. Nat. Acad. Sci. USA 72:1768, 1975.
4. Sheldrick, P. and Berthelot, N.: Cold Spring Harbor Symp. Quant. Biol. 39:667, 1975.
5. Hayward, G.S., Jacob, R.J., Wadsworth, S.C. and Roizman, B: Proc. Nat. Acad. Sci USA 72:4243, 1975.
6. Honess, R.W. and Roizman, B.: J. Virol. 12:1346, 1973.
7. Morse, L.S., Pereira, L., Roizman, B., and Schaffer, P.A.: J. VIrol. 26:389, 1978.
8. Marsden, H.S., Stow, N.D., PReston, V.G., Timbury, M.C., and Wilkie, N.M. J. Virol. 28:624, 1978.
9. Wagner, E.K.: In: Individual HSV Transcripts in Herpesviruses, Vol 3, B. Roizman (ed.), Plenum Press N.Y., 1984, in press.
10. Honess, R.W. and Roizman, B.: J. Virol. 14:8, 1974.
11. Honess, R.W. and Roizman, B.: Proc. Nat. Acad. Sci. USA 72:1276, 1975.
12. Knipe, D.M., Ruyechan, W.T., Roizman, B., and Halliburton, I.W.: Proc. Nat. Acad. Sci. USA 75:3896, 1978.
13. Dixon, R.A.F. and Schaffer, P.A.: J. Virol. 36:189, 1980.
14. Watson, R.J. and Clements, J.B. Nature 285:329, 1980.
15. Post, L.E. and Roizman, B.: Cell 25:227, 1981.
16. Poffenberger, K.L., Tabares, E., and Roizman, B.: J. Virol. 46:103, 1983.
17. Mackem, S. and Roizman, B.: Proc. Nat. Acad. Sci. USA 77:7122, 1980.
18. Dutia, B.M.: J. Gen Virol. 64:513, 1983.
19. Huszar, D. and Bacchetti, S.: Nature 302, 76, 1983.
20. Ruyechan, W.T., Morse, L.S., Knipe, D.M., and Roizman, B.: J. Virol. 29:677, 1979.
21. Roizman, B., Norrild, B., Chan, C., and Pereira, L.: Virology, 1984, in press.

22. Roizman, B.: Cell 16:481, 1979.
23. Grafstrom, R.H., Alwine, J.C., Steinhart, W.L., Hill, C.W., and Hyman, R.W.: Virology 67:144, 1975.
24. Mocarski, E.S. and Roizman, B.: Proc. Nat. Acad. Sci. USA 78:7047, 1981.
25. Davidson, A.J. and Wilkie, N.: J. Gen. Virol. 55:315, 1981.
26. Wagner, M.M. and Summers, W.C.: J. Virol. 27:374, 1978.
27. Locker, H. and Frenkel, N.: J. Virol. 32:429, 1979a.
28. Mocarski, E.S. and Roizman, B.: Cell 31:89, 1982.
29. Frenkel, N., Locker, H., and Vlazny, D.: In: Herpesvirus DNA: Recent Studies of the Viral Genome, Y. Becker (ed.), Martin Nijihoff Pub., The Hague, pp. 149, 1981
30. Mocarski, E.S., Post, L.E., and Roizman, B.: Cell 22:243, 1980.
31. Smiley, J.R., Fong, B.S., and Leung, W.-C.: Virology 11, 345, 1981.
32. Mocarski, E.S. and Roizman, B.: Proc. Nat. Acad. Sci. USA 79:5626, 1982.
33. Frenkel, N., Locker, H., Batterson, W., Hayward, G.S., and Roizman, B.: J. Virol. 20:527, 1976.
34. Stow, N.D.: EMBO J. 1, 863, 1982a.
35. Stow, N.D. and McMonagle, E.C.: Virology 130:427, 1983.
36. Spaete, R.R. and Frenkel, N.: Cell 30:295, 1982.
37. Locker, H., Frenkel, N., and Halliburton, I.: J. Virol. 43:574, 1982.
38. Murchie, M.J. and McGeoch, D.J.: J. Gen. Virol. 62:1, 1982.
39. Watson, R.J. and VandeWoude,G.F.: Nucleic Acid Res. 10:979, 1982.
40. Anderson, K.P., Costa, R., Holland, L., and Wagner, E.: J. Virol 34:9, 1980.
41. Watson, R.J., Preston, C.M., and Clements, J.B.: J. Virol. 31:42, 1979.
42. Mackem, S. and Roizman, B.: Proc. Nat. Acad. Sci. USA 79:4917, 1982.
43. Mackem, S. and Roizman, B.: J. Virol. 44:939, 1982.
44. Cordingley, M.G., Campbell, M.E.M., and Preston, C.M.: Nucleic Acids Res. 11:2347, 1983.
45. Herz, C. and Roizman, B.: Cell 33:145, 1983.
46. Post, L.E., Mackem, S., and Roizman, B.: Cell 24:555, 1981.
47. Post, L.E., Norrild, B., Simpson, T., and Roizman, B.: Molecular and Cellular Biology 2:233, 1982.
48. Mackem, S. and Roizman, B.: J. Virol. 43:1015, 1982.
49. Batterson, W. and Roizman, B.: J. Virol., 46:371, 1983.

50. Roizman, B., Kristie, T., Batterson, W., and Mackem, S.: In: Proceedings of The P & S Biomedial Sciences Symposium on Transfer and Expression of Eukaryotic Genes, H. Vogel and H.S. Ginsberg (eds.), Columbia Univ. Press, 1983, in press.
51. Leiden, J.M., Buttyan, R., and Spear, P.G.: J. Virol. 20:413, 1976.
52. Kit, S., Dubbs, D.R., and Schaffer, P.A.: Virology 85:456, 1978.
53. McKnight, S.L.: Cell 31:355, 1982.
54. McKnight, S.L. and Kingsbury, R.: Science 217:316, 1982.
55. McKnight, S.L., Gavis, E.R., Kingsbury, R., and Axel, R. Cell 25:385, 1981.
56. Zipser, D., Lipsich, L., and Kwoh, J.: Proc. Nat. Acad. Sci. USA 78: 6276, 1981.
57. Smiley, J.R., Swan, H., Pater, M.M., Pater, A., and Halpern, M.E.: J. Virol 47:301, 1983.
58. Conley, A.F., Knipe, D.M., Jones, P.C., and Roizman, B.: J. Virol. 37:191, 1981.

The Role of Viruses in Human Cancer. Volume II
G. Giraldo and E. Beth editors

HERPES SIMPLEX VIRUS AND OCULAR DISEASE

YSOLINA M. CENTIFANTO-FITZGERALD

Lions Eye Research Laboratories, LSU Eye Center, Louisiana State University Medical Center School of Medicine, 136 S. Roman St., New Orleans, LA 70112 (U.S.A.)

INTRODUCTION

Herpes simplex virus type 1 (HSV-1) is one of the most dangerous pathogens for the eye, as it produces severe disease that can lead to blindness. For the past several years we have been concerned with the differences that exist among HSV-1 isolates from patients with herpetic ocular disease. A study of the disease manifestations in our patients indicated that severity varies from patient to patient, and more importantly that the type of disease varies from patient to patient. In severe cases, there is deep stromal involvement with tissue necrosis and opacity, which leads to corneal scarring and blindness. In mild cases, there may be only an annoying conjunctivitis, which is self-limiting and resolves within a few days. Between these two extremes, we see still other disease manifestations, including epithelial dendritic ulcers, geographic and ameboid epithelial ulcers, disciform edema, and iritis.

We were interested in looking into the basis for such diversity, as there appeared to be absolutely no correlation between disease pattern and age, health, or immune status of the host. These observations led us to speculate that perhaps some part of the character of the disease may be dependent not on the host, but on the infecting virus strain. This speculation was reinforced by a number of observations.

1) It is well known that, among laboratory strains, some viruses are very virulent and some are not.
2) In the case of primary infection, the outcome is not predictable. Some patients will have occasional recurrent episodes of the disease, others may have frequent recurrences, and still others may never have a recurrence.
3) The majority of the population can be shown to have been exposed to the virus; this is demonstrable by antibody levels to HSV in the population at large. Yet only a fraction, and a very small fraction at that, ever have overt disease.

Based on these observations, we formulated the hypothesis that the different clinical patterns of herpetic ocular disease may be attributed at least partially to the characteristics of the virus strain and that these differences may indeed represent specific viral functions. Therefore, we proceeded to undertake an in-depth study of HSV-1 strains and their pathogenicity in the eye.

HERPETIC OCULAR DISEASE

Strain specificity

The New Zealand white rabbit has been used extensively in our laboratory as a model for herpetic ocular disease. In the first part of this work, incisions were made in the superficial epithelium of each cornea, 50 microliters of a virus suspension were applied topically, the lid was closed, and the eye rubbed twice. The animals were examined daily by means of the slit lamp; on the third day after infection, disease was evident in all eyes.

Two groups of viruses were used. One group consisted of isolates from our clinic population, which were tested at an early passage (P-1) in the rabbit cornea. We classified the resulting disease as to severity and kind, and we found that we were able to reproduce, in the rabbit, the majority of the disease manifestations that we have seen in our clinic population.

Of importance to us was the realization that a few isolates produced extremely mild disease, i.e., the strain was practically avirulent. Initially, we tried to determine whether there was a defect in the mechanism of infection from these isolates, that is to say, whether their observed avirulence was due to receptor-site deficiencies and/or non-permissiveness of the corneal cells. We grew the virus in corneal tissue both in vitro, in excised pieces of cornea, and in vivo, in the rabbit eye. After infection, we followed these animals for 40 days and observed virus shedding in the tears. We also looked at the latent period, i.e., virus colonization of the ganglia, and we were able to recover the latent virus from the trigeminal ganglia.

From these findings, we concluded that in terms of the mechanism of infection, the avirulent strains are not different from the wild-type strains. They can grow in the corneal tissue in the animal and outside the animal. They persist in the cornea postinfection in the absence of disease, they colonize the ganglia, and they are shed in the tears. In essence, they behave exactly like the wild-type strains, except that they do not produce overt disease.[1] The implications of this statement are discussed below.

The second group of viruses consisted of laboratory strains with well-known disease characteristics. In this group, there were strains that produce epithelial disease, stromal disease, stromal necrosis with corneal vascularization, and geographic defects. In essence, we found several disease parameters that are also seen in the human population.

For this experiment, the virus strains were grown in the same cell line to preclude variation, and several viral inoculum sizes were used for each strain. All the eyes were examined by means of the slit lamp three times a week for at least 90 days. The results established that the disease pattern remains true. Epithelial-disease-producing virus failed to produce stromal disease regardless of titer and stromal-disease-producing virus produced stromal disease even at lowest titers. There was a relationship between titer and incidence of disease, but the type of disease was not affected. Because these experiments were done in randomly chosen and presumably immunocompetent rabbits, differences in type of disease could not be attributed to immune deficiencies in the host. We concluded that there must be some differences among the virus strains, and that the types and severity of ocular disease produced in the rabbit eye are inherent properties of the infecting strain, i.e., strain-specific differences can account for variability in ocular herpetic disease.[2,3]

Stromal disease

Stromal keratitis is an entirely different entity from epithelial disease, although both are caused by infecting herpesviruses. In epithelial disease, the epithelial layer of the cornea shows the formation of typical dendritic ulcers with a characteristic shape. The ulcers may grow and coalesce to form geographic defects, which cover a large area of the cornea. The disease usually reaches a peak in about seven days and subsides thereafter, and tissue destruction appears to be a result of the growth of the virus in the epithelial layer of the cornea.

In contrast, stromal keratitis is a two-step phenomenon. First, the virus multiplies in the epithelial layer. Disciform edema appears by the second week and can persist for several weeks, leading to, in some cases, granular opacities. By this time, the stroma is infiltrated by lymphocytes, macrophages, and polymorphonuclear leukocytes.[4-6]

This clinical picture suggested to us that the development of stromal keratitis might be an immune reaction of the host to the presence of HSV antigens in the stromal layer. Also, Metcalf et al.[7] studied the role of T cells in severe stromal disease with tissue necrosis and corneal vascularization in athymic mice and their results imply an immunopathologic basis for this disease. However, we do not know which and to what extent specific antigens are involved in stimulating the immune system.

We approached the identification of the proteins that are involved in stromal keratitis in two ways. The HSV-specific proteins that are the reacting moieties that trigger the host immune response during HSV infection appeared to be likely candidates. Work from our laboratory showed that the three major groups of HSV glycoproteins are secreted in vitro by the herpesvirus strains studied and that glycoprotein C is the major constituent of these secretions, a finding in agreement with the results of other investigations.[8-10] We also demonstrated that the stromal-disease-producing viruses secrete larger amounts of glycoproteins than the epithelial-disease-producing viruses.[11] Our data indicated a correlation between the appearance of stromal or epithelial disease and the amount of glycoprotein secreted. In addition, we found that clinical manifestations of virus-induced stromal disease could be reproduced by the injection of these secreted glycoproteins into the corneas of immune animals. It is possible that the presence of these glycoproteins in the stromal layer at the time of viral growth in the epithelium is a determinant in the induction of herpetic stromal disease. In a second approach, we identified several of the viral proteins present in infected rabbit corneas by means of SDS-polyacrylamide gel electrophoresis. Thus, we documented that HSV glycoproteins and other proteins are involved in the immunopathology of stromal disease.

However, the critical difference between stromal- and epithelial-disease-producing strains may be somewhat more subtle. We compared the antigenicity of each individual stromal-disease-producing strain to the antigenicity of the strains that produced only epithelial disease. But we were also interested in whether the relative capacity to elicit the immune response may be a co-determinant in the appearance of stromal keratitis.

GENETICS OF OCULAR DISEASE

The relationship of characteristic infecting strains and the disease produced was documented further.[12] In this study, we used the HSV-1(MP) strain, which produces stromal disease; HSV-1(F), which produces epithelial disease; and six recombinants derived by transfer of a specific DNA region of the donor HSV-1(MP) strain into the recipient HSV-1(F) strain. We showed that herpetic ocular disease characteristics, such as morphology of the dendritic ulcers and severity of epithelial disease, as well as incidence and duration of stromal disease, are genetically determined by the virus strain. In essence, we showed that: a) the morphology of the dendrites in epithelial disease is genetically determined; b) the viral functions are found between 0.70 and 0.83 map units of the HSV-1 genome and within the BglII F DNA fragment; and c) these functions do segregate but are closely linked and are independent of the production of glycoprotein C or fusion of HEp-2 infected cells.

These results confirm experimentally our initial observation that the ocular disease patterns produced by the wild-type isolates are truly characteristic of each individual isolate. This study represents the first attempt to define the viral functions responsible for the disease patterns caused by herpes simplex virus.

LATENCY

The preceding findings are a clear demonstration that several herpetic disease parameters are determined by the viral DNA. From these results, we can infer that strain differences may also be important in determining the rate of clinical recurrence. The rationale for such a premise is as follows: 1) herpes infections are common in the natural host, man; 2) infectious virus has been readily isolated from the ganglia of individuals with no history of disease; and 3) although two virus strains have been isolated from one individual at different sites, in the majority of individuals with no history of disease, only one strain has been isolated from multiple cervical ganglia.[13] Based on these observations, one can speculate that either these individuals were infected with a relatively avirulent strain that did not produce overt disease, or that they were somehow resistant to superinfection.

To investigate this apparent protection, we infected a large number of rabbits with the benign E-43 strain, which resulted in mild disease and ganglionic colonization. When these animals were challenged with a neurotropic virus strain (McKrae), the resulting disease was less severe than expected. In some animals, the challenging strain remained in the ocular tissues for weeks, but the only virus isolated from the trigeminal ganglia 6 weeks after challenge was the primary infecting strain (E-43). Similar results were seen when we challenged the animals with the MP strain.[14] The factors involved in this type of protection are not understood, but this sort of mechanism may play a role in mitigating overt disease in the majority of the population, most of whom have been exposed to the virus early in life.

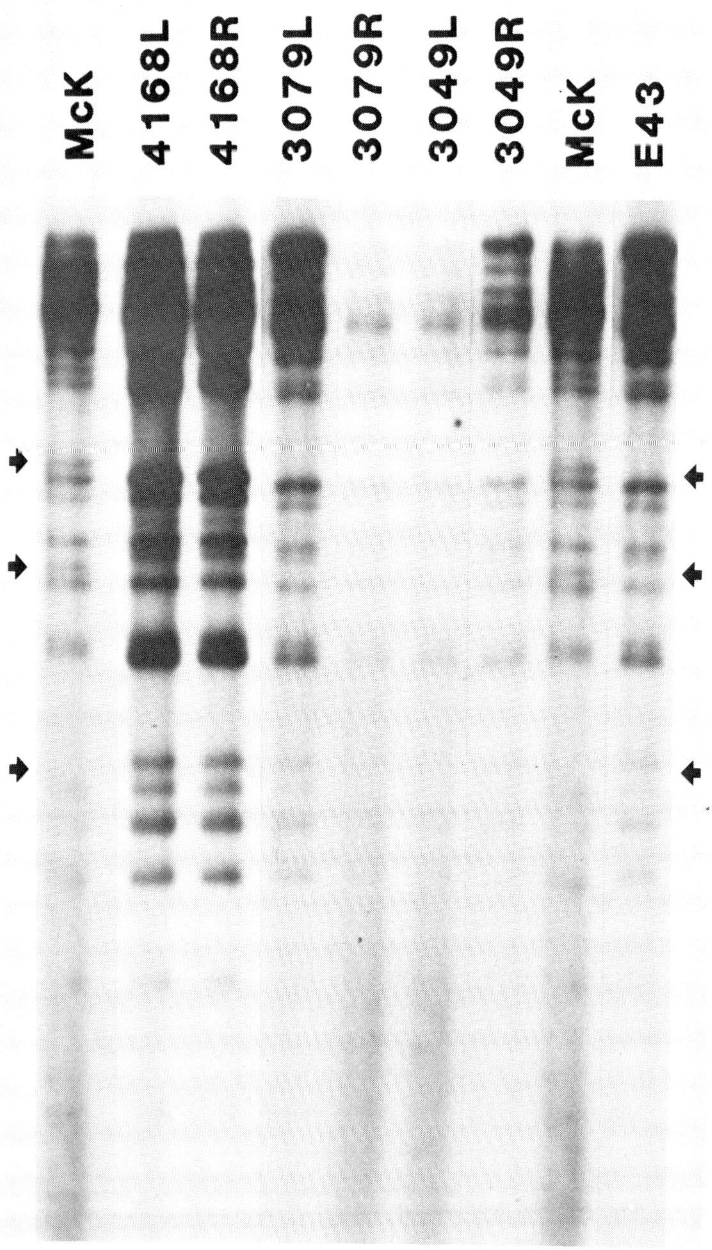
McK
4168L
4168R
3079L
3079R
3049L
3049R
McK
E43

IMMUNITY

The next step in this investigation involved the attempt to ascertain whether this apparent protection is a result of immunity, produced either by active infection or by immunization with live virus. In one set of experiments, animals infected in each eye with a standard inoculum (10^5 PFU) of E-43 strain were challenged with the McKrae strain six months after primary infection. After challenge, all animals showed ocular disease and shed virus at a rate similar to that characteristic of the McKrae strain, but restriction endonuclease analysis showed that the virus recovered from the ganglia was the original E-43 strain (Figure).

In another set of experiments, animals were immunized systemically with live virus and topically with UV-inactivated virus, producing a blood antibody titer of 1:256 and a tear antibody titer of 1:16. This level of immunity failed to protect the animals against colonization of the ganglia upon subsequent challenge, i.e., the challenging virus could be isolated from the ganglia. These findings indicated that antibodies, either humoral or secretory, provide some minimal protection from herpetic ocular disease, but do not prevent ganglionic colonization.[15]

To study the protective mechanisms involved in ganglionic superinfection, we felt it necessary to document all pertinent events in the E-43 system. Initially, we asked how much virus must be present in the cornea to colonize the ganglia in sufficient amounts to preclude superinfection. To quantitate our standard model, the _in vivo_ titer of the E-43 strain in the rabbit cornea was then determined. A 50 microliter inoculum containing 10^5 PFU was used to infect the rabbit eyes. Every day from day 3 through day 12 postinfection, rabbits were sacrificed and the eyes enucleated. Corneal scrapings were taken from the entire surface of each extirpated cornea, and the HSV titer of each individual cornea was determined. Disease was evident at day 3 postinfection in all animals. The amount of virus recovered increased thereafter, reaching a peak by day 7. The titer subsided over the next 5 days; by day 12, the eyes were clear but virus could still be recovered from the cornea. This degree of infection at the point of entry resulted in 100% protection from ganglionic superinfection.

FIGURE (facing page). Electrophoretic pattern of viral DNA digested with restriction endonuclease enzyme (KpnI). The digestion profile of the viral DNA of the isolate is similar to that of the E-43 strain and distinctly different from that of the McKrae strain. Arrows point to specific cleavage differences in the McKrae strain.

In the trigeminal ganglia, the virus produced an acute infection, with tissue necrosis of the neuronal tissue as the production of virus particles increased. At 6 weeks postinfection, the virus was in the latent state, and less than 1% of the total neurons stained positively for HSV antigens; this percentage is in agreement with findings by other investigators.[16]

Next, inocula above and below the E-43 standard dosage were used to determine the effect of inoculum size on ocular disease and ganglionic superinfection. Primary infection with inocula of 10^5, 10^4, and 10^3 PFU produced evident ocular disease, and protected the animals from subsequent ocular disease and mortality upon challenge with the McKrae strain (Table). In con-

TABLE

EFFECT OF INOCULUM SIZE ON OCULAR DISEASE AND GANGLIONIC SUPERINFECTION

INOCULUM SIZE (PFU)	RABBIT NUMBER	OCULAR DISEASE 5 DAYS POSTINFECTION OD	OCULAR DISEASE 5 DAYS POSTINFECTION OS	OCULAR DISEASE 7 DAYS POSTINFECTION OD	OCULAR DISEASE 7 DAYS POSTINFECTION OS	OCULAR DISEASE 5 DAYS POSTCHALLENGE OD	OCULAR DISEASE 5 DAYS POSTCHALLENGE OS	VIRUS RECOVERED FROM GANGLIA[a]
10^5	3896	1/2[b]	1	1+	2S[c]	0	0	E-43
	3897	1	1	1	+2S	0	0	E-43
	3898	1/2	1 1/2	1 1/2	3			
10^4	3899	1 1/2	1/2	3	2S	0	0	E-43
	3900	1/2	1	1/2	2S	0	0	E-43
	1557	1/2	1	1/2	1S	0	0	E-43
10^3	1558	1/2	1/2	1	1	0	0	*[d]
	1559	1/2	1/2	1/2	1/2	0	0	E-43
	1560	0	0	0	0	0	0	E-43
10^2	1555	0	1/2	0	0			
	1551	0	0	0	0	3	3	*
	1552	0	0	0	0	1/2	2	*
10^1	1556	0	0	0	0			
	1553	0	0	0	0	1/2	1	*
	1554	0	0	0	0	1/2	2	*

[a]Six weeks postchallenge
[b]Severity of ocular disease graded from 1 to 4, with 1 = mild; 4 = severe.
[c]Stromal disease
[d]Rabbit died before 6 weeks postchallenge

trast, very low inocula produced no discernible disease after primary infection, and the animals were not protected from ocular disease or death.

The question of which strains are found in the ganglia of the doubly-infected animals was then investigated. The rabbits were sacrificed 6 weeks postchallenge and the DNA recovered from the virus was analyzed by restriction endonuclease analysis. The ganglia from rabbits that received one of the large inocula (10^5, 10^4, or 10^3) and that displayed definite ocular herpetic disease yielded only the initial infecting E-43 strain. Thus, it is apparent that in this model, which uses the E-43 strain to produce subsequent protection from challenge with the McKrae strain, the inoculum must be of sufficient size to produce both ocular disease and ganglionic colonization.

CAN TWO DIFFERENT STRAINS COLONIZE THE SAME GANGLION?

We were interested to determine if the number of virus-infected neurons in the trigeminal ganglia varied with the size of the inoculum. To this end, the number of neurons positive for HSV antigens was determined by immunofluorescence staining. Three groups of animals were infected with 10^7, 10^5, or 10^3 PFU, and sacrificed 6 weeks postinfection. The ganglia were removed, divided into anterior, middle, and posterior sections, quick-frozen, and processed for immunofluorescence staining. The number of positive neurons was similar in all three groups (0.05% - 0.1%); even a thousandfold difference in the size of the inoculum did not strikingly alter the number of latently-infected neurons. This suggests that the number of susceptible neurons that can harbor the latent virus is small.

Since the number of positive neurons did not vary appreciably in animals infected with inocula of different sizes, but all animals showed clearly defined ocular disease, it may be that in animals with overt disease, all the susceptible neurons are infected. If the number of available, susceptible neurons controls the feasibility of superinfection, then it follows that superinfection is not possible in ganglia where all the neurons are "occupied" by an earlier infecting strain. Thus experimental animals in whom overt disease is induced by a primary infection would not be subject to superinfection by a challenging virus strain. Experimentally, we found in this study that animals that were infected with the E-43 strain and that showed mild ocular disease could be challenged with the McKrae strain, but yielded only E-43 strain upon virus recovery from the ganglia.

It appears that there is no simple relationship between relative size of the inoculum and increasing or decreasing protection; the effect seems to be more a threshold phenomenon, where the lower limit is the amount of virus required to produce clinical disease, which presumably correlates with a specific amount of virus in the ganglia. In our study, we saw a relationship between overt disease at the port of entry and protection against ganglionic colonization. Those animals in which ocular disease could be documented demonstrated protection against superinfection by a challenging virus strain, and finally, showed measurable numbers of virus-occupied neurons.

Infections produced with smaller inocula (less than 10^3 PFU) were studied by means of a different approach. Because the initial amount of virus in the ganglia was small, virus recovery and the differentiation of the primary virus from the challenging virus were difficult. For these reasons, viruses that differ in their cytopathogenic effects, e.g., rounded cells versus polykaryocyte formation, were chosen to facilitate identification in possible mixed infections.

In one set of experiments, F(MP)F, a recombinant strain that causes no apparent disease in the rabbit eye, was used as the primary infecting virus, and MP, a strain that is fatal in rabbits, was used as the challenging virus. When the trigeminal ganglia of these animals were examined for latent virus after primary infection, the results were always negative, perhaps due to the low amount of virus.

After challenge with the MP strain, the results varied according to the time of challenge. Briefly, if the challenge occurred 7 to 14 days after primary infection, the animals were protected from death caused by the MP strain, and the virus recovered from the ganglia was the challenging strain (MP). If the challenge was performed 30 days postinfection, some ganglia yielded the primary F(MP)F strain and other ganglia produced the MP strain.

From what is known of the events in the ganglia, we can infer the following. Infectious virus placed in scarified corneas reaches the trigeminal ganglia. With wild-type virus, the acute infection in the ganglia follows the normal course. At 5 to 7 days, virus is evident in the ganglia, and lingers for at least 7 to 10 days more. So, it is conceivable that in the 30 to 45 days after primary infection, the virus may spread throughout the ganglia and perhaps set up the so-called "state of resistance." There is no doubt that in the laboratory, varying experimental conditions can enable the delivery of more than one virus to the same ganglion, as has been recently reported in mice.[17] Our studies, however, employ wild-type clinical isolates that have purposely not been plaque-purified, so that they may be as close to the original human disease-producing pathogen as possible.

CONCLUSIONS

Our studies of strain variability and its relationship to disease clearly indicate that the type and severity of herpetic disease are determined by the characteristics of the infecting strain and these disease parameters are determined by the viral DNA. We have also shown that relatively benign or avirulent strains do exist in the wild-type population, and many of these strains can colonize the ganglia and protect against subsequent infection with more virulent strains. This may be one mechanism for determining not only the presence or absence of clinical disease, but also the severity of disease in individuals with an apparently functional immune system and no perceptible increase in susceptibility to other diseases.

ACKNOWLEDGEMENTS

This work was supported in part by PHS grants EY02389 and EY02377 from the National Eye Institute, National Institutes of Health, Bethesda, Maryland.

REFERENCES

1. Centifanto-Fitzgerald, Y.M. and Kaufman, H.E.: In: The Human Herpesviruses, an Interdisciplinary Perspective. A.J. Nahmias, W.R. Dowdle, and R.F. Schinazi (eds.), Elsevier, New York, 1981, p. 595.

2. Wander, A.H., Centifanto, Y.M., and Kaufman, H.E.: Arch. Ophthalmol. 98:1458, 1980.

3. Centifanto-Fitzgerald, Y.M., Fenger T., and Kaufman, H.E.: Exp. Eye Res. 35:425, 1982.

4. McNeill, J.I. and Kaufman, H.E.: Arch. Ophthalmol. 97:727, 1979.

5. Metcalf, J.F. and Kaufman H.E.: Am. J. Ophthalmol. 82:827, 1976.

6. Metcalf, J.F., McNeill J.I., and Kaufman, H.E.: Invest. Ophthalmol. 15:979, 1976.

7. Metcalf, J.F., Hamilton, D.S., and Reichert, R.W.: Infect. Immun. 26:1164, 1979.

8. Randall, R.E., Killington, R.A., and Washington, D.H.: J. Gen. Virol. 48:297, 1980.

9. Kaplan, A.S., Erickson, J.S., and Ben-Porat, T.: Virology 64:132, 1975.

10. Chen, A.B., Ben-Porat, T., Whitley, R.J., and Kaplan, A.S.: Virology 91:234, 1978.

11. Smeraglia, R., Hochadel, J., Varnell, E.D., Kaufman, H.E., and Centifanto-Fitzgerald, Y.M.: Exp. Eye Res. 35:443, 1982.

12. Centifanto-Fitzgerald, Y.M., Yamaguchi, T., Kaufman, H.E., Tognon, M., and Roizman, B.: J. Exp. Med. 155:475, 1982.

13. Lonsdale, D.M., Brown, S.M., Subak-Sharpe, J.H., Warren, K.G., Koprowski, H.: J. Gen. Virol. 43:151, 1979.

14. Centifanto-Fitzgerald, Y.M., Varnell, F.D., and Kaufman, H.E.: Infect. Immun. 35:1125, 1982.

15. Centifanto-Fitzgerald, Y.M.: Abstract, Eighth International Herpesvirus Workshop, Oxford, England, July 31 - August 5, 1983, p. 172.

16. Tenser, R.B., Dawson, M., Ressel, S., and Dunstan, M.: Ann. Neurol. 11:285, 1982.

17. Meigner, B., Norrild, B., and Roizman, B.: Infect. Immun. 41:702, 1983.

© 1984 Elsevier Science Publishers B.V.
The Role of Viruses in Human Cancer. Volume II
G. Giraldo and E. Beth editors

ROLE OF THE PROMOTER SEQUENCE OF THE THYMIDINE KINASE (TK) GENE OF HSV-1 IN BIOCHEMICAL TRANSFORMATION OF CELLS, GENE EXPRESSION AND NEUROVIRULENCE IN MICE

YECHIEL BECKER, YEHUDA SHTRAM, AVNER BARASOFSKY, MICHELLE HABER*, ANNIE SCEMAMA, YAEL ASHER, EYNAT TABOR, TAMIR BEN-HUR, DONALD GILDEN** AND JULIA HADAR
Department of Molecular Virology, Faculty of Medicine, Hebrew University of Jerusalem, Jerusalem, Israel.

TRANSFORMATION OF CELLS BY HERPES SIMPLEX VIRUSES

The role of herpes simplex virus type 1 (HSV-1), and even more so type 2 (HSV-2) in transformation of cells in-vitro and in-vivo remains unclear, in spite of the large number of experiments and published reports. Although it appears that cells under in-vitro conditions are morphologically transformed by the BglII/N DNA fragment of HSV-2 (1, 2; see also Abstracts International Herpesvirus Workshop, Oxford, 1983, pp. 70, 187) and more specifically by a sequence of 700 bp in this DNA fragment, the sequence does not appear to remain in the affected cells, and the exact effect on the cell DNA is unknown. A "hit-and-run" mechanism to explain the effect of HSV-2 on cellular DNA was suggested by D.A. Galloway and colleagues (Abstracts International Herpesvirus Workshop, Oxford, 1983, p. 70). Huszar and Bacchetti (3) proposed that the HSV-2-coded ribonucleotide reductase, which shares sequences with the HSV-2 BglII/N and C fragments,might be involved in the transformation of cells by providing an enzyme for the possible "hit-and-run" effect of the viral DNA by acting as a mutator. The role of HSV-2 DNA in cervical carcinoma is still unclear, in spite of numerous attempts to demonstrate the presence of HSV-2 DNA sequences in the cancerous tissues (4). However, transformation of rodent cells with inactivated HSV-2 (5) did result in the presence of viral DNA sequences in the cell (6). This may indicate that certain sequences of the viral DNA are able to recombine and most probably integrate into the cellular DNA, with or without affecting the ability of the cell to proliferate. Attempts to link HSV-2 DNA with a specific chromosome in transformed cells resulted in contradictory findings. McKinlay et al. (7) reported that the HSV-2 thymidine kinase (TK) gene was integrated into the DNA of chromosome 18, whereas Kit et al. (8) did not find any association between the HSV-2 TK gene and a particular human chromosome. Smiley et al.(9)

Present address:
* School of Pathology, The University of New South Wales, P.O. Box 1 Kensington, N.S.W., Australia 2033
** Department of Neurology Gl, University of Pennsylvania, Philadelphia, Pa. 19104, USA

reported that transformation experiments with uv-inactivated HSV-1 yielded stable and unstable classes of transformed cells and suggested that the viral DNA is integrated in murine cells in a specific chromosome.

The ability of the HSV TK gene to become integrated into the cellular DNA, as shown by Pellicer et al. (10), seems to be unrelated to cell transformation. Reyes et al. (11) reported that morphologically transforming HSV-1 and 2 DNA fragments do not correlate with the DNA fragments which biochemically transform L(TK⁻) cells. Hampar et al. (12) reported that cells are not transformed morphologically by the viral TK gene and suggested that "morphological and biochemical transformation are independent events". However, an indication that the integration of HSV-1 TK DNA into the cellular DNA has an effect on the host cell DNA was reported by Reyes et al. (13).

Most cell lines transfected with DNA fragments from recombinant plasmids or viral DNA fragments carrying the HSV-1 TK gene contained only one copy of the viral TK gene, although a few cell lines contained multiple copies. Analysis of such cell lines revealed that the viral DNA sequences were amplified to create up to 20 tandem duplications. Such amplified sequences were lost on subsequent passaging of cells, and the number of TK gene copies fell to 3-4 copies per haploid cell. Local amplification of the viral TK gene integrated into cell DNA might resemble the effect that carcinogens have on SV40 DNA integrated into cell DNA reported by Lavi (14). It was shown that treatment of SV40-transformed cells with the carcinogen 7,12-dimethylbenz[a]anthracene resulted in the tandem amplification of the integrated SV40 DNA. Restriction enzyme analysis of the amplified DNA molecules present in the Hirt supernatant extracted from the treated cells revealed that not all the integrated SV40 sequences were represented in amplified DNA. Recently Matz et al. (Abstracts International Herpesvirus Workshop, Oxford, 1983, p. 71) reported that infection of SV40-transformed cells with HSV-1 and HSV-2 resulted in the amplification of the SV40 DNA sequences in the cells.

The ability of the HSV-1 TK gene from viral DNA (15) or from fragmented DNA (16-18) to integrate into cellular DNA, although lacking the ability to transform cells morphologically (12), has been studied. Studies on the expression of the integrated viral TK gene in biochemically transformed cells might shed light on the mechanism of cell transformation mediated by HSV-1 and HSV-2 DNA sequences and the role of the viral gene promoters in the cells.

CONSTRUCTION OF RECOMBINANT PLASMIDS CONTAINING THE HSV-1 TK GENE

Numerous studies on HSV-1, and more recently on HSV-2, TK genes have been reported in recent years (reviewed in ref. 19). The nucleotide sequence of the HSV-1 TK was reported by McKnight (20) and Wagner et al. (21), and the ability of the recombinant plasmids to express the viral TK gene in *E. coli* was demonstrated (22-24). Studies on biochemical transformation of cells by recombinant plasmids containing the HSV-1 TK gene (19) provided an insight into the organization of the TK gene and the upstream promoter sequences involved in regulating the expression of the gene.

Table 1 describes the recombinant plasmids constructed in our laboratory to study expression of the HSV-1 TK gene. The TK gene was cloned from different HSV-1 strains, and some plasmids also contained unrelated HSV-1 DNA sequences in addition to the TK gene. The viruses included the HSV-1(F) strain and isolates derived in our laboratory from an NIH strain (25). Our large plaque (LP) TK^{+} virus produces a high level of TK activity, and the small plaque (SP) $TK^{1/4}$ mutant has low TK activity (25% of that of the LP TK^{+} virus). We also isolated an LP TK^{-} mutant with no TK activity. The TK gene from HSV-1 (F) was cloned in pBR322 as a BamHI/Q' fragment (e.g. pBRTK), or as a BglII-BamHI/Q fragment (pBY718; ref. 24) and also in pSVOd (Fig. 1), a plasmid that lacks the toxic sequences of pBR322 and contains the origin of replication (designated ORI) of SV40 (26). Fig. 2 describes the transfer of the HSV-1 TK gene from plasmid pBY513-1 (TK gene lacking part of the promoter present in pSVOd) to plasmid pBY19 which contains the BamHI/G and V fragments of HSV-1 (F) DNA. Fig. 3 shows the transfer of the HSV-1 TK gene (lacking part of its promoter sequence) to plasmid pBY40-2 (containing the BglII/I fragment of HSV-1 LP TK^{+} DNA) which is the cloning vehicle to produce plasmid pBY42-28. As in Fig. 1, the SV40 ORI and the TK gene were transferred as described by Becker et al. (24, 27). These plasmids were used to study transformation of L(TK^{-}) cells and TK gene expression in biochemically transformed cells and in *E. coli*.

THE PROMOTER SEQUENCE OF HSV-1 TK IS REQUIRED FOR EFFICIENT BIOCHEMICAL TRANSFORMATION OF L(TK^{-}) CELLS

L(TK^{-}) mutant mouse cells were transformed by the plasmid preparations (3 µg DNA/plate), usually three plates for each sample done at least twice. The transformation technique was as previously reported (28). The cultures were stained and the number of colonies/plate was counted. Intact pBRTK plasmids efficiently transformed L(TK^{-}) cells, but cleaving the plasmid

TABLE 1

RECOMBINANT PLASMIDS CONTAINING CLONED HSV-1 DNA

Plasmid	Cloning vector	Cloned HSV-1 TK gene	Second DNA fragment cloned in plasmid
pBRTK	pBR322	BamHI-BamHI Q fragment of HSV-1 (F)	--
pBY16	pBR322	BamHI-BamHI Q fragment of HSV-1(LP)TK$^+$	--
pHA101	pHA10	BamHI-BamHI Q fragment of HSV-1 (F)	--
pBY718	pBR322	BglII-BamHI from Q fragment HSV-1 (F) (lacking the promoter)	--
pBY718/pBY10(1)	pBR322	" "	Bam HI fragment N of HSV-1
pBY718/pBY10(5)	pBR322	" "	as above, but in the opposite orientation
pBY718/pBY7(17)	pBR322	" "	BamHI fragment J of HSV-1 DNA
pBY703	pBR322	BamHI-BglII containing the viral TK gene promoter sequence	--
pBY4	pBR322	BamHI-BamHI fragment from HSV-1 (LP) TK$^+$	--
pBY10	pBR322	BamHI-BamHI fragment N from HSV-1 (LP) TK$^+$	--
pBYAB53-12	pBR322	BamHI-BamHI Q fragment from HSV-1 (SP) TK¼	
pBYAB53-12-16	pBR322	BglII-BamHI from a fragment from HSV-1 (SP) TK¼	
pBY510-4	pSV0d	BamHI-BamHI Q fragment HSV-1 (LP) TK$^+$	SV40 origin of replication (ORI)
pBY511-6	pSV0d	BglII-BamHI Q fragment HSV-1 (LP) TK$^+$	SV40 ORI
pBY513-1	pSV0d	BglII-BamHI Q fragment HSV-1 (F)	SV40 ORI
pBY24-4	pSV0d	BglII-BamHI from Q fragment HSV-1 (F)	SV40 ORI
pBY24-18		" (two copies)	SV40 ORI
pBY40-2	pBR322	BglII/I HSV-1 (LP) TK$^+$	
pBR42-28	pBY40-2	EcoRI-EcoRI from pBY513-1	

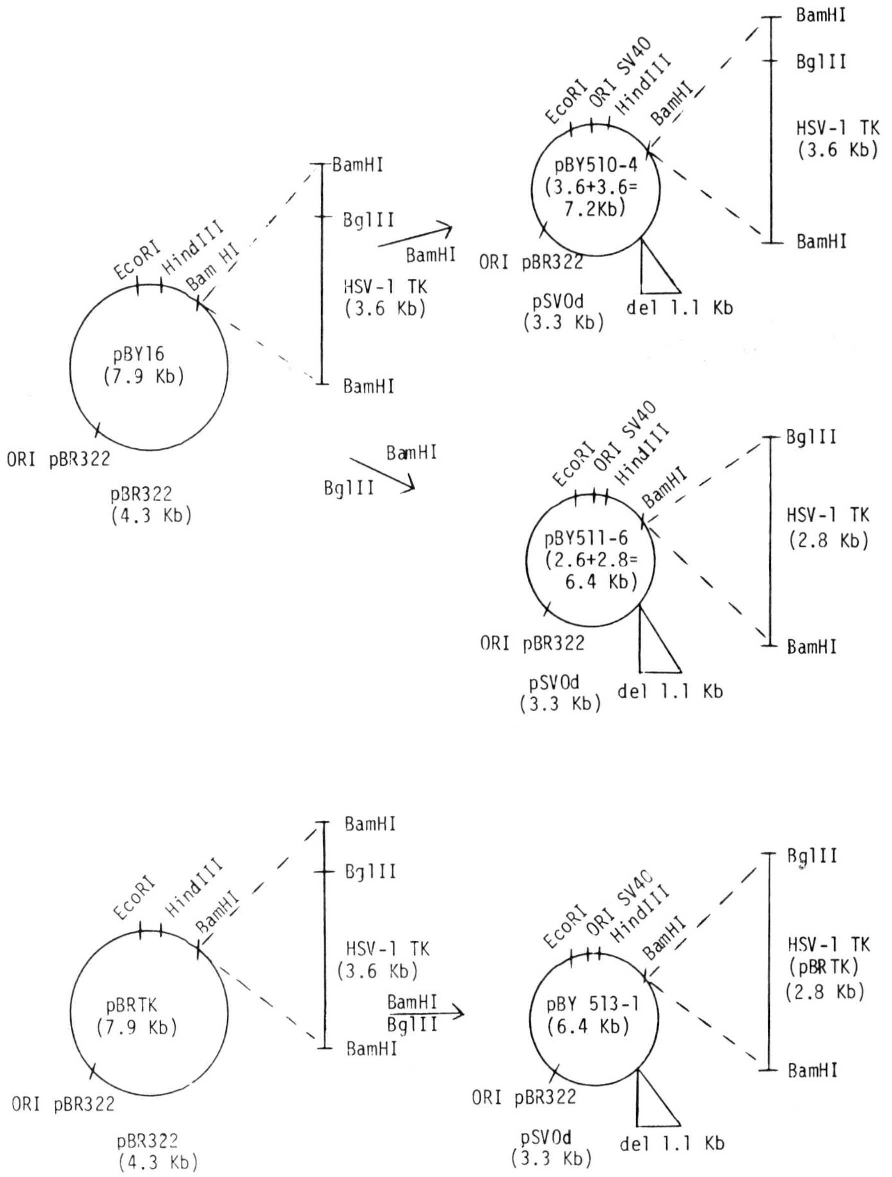

Fig. 1. Steps in the construction of recombinant plasmids containing the HSV-1 TK gene in pBR322.

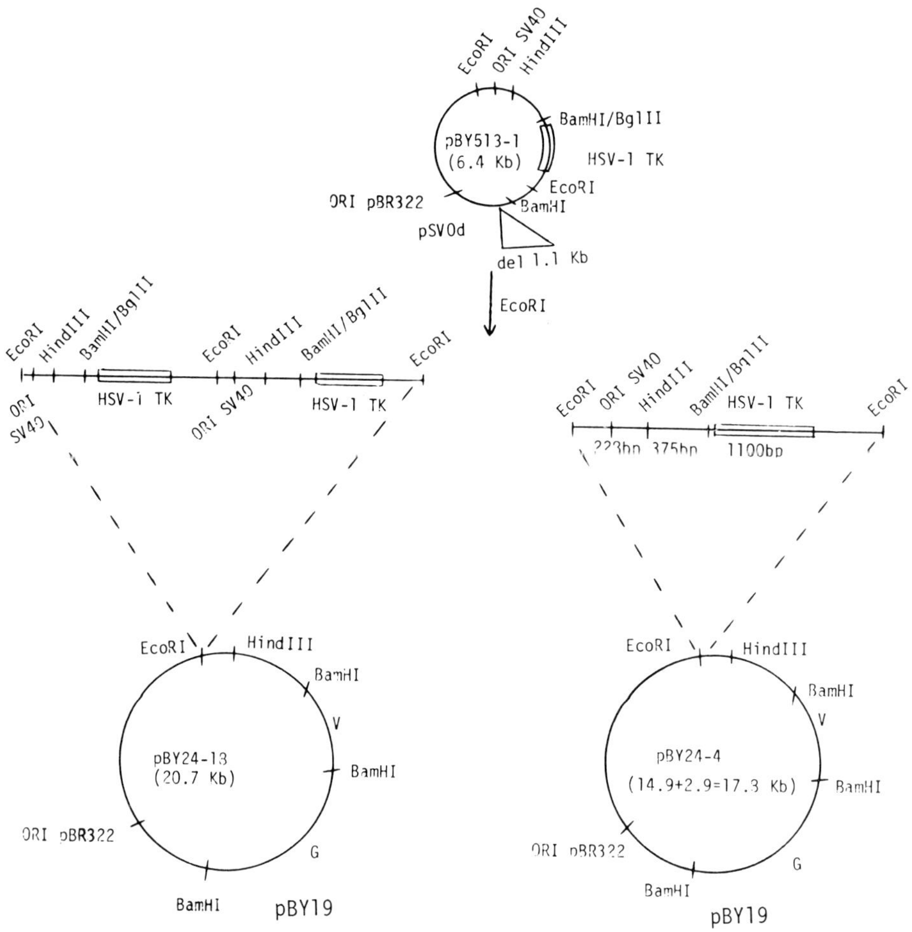

Fig. 2. Insertion of the HSV-1 TK gene and SV40 sequences which contain the SV40 origin of replication (ORI SV40) into recombinant plasmid pBY19.

with BamHI (that released the BamHI-BamHI DNA fragment containing the viral TK gene and its promoter) enhanced transformation threefold (Table 2). However, cleavage of pBRTK with the restriction enzyme EcoRI resulted in a marked decrease in the ability of the DNA to transform L(TK$^-$) cells. Other plasmids (pBY16, pHA101) that carry the BamHI-BamHI DNA fragment containing the viral TK gene from HSV-1 LP TK$^+$ also transformed L(TK$^-$) cells.

Transformation of L(TK$^-$) cells with plasmid pBY718 that contains the HSV-1 TK gene without part of its promoter sequence yielded fewer than ten colonies of transformed cells per plate (about 1% of cells transformed by

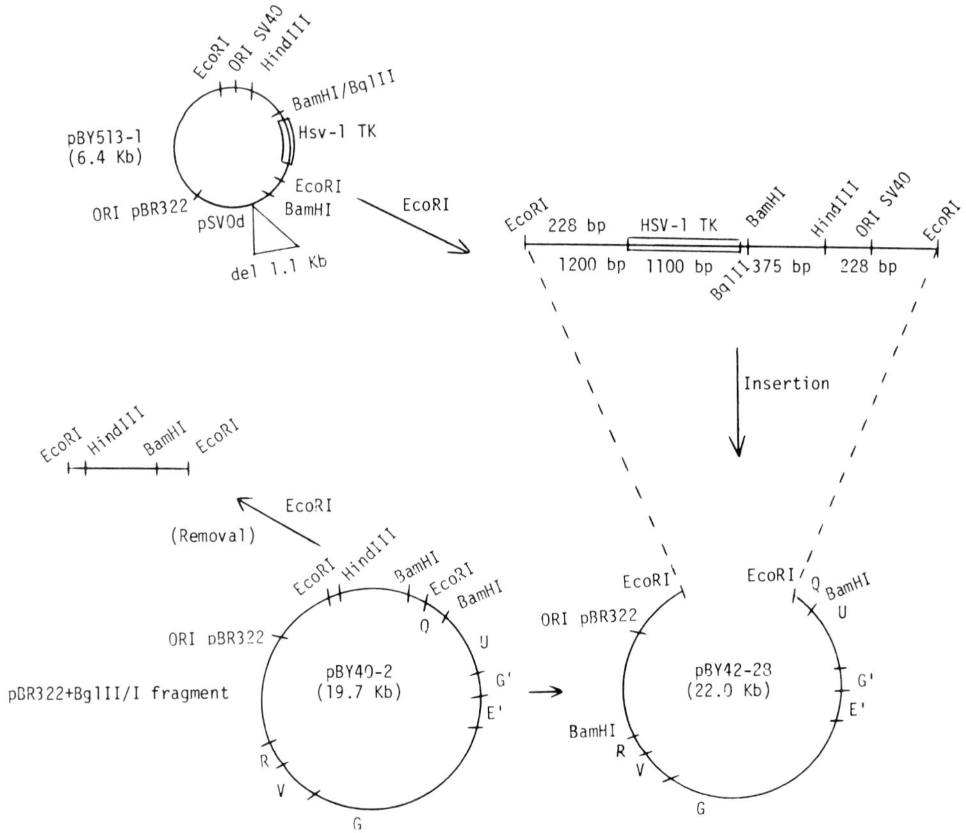

Fig. 3. Construction of a recombinant plasmid which contains the HSV-1 (LP) BglII/I fragment, HSV-1 TK and ORI sequences in pBR322.

pBRTK). This is a marked reduction in transforming ability, as compared to pBRTK that contains the intact TK gene and its promoter. In their study, Colbere-Garapin et al. (29) obtained no TK^+ colonies whatsoever in an HSV-1 TK-containing plasmid cleaved with BglII. Cloning of additional viral DNA sequences in pBY718 did not change the extent of $L(TK^-)$ transformation. No transformed colonies were found in cells transfected with pBY703 (carrying the viral promoter without the TK gene) or pBY4 and pBY10 containing HSV-1 DNA from other sites in the viral genome. Thus, removal of the promoter sequence from the viral TK gene markedly affects the ability of the gene to transform $L(TK^-)$ cells.

TABLE 2

TRANSFORMATION OF L(TK$^-$) CELLS WITH RECOMBINANT PLASMIDS CONTAINING HSV-1 TK GENE LACKING ITS PROMOTER

Plasmid	Cleavage with restriction enzyme	No. of colonies/ 3 μg plasmid DNA
Plasmids containing Q fragment (viral promoter and TK gene)		
pBRTK	--	360
pBRTK	BamHI	1100
pBRTK	EcoRI	13
pBY16	--	52
pHA101	--	155
pHA101	BamHI	136
Plasmids containing TK gene lacking its promoter		
pBY718	--	6
pBY718	BamHI	8
pBY718/pBY10(1)	--	2
pBY718/pBY10(5)	--	4
pBY718/pBY7 (17)	--	6
Plasmids not containing HSV-1 TK gene		
pBY703	--	0
pBY4	--	0
pBY10	--	0

PROMOTER SEQUENCE OF THE TK GENE IS REQUIRED FOR GENE EXPRESSION IN BIOCHEMICALLY TRANSFORMED CELLS

The cell lines resulting from the biochemical transformation of L(TK$^-$) cells (designated by the number of the recombinant plasmid used for the transformation) were studied for their constitutive expression of the viral TK gene. This was compared with induced expression of the TK gene after infection of these cells with a TK$^-$ mutant of HSV-1 Such an infection leads to the synthesis of the immediate early proteins by the viral alpha genes. These, in turn, interact with sequences in the promoter of the TK gene (a beta viral gene) leading to enhanced expression of the TK gene (30). As shown in Fig. 4, the various cell lines differed in their response to infec-

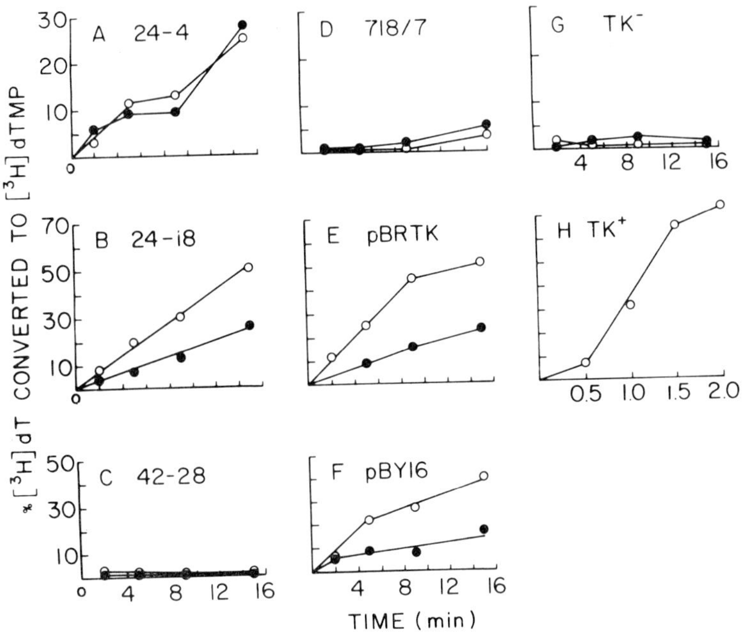

Fig. 4. Induction of HSV-1 TK genes in cells biochemically transformed by recombinant plasmids carrying the HSV-1 TK gene. The transformed colony was selected after transformation by each plasmid and further propagated in HAT medium. The colonies were designated with the number of the recombinant plasmid used for transformation. Parallel cultures were prepared for studying constitutive expression of the HSV-1 TK present in the cells (●—●) and viral TK gene expression after infection with a TK⁻ mutant of HSV-1 (LP) (o—o). At 20 hr p.i., the cultures were harvested, homogenized, and the TK activity was determined by incubating the cell homogenates with ^{3}H-thymidine. Samples were removed at 2 to 16 min after incubation at 37°C. The phosphorylation of ^{3}H-thymidine was determined as previously described (24).

tion with the HSV-1 TK⁻ mutant. In the control (Fig. 4G), it can be seen that, when infected with the TK⁻ mutant, L(TK⁻) cells showed no TK activity, whereas the TK⁺ HSV-1 parent strain (LPTK⁺) did induce TK activity in the cells (Fig. 4H). Cell lines biochemically transformed by plasmids pBRTK (Fig. 4E) and pBY16 (Fig. 4F) that contain the BamHI/Q fragment of HSV-1(F) and HSV-1(LP) virus strain, respectively, showed enhanced TK activity after infection with the TK⁻ virus mutant. However, all the recombinant plasmids that lack part of the promoter sequence and contain the BglII-BamHI fragment did not respond with enhanced expression of the TK gene after infection with the TK⁻ virus (Fig. 4D, 4C and 4A). The exception was the cell line transfected by plasmid pBY24-18, in which the ability of the TK gene to respond to TK⁻ virus infection was restored. It is of interest that this plasmid (see Fig. 2) contains two TK genes and two origins of SV40 DNA replication. The role of the SV40 origin sequence in the restoration of the response of the HSV-1 TK gene to the viral alpha proteins is still to be clarified.

Table 3 provides additional information on the expression of the HSV-1 TK gene in L(TK⁻) cells biochemically transformed by different plasmids. In cell line pBYAB720 transformed with a plasmid resembling pBY718 (containing the BglII-BamHI/Q fragment, namely the TK gene without part of the promoter) the TK activity was low (background level), and infection with the TK⁻ virus mutant had no effect on the expression of the TK gene. However, cells biochemically transformed with pBRTK (the BamHI/Q fragment) responded with enhanced TK expression after TK⁻ virus infection. Addition of phosphonoacetic acid (PAA), an inhibitor of the viral DNA polymerase, led to an additional increase in TK gene expression. This is due to inhibition of the replication of the viral DNA and the continued expression of the viral immediate early genes (alpha) needed for TK (beta) gene expression.

Cells transformed with plasmid pBYAB53-12 cleaved with BamHI responded remarkably well to TK⁻ virus infection, with even higher expression in the presence of PAA. Cells transformed by pBRTK cleaved with EcoRI also responded with enhanced TK expression after TK⁻ virus infection that increased after PAA treatment.

Removal of the promoter sequence of the TK gene as far as the BglII site thus prevented the gene from responding to the immediate early proteins, products of the alpha genes. Removal of sequences upstream to the EcoRI site did not affect the response of the TK gene to the alpha proteins. It is concluded that the TK promoter sequence needed for enhanced expression of the TK gene is between the BglII and the upstream EcoRI restriction site. This observation is in agreement with the findings of Hayward et al. (19), McKnight (20), and Zipser et al. (31).

EXPRESSION OF HSV-1 TK GENE IN *E. coli*

In our studies (24, 27), the TK gene of the HSV-1 TK⁺ virus strain was cloned and expressed in *E. coli*, using the tetracycline resistance (tet^r) plasmid promoter for the expression of the viral gene. Removal of the 800 bp sequence upstream of the BglII site in the BamHI Q fragment enhanced

TABLE 3

EFFECT OF HSV-1 TK^- VIRUS INFECTION ON TK GENE EXPRESSION IN $L(TK^-)$ CELLS BIOCHEMICALLY TRANSFORMED WITH RECOMBINANT TK PLASMIDS

Cell line from plasmid		% Conversion of [^{3}H]dt to [^{3}H]dTMP			
	Time (min)	0.5	1	1.5	2
pBRTK EcoRI		6.0	7.4	7.6	7.6
"	+ PAA	6.5	7.4	7.7	7.9
"	+ HSV-1(TK^-)	7.5	10.0	15.0	18.0
"	+ HSV-1(TK^-) + PAA	10.0	17.0	24.0	27.5
pBYAB53-12 BamHI		5.0	5.0	5.0	7.0
"	+ PAA	5.0	5.0	7.0	9.0
"	+ HSV-1(TK^-)	9.5	17.5	24.0	32.0
"	+ HSV-1(TK^-)+PAA	13.0	22.0	30.0	38.0
pBRTK BamHI			ND*		4.0
"	+ PAA				4.0
"	+ HSV-1(TK^-)				8.0
"	+ HSV-1(TK^-) + PAA				15.0
pBYAB720			ND*		5.0
"	+ PAA				5.0
"	+ HSV-1(TK^-)				5.0
"	+ HSV-1(TK^-) + PAA				5.0

* Not determined

For experimental details, see legend to Fig. 4

the expression of the TK gene fourfold in *E. coli* (27). These results indicated that the 800 bp fragment (containing the TATTA box) includes a nucleotide sequence that reduces the expression of the viral TK gene.

The reason for the reduced expression of the TK gene in the BamHI/Q fragment was investigated by cloning the TK gene from TK¼ virus DNA. Cloning of the BamHI/Q fragment of the TK¼ gene in pBR322 (BamHI site: plasmid pBYAB53-12, Table 1) resulted in no expression of the TK gene in *E. coli* (Table 4). The TK gene from the HSV-1 TK¼ virus thus contains a sequence that inhibits expression of the TK gene in *E. coli*. Removal of the 800 bp upstream to the

TABLE 4

EXPRESSION IN *E. coli* OF THE HSV-1 TK½ TK GENE WITH[a] AND WITHOUT[b] THE BamHI-BglII UPSTREAM NUCLEOTIDE SEQUENCE

Plasmid	[^{3}H]dT incorporation				% conversion of [^{3}H]dT to [^{3}H]dTMP			
Time (min)	5	10	20	40	5	10	20	40
pBYAB53-12[a]	822	898	1172	1438	3.1	2.64	3.08	2.54
pBYAB53-12-16[b]	12470	23178	29962	30852	7.8	11.8	20.3	18.3
pBY718[b]	5846	12632	20200	24832	8.8	12.2	15.9	16.5
pBR322[a]	1280	1130	1300	1350	2.04	1.8	1.63	1.5

For experimental details, see legend to Fig. 6.

BglII site in the TK gene of the TK½ virus (plasmid pBYAB53-12-16) resulted in enhancement of TK expression to a level slightly higher than that of pBY718 (Table 4).

Analysis of the sequence of the TK gene and its upstream promoter sequence (20,21) revealed a palindrome spanning both sides of the BglII sequence (Fig. 5). The palindromic sequence resembles an attenuator sequence present in the trp gene of EcoRI (32). The attenuator sequence is centrally located between the mRNA start (cap) site, designated "0", and the methionine codon (ATG) in the TK gene (designated 107). It is hypothesized that mutations of C to T in the attenuator sequence might be responsible for increased homology between bases in the palindrome; increased matching between base pairs could result in reduced expression of the TK gene not only in *E. coli*, but possibly also in mammalian cells. This might happen by the formation of a stem and loop in the TK mRNA. On the other hand, mutations from T to C might reduce the sequence homology in the palindrome and might lead to an increase in expression of the TK gene. Further studies are in progress to determine the role of the attenuator sequence in the expression of the TK gene. It is of interest that a palindrome resembling an attenuator sequence could be identified in the sequence of the glycoprotein C (gC) gene (33) (around nucleotide +60; see Fig. 8). This palindrome is centrally located between the mRNA start and the ATG methionine codon of the gene. The palindrome sequence is missing from the spliced mRNA transcribed from this gene, indicating a mechanism of removal from mRNA of a palindromic sequence which might inhibit translation of the mRNA (33).

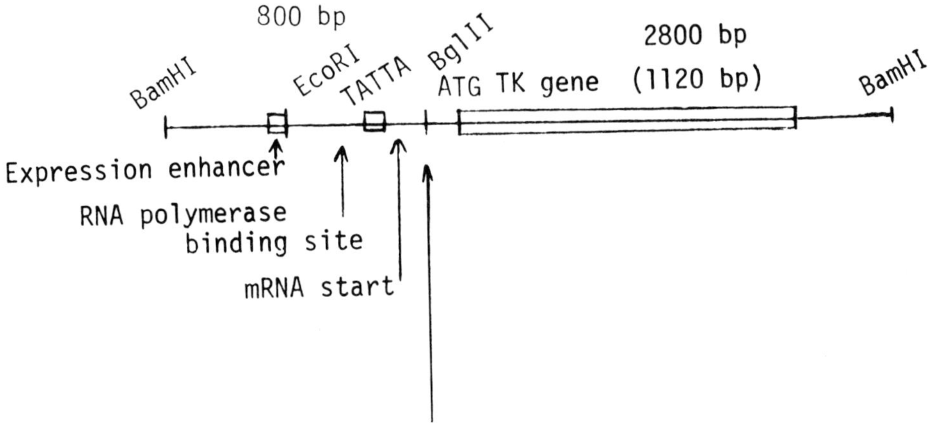

ATTENUATOR

BglII

CGTGCCGCAGATCTTGGTGGCGTG

trp (E. coli) AGCCCGCCTAATGAGCGGGCT*

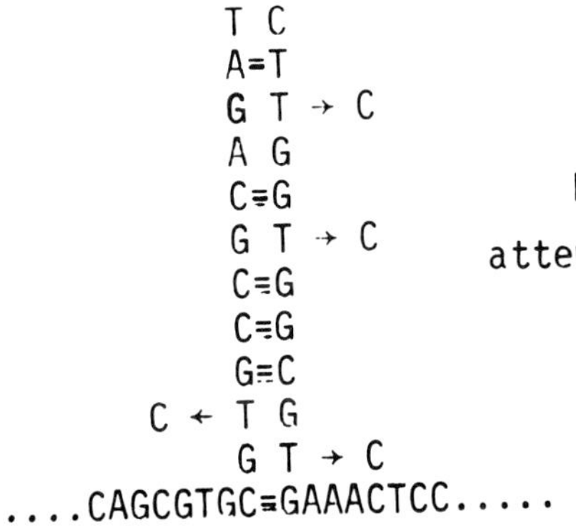

HSV-1 TK attenuator sequence

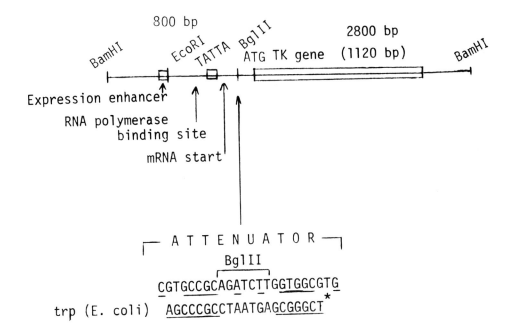

Fig. 5. Schematic representation of the promoter region of the HSV-1 TK gene and the possible changes in the nucleotides of the proposed attenuator.

* Reviewed in ref. 32.

A CHIMERIC PROMOTER (SV40 + tet^r) REDUCES TK GENE EXPRESSION

Plasmid pSVOd contains 228 nucleotides from the EcoRII/G fragment of SV40 DNA (26). The latter contains the SV40 origin of replication fused to the HindIII site in the promoter sequence of the tet^r gene of pBR322. As a result, expression of the HSV-1 TK (in plasmid pBY513-1) is reduced by a factor of 2 (Fig. 6), as compared to expression of the TK gene in *E. coli* transfected with pBY718. Thus, the SV40 nucleotide sequence (from a mammalian virus) partially replaces the promoter sequence of the promoter of the bacterial tet^r gene. Analysis of the nucleotide sequences both upstream to the BamHI site of pSVOd (originating from pBR322) and downstream from the BglII sequence of HSV-1 DNA (as reported by McKnight (20) and Wagner et al. (21) for HSV-1 strain F) revealed a palindrome with four G=C pairs and one A=T pair, as shown in the sequence:

C A C A C C C G T C C T G T G↓G A T C T T G G T G G C G T C
G T G T G G G C A G G A C A C C T A G↑A A C C A C C G C A G

pBR322 DNA HSV-1 DNA

Under the appropriate conditions, these sequences could form a stem and loop structure that might prevent or decrease transcription and/or translation of the mRNA. It is suggested that this palindromic sequence may influence the level of expression of the TK gene also in mammalian cells.

ORGANIZATION OF THE PROMOTER SEQUENCE OF THE HSV-1 GENE

Comparison of HSV-1 TK promoter to SV40 early gene promoter. Studies in recent years on the organization of the promoter sequences of the SV40 early gene coding for T and t antigens (34, 35) have helped to elucidate how transcription of the early gene takes place. To obtain further insight into the promoter of the HSV-1 TK gene, its sequences were compared with those of the promoter of the SV40 early gene, using the TATTAA sequence for alignment (Fig. 7).

The following conclusions can be drawn from the comparison: (a) the mRNA start in both genes is almost in the same position (difference of one nucleotide); (b) near the mRNA start of the SV40 gene, there is a palindrome, while in the TK sequence such a palindrome is not evident; (c) the stretch of T sequences in SV40 upstream from the TATTA box is not present in the TK sequence; (d) in the SV40 DNA sequence, there are three groups of 21 bp repeats, each containing two CCGCCC sequences, while in the TK there are only two similar sequences: one sequence (GGGGCGGC) is near the right hand side

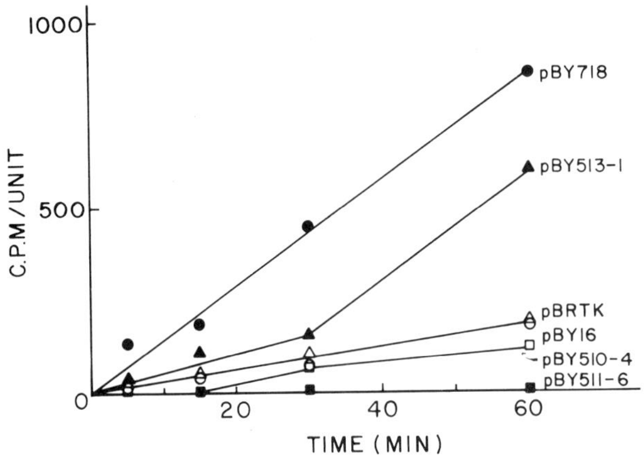

Fig. 6. Expression of HSV-1 TK gene in *E. coli* K-12. Recombinant plasmids carrying the HSV-1 TK gene were transfected to *E. coli* K-12 tdk mutant (Ky895). The bacterial cultures were labeled with ^{3}H-thymidine, and at different time intervals the density of the cultures was determined and a cell sample was removed, precipitated with trichloracetic acid, and the insoluble radioactivity in the DNA was determined. The total radioactivity in each sample was normalized by dividing by the Klett units (CPM/unit).

of the 21 bp repeat I of SV40, and the second sequence (CCCCGCCC) is located near the left hand side of the 21 bp repeat III; (e) following the 21 bp repeats, there are two sequences each of 72 bp in the SV40 promoter (the 72 bp repeat II is not shown). The promoter region of the TK gene is very similar when aligned with the 72 bp repeat I of SV40.

The 21 bp repeats in the SV40 early gene promoter with the two repeating sequences CCGCCC serve as binding sites for the virus-coded T-antigen molecules. The regions of the 21 bp sequence might be compatible with the structure of DNA to which protein molecules can bind. The sequence GGGGCGGC in the TK promoter may be responsible for the attachment of the alpha immediate-early protein. Indeed, a change in this sequence prevented the stimulation of TK expression by the viral alpha protein (31).

Comparison between SV40 early gene promoter and promoters of the HSV-1 alpha, beta and gamma genes. To further analyze the promoter sequences in the different HSV-1 genes, the same method that was used to compare the HSV-1 TK

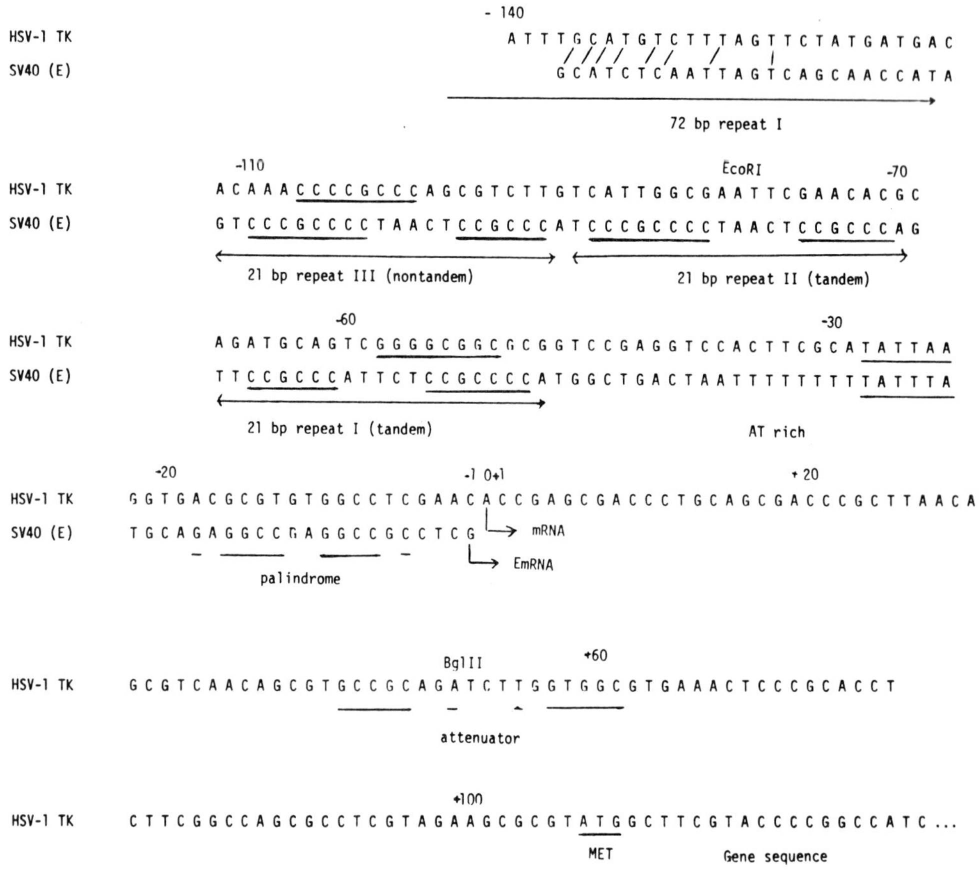

Fig. 7. Comparison between the nucleotide sequence in the promoter region of the SV40 early gene and the nucleotide sequence of the HSV-1 TK promoter. The alignment is from the TATTAA boxes of the two promoters.

promoter and the SV40 early gene promoter (Fig. 8) was applied. The promoters of the HSV-1 alpha gene 4 (30), as well as the HSV-1 late gamma genes, glycoprotein C (gC) (33) and glycoprotein D (gD) (36) were compared with the SV40 early (E) gene promoter (35). The HSV-2 TK gene promoter (37) was also included, as well as the Epstein-Barr virus (EBV) L_1 gene promoter (38). The HSV-1 gene TK sequence was taken from McKnight (20) and Wagner et al. (21). Fig. 8 shows that the HSV-1 gene 4 (an alpha gene) has more GC-rich sequences when compared with the 21 bp repeats I, II and III of SV40. The TK genes (beta genes) have fewer GC-rich sequences than are present in each

Fig. 8. Comparison between the promoter sequence of the SV40 early gene and HSV-1 alpha, beta and gamma gene promoter sequences. The promoter of the EBV L_1 gene was also included for comparison.

of the 21 bp repeats of SV40. The glycoprotein genes gD and gC (alpha genes) have lost the GC-rich sequences. In addition, the alpha gene 4 has GC-rich sequences on either side of the 0 nucleotide position. An attenuator sequence found in the TK genes upstream to the ATG codon (between nucleotides 46 and 60) can be seen in the gamma genes gC (between nucleotides 45 and 73) and gD (between nucleotides 51 and 60), but not in the alpha gene 4.

Functional domains in the promoter of the HSV-1 TK gene. Fig. 9 illustrates a proposal for the organization of the promoter of the TK gene based on the comparison of similar sequences in the promoter of the SV40 early gene. Between the mRNA start (cap site: position nucleotide 0) and the ATG codon of the TK gene (nucleotide +107) there is an attenuator sequence which may reduce or enhance expression of the TK gene, depending on the homology in the palindrome. The 28 nucleotide sequences upstream of the mRNA initiation codon are involved in accurate transcription of the mRNA (Fig. 9). Zipser et al. (31) reported that nine bp upstream of the cap site are involved in constitutive expression of the TK gene, since this is prevented by a mutation in these sequences. However, induction of the gene is not affected, although changes in the 50 bp upstream from the cap site do affect the expression of this gene. Around position 51 (Fig. 9) the GGGGCGGCGCGG sequence resembling the SV40 early gene promoter, to which the T antigen binds, is located. It is suggested that the immediate early alpha protein binds to this sequence. A similar (opposite) sequence is found around position -100 with a 63 bp sequence similar to that of SV40. The RNA polymerase II binding site may, therefore, be situated between -112 bp to -184 bp (and possibly -256 bp), if the comparison with the SV40 early gene promoter is accepted.

ARE NUCLEOTIDE SEQUENCES FLANKING THE HSV-1 TK GENE INVOLVED IN RECOMBINATION WITH CELL DNA?

Numerous studies (8, 10, 18, 19, 21, 23, 28) have demonstrated that the HSV-1 TK sequences efficiently integrate into cell DNA leading to the biochemical transformation of TK^- cells. Integration of the viral DNA into the cellular DNA is most likely carried out by the cellular recombination enzymatic system, provided that the TK gene is flanked by sequences that can efficiently recombine with similar cellular recombination sequences. Recently, Kataoka et al. (39) reported on *E. coli* extract-catalyzed recombination in switch regions of mouse immunologulin genes. The nucleotide sequences surrounding the recombination sites were determined and five different sequences involved in recombination were found. It is of interest that a recombination

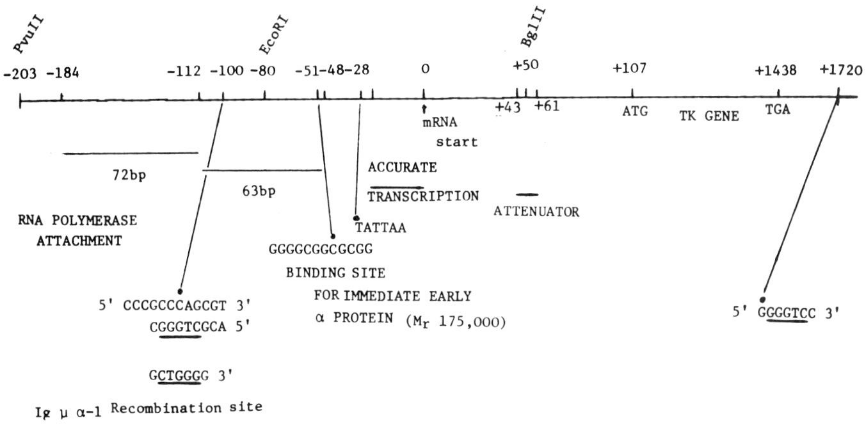

Fig. 9. A schematic organization of the HSV-1 TK gene promoter with its functional domains.

sequence similar to CTGGG of the Igμa-1 gene is present at position -100 bp and +1720 bp of the TK gene in the form of GGGTC (Fig. 9). Indeed, cleavage of the HSV-1 TK promoter with EcoRI removed one of the two recombination-like sequences from the DNA, resulting in a marked reduction in biochemical transformation of L(TK$^-$) cells (see Table 2).

PATHOGENICITY OF HSV-1 IN MICE DEPENDS ON LEVEL OF TK GENE EXPRESSION

A study on the organization and function of HSV-1 TK gene is incomplete, without an analysis of the role of the TK gene in neurovirulence of HSV-1 for mice. Our study on TK$^+$, TK¼ and TK$^-$ mutants of HSV-1 in mice (25). showed that in 4-week-old mice the TK$^+$ virus was highly pathogenic when inoculated into the eyes, while TK¼ was less pathogenic and became latent in the trigeminal ganglia of infected mice. The TK$^-$ was nonpathogenic and did not cause virus latency. It was of interest to study the neurovirulence of these three virus strains in mice of different ages, in which the central nervous system (CNS) TK gene still functions until age 10-15 days before it is regulated. The pathogenicity of the three virus strains in mice of different ages is presented in Fig. 10. The TK$^+$ and TK¼ virus strains were pathogenic in mice of all ages. The TK$^-$ mutant was highly pathogenic for mice aged up to seven days, but mice above ten days of age were resistant to this virus.

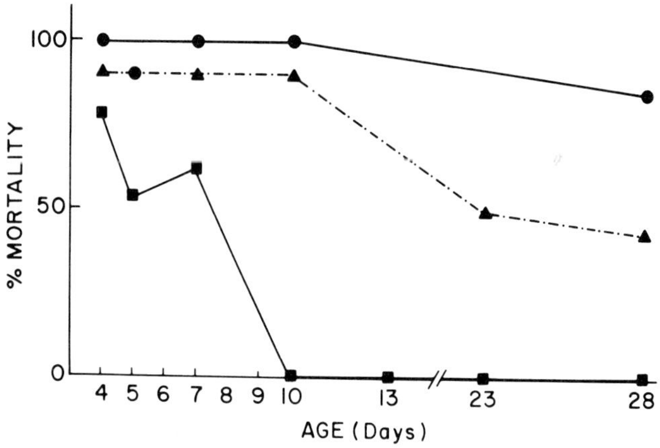

Fig. 10. Neurovirulence of HSV-1 strains LP TK^+ (●), SP TK¼ (▲) and LP TK^- (■) in mice of different ages. The virus at a dose of 10^7 pfu/ml was inoculated onto scarified corneas and the mortality of the mice was recorded over a period of three weeks.

It is known that infant mice are highly susceptible to infection with HSV, but it must be remembered that TK^- HSV-1 strains cannot replicate in neurons in the absence of TK activity. For this reason, we claim that in growing young mice TK activity is provided by the cellular enzyme, and this allows the TK^- herpes simplex virus, that is not pathogenic in adult mice, to become pathogenic in infant mice. The high neurovirulence of the HSV-1 TK^- mutant for young mice which have an immature nervous system precludes its use as a vaccine virus and indicates the potential danger of HSV-1 TK^- mutants for children.

REFERENCES

1. Aurelian, L., Manak, M.M., and Ts'o, P.O.P.: In: *Biological Carcinogenesis*, M.A. Rich and P. Furmanski (eds.), Manuel Dekker, Inc., New York and Basel, 1982, p. 229.

2. Jariwalla, R.J., Aurelian, L., and Ts'o, P.O.B.: *Proc. Natl. Acad. Sci. U.S.A.* 77:2279, 1980.

3. Huszar, D., and Bacchetti, S.: *Nature* 302:76, 1983.

4. Maitland, N.J., Kinross, J.H., Busuttil, A., Ludgate, S.M., Smart, G.E. and Jones, K.W.: *J. Gen. Virol.* 55:123, 1981.

5. Duff, R., and Rapp, F.: *Nature (London)* 233:48, 1971.

6. Leiden, J.M., and Frenkel, N.: In: *Herpesvirus DNA*, Y. Becker (ed.), *Developments in Molecular Virology, Vol. 1*, Martinus Nijhoff Publishers, The Hague, Boston, London, 1981, p. 197.

7. McKinlay, M.A., Wilson, D.E., Harrison, B., and Povey, S.: *J. Natl. Cancer Inst.,* 64:241, 1980.

8. Kit, S., Teitz, Y., Hazen, M., and Qavi, H.: *Int. J. Cancer* 23:846, 1979.

9. Smiley, J.R., Steege, D.A., Juricek, D.K., Summers, W.P., and Ruddle, F.H.: *Cell* 15:455, 1978.

10. Pellicer, A., Wigler, M., Axel, R., and Silverstein, S.: *Cell* 14:133,1978.

11. Reyes, G.R., LaFemina, R., Hayward, S.D., and Hayward, G.S.: *Cold Spring Harbor Symp. Quant. Biol.* 44:629, 1979.

12. Hampar, B., Derge, J.G., Boyd, A.L., Tainsky, M.A., and Showalter, S.D.: *Proc. Natl. Acad. Sci. U.S.A.* 78:2616, 1981.

13. Reyes, G.R., McLane, M.W., and Hayward, G.S.: *J. Gen. Virol.* 60:209, 1982.

14. Lavi, S.: *Proc. Natl. Acad. Sci. U.S.A.* 78:6144, 1981.

15. Munyon, W., Kraiselbund, E., Davis, D., and Mann, J.: *J. Virol.* 7:813,1971.

16. Bacchetti, S., and Graham, F.L.: *Proc. Natl. Acad. Sci. U.S.A.* 74:1590, 1977.

17. Maitland, N.J., and McDougall, J.K.: *Cell* 11:233, 1977.

18. Wigler, M., Silverstein, S., Lee, L.S., Pellicer, A., Cheng, Y.C., and Axel, R.: *Cell* 11:223, 1977.

19. Hayward, G.S., Reyes, G.R., Gavis, E.R., and McKnight, S.L.: In: *Herpesvirus DNA,* Y. Becker (ed.), *Developments in Molecular Virology, Vol. 1,* Martinus Nijhoff Publishers, The Hague, Boston, London, 1981, p. 271.

20. McKnight, S.L.: *Nucl. Acids Res.* 8:5949, 1980.

21. Wagner, M.J., Sharp, J.A., and Summers, W.C.: *Proc. Natl. Acad. Sci. U.S.A.,* 78:1441, 1981.

22. Garapin, A.C., Colbere-Garapin, F., Cohen-Solal, M., Horodniceanu, F., and Kourilsky, P.: *Proc. Natl. Acad. Sci. U.S.A.* 78:815, 1981.

23. Kit, S., Otsuka, H., Qavi, H., Trkula, D., Dubbs, D.R., and Hazen, M.: *Nucl. Acids Res.* 8:5233, 1980.

24. Becker, Y., Shtram, Y., Honigman, A., Laban, A., and Cohen, A.: *Gene* 21:51, 1983.

25. Gordon, Y., Gilden, D.H., Shtram, Y., Asher, Y., Tabor, E., Wellish, M., Devlin, M., Snipper, D., Hadar, J., and Becker, Y.: *Arch. Virol.* 76:39, 1983.

26. Mellon, P., Parker, V., Gluzman, Y., and Maniatis, T.: *Cell* 27:279, 1981.

27. Becker, Y., Shtram, Y., Snipper, D., Asher, Y., Tabor, E., Gordon, Y., Gilden, D., Wellish, M., Hadar, J., Becker, O., Cohen, A., Laban, A., and Honigman, A.: In: *Herpesvirus: Clinical, Pharmacological and Basic Aspects,* Y-C. Cheng and W.H. Prusoff (eds.), Int. Congress Series 571, Excerpta Medica, Amsterdam-Oxford-Princeton, 1982, p. 57.

28. Wigler, M., Sweet, R., Sim, G.K., Wold, B., Pellicer, A., Lacy, E., Maniatis, T., Silverstein, S., and Axel, R.: *Cell* 16:777, 1979.

29. Colbere-Garapin, F., Chousterman, S., Horodniceanu, F., Kourilsky, P., and Garapin, A-C.: *Proc. Natl. Acad. Sci. U.S.A.* 76:3755, 1979.

30. Mackem, S., and Roizman, B.: *Proc. Natl. Acad. Sci. U.S.A.* 77:7122, 1980.

31. Zipser, D., Lipsich, L., and Kwoh, J.: *Proc. Natl. Acad. Sci. U. S. A.* 78:6276, 1981.

32. Rosenberg, M., and Court, D.: *Ann. Rev. Genet.* 13:319, 1979.

33. Frink, R.J., Eisenberg, R., Cohen, G., and Wagner, E.K.: *J. Virol.* 45:634, 1983.

34. Bergsma, D.J., Olive, D.M., Hartzell, S.W., and Subramanian, K.N.: *Proc. Natl. Acad. Sci. U.S.A.* 79:381, 1982.

35. Byrne, B.J., Davis, M.S., Yamaguchi, J., Bergsma, D.J., and Subramanian, K.N.: *Proc. Natl. Acad. Sci. U.S.A.* 80:721, 1983.

36. Watson, R.J., Weis, J.H., Salstrom, J.S., and Enquist, L.W.: *Science* 218:381, 1982.

37. Swain, M.A., and Galloway, D.A.: *J. Virol.* 46:1045, 1983.

38. Farrell, P.J., Deininger, P.L., Bankier, A., and Barrell, B.: *Proc. Natl. Acad. Sci. U.S.A.* 80:1565, 1983.

39. Kataoka, T., Takeda, S-I, and Honjo, T.: *Proc. Natl. Acad. Sci. U.S.A.* 80:2666, 1983.

© 1984 Elsevier Science Publishers B.V.
The Role of Viruses in Human Cancer. Volume II
G. Giraldo and E. Beth editors

MOLECULAR BIOLOGY OF THE RELATIONSHIP BETWEEN HERPES SIMPLEX VIRUS-II AND CERVICAL CANCER

JAMES K. MCDOUGALL[1,2], PATRICIA SMITH[1], HISHAM K. TAMIMI[2], ERNEST TOLENTINO[1] AND DENISE A. GALLOWAY[1,2]

[1]Fred Hutchinson Cancer Research Center, 1124 Columbia Street, Seattle, WA 98104 and [2]University of Washington, Seattle, WA 98105 (U.S.A.)

The increasing prevalence of herpes genitalis as a sexually transmitted disease (1) and an increasing incidence of pre-malignant changes in the exposed tissues of younger women (2) have stimulated attempts to establish an association between previous infection with herpes simplex virus (HSV) and the development of uterine cervical carcinoma. Many epidemiological and serological studies (3-7) have indicated a strong correlation between neoplastic changes in the cervix and earlier infection with HSV-2 and it has furthermore been shown that virus-specific proteins (8-12) and ribonucleic acid (RNA) (12-17) can be detected in the neoplastic tissues. Frenkel et.al. (18) described solution hybridization studies on one cervical tumor in which a portion of the HSV-2 genome was detected, but subsequent similar experiments on a large number of cervical tumors provided only negative results (19,20). Those experiments while capable of detecting 1 viral genome per diploid cell, would not have resolved small fragments of viral DNA although a transforming gene could be wholly represented within less than 2 kb of viral DNA (21,22).

HSV-1 and HSV-2 are able to transform rodent cells to a malignant phenotype (23,24). Several laboratories have concentrated on identifying the viral genes responsible for morphological transformation and the viral products that persist in transformed and tumor cells. Exposure to specific sequences of HSV DNA can produce changes in rodent cells that cause them to be tumorigenic. Also, certain regions of the genome are frequently expressed and translated into recognizable viral antigens in both experimentally transformed cells and in human cervical carcinoma cells. Although these viral footprints can be catalogued there is no certainty that the continued expression of any particular herpesvirus antigen is necessary to maintain the transformed state nor that sequences of HSV DNA are always retained or expressed.

We have approached the problem of attempting to establish a molecular basis for a role for HSV genes in transformation *in vitro* and neoplasia *in vivo* by a strategy employing three separate but related tactics. Firstly, we have examined the viral DNA sequences retained in cells transformed *in vitro* by HSV-2 and asked whether those sequences can transform cells independently of their role in the viral genome. Secondly, we have hybridized labelled sub-genomic viral DNA probes *in situ* to sections from normal and neoplastic cervical tissues to determine whether differences in hybridization may indicate the presence of viral RNA in the neoplastic tissues and, thirdly, we have used the cloned sub-genomic probes in filter hybridization experiments in attempts to detect viral-specific sequences in DNA extracted from cervical tumors.

HSV-2 TRANSFORMATION

Our initial experiments utilized a number of cloned cell lines derived from a cell line originally isolated by Duff and Rapp (23,24). The 333-8-9 cell line, which was derived from primary hamster cells exposed to UV-inactivated HSV-2, strain 333, exhibits unlimited *in vitro* growth, is of predominantly fibroblastic morphology and cannot be induced to release whole infectious virus. The cell line is oncogenic in new born syngeneic hamsters, producing undifferentiated fibrosarcomas and metastases.

In order to determine which viral sequences might be retained we carried out the following experiments. Herpes simplex virus type 2 (HSV-2) DNA was treated with four restriction endonucleases (*Eco*RI, *Hind*III, *Bgl*II, or *Xba*I) and eight fragments were purified and labeled with ^{32}P *in vitro*. The kinetics of renaturation of each of the fragments was measured in the presence of DNA extracted from 333-8-9 and from a series of cloned derivatives and their tumor lines. All of the lines examined contained a partial set of viral sequences present at only a few copies per cell. Passage of the cell lines in tissue culture or in animals resulted in partial loss of viral DNA. Two blocks of sequences were present in most of the lines examined; those mapping at positions 0.21-0.33 of the HSV-2 genome were detected in seven of seven cell lines tested and those at positions 0.60-0.65 were detected in six of eight. Other sequences from the L component were also present in the DNA of HSV-2 transformed hamster cells, but none of these sequences was consistently detected (25). From these results it was possible to reach the following conclusions about the 333-8-9 cell lines and these conclusions when taken together with results from Frenkel and colleagues (26,27) may provide a general picture for HSV transformants. (i) All of the cell lines examined contain HSV-2 DNA. (ii) The sequences retained are present in few copies per cell. (iii) Only a subset of the viral DNA sequences is retained. (iv) While some lines have an extensive set of sequences, others have small amounts. (v) Sequences continue to segregate upon cloning and passage. (In the original cloning of the parental line there was selection for sub-populations of cells). (vi) No particular set of viral sequences is correlated with degree of tumorigenicity. These results go part of the way towards defining a minimal set of viral sequences which may contain essential information for transformation.

For further definition of the transforming sequences the calcium phosphate transfection method (28) was used. HSV-2 DNA was digested with several restriction endonucleases, and the entire collection of fragments was used in transformation experiments. Transformants were produced with DNA digested with *Hind*III, *Bgl*II and *Xba*I. Fragments of *Bgl*II-cleaved DNA were chosen for further study because at least one produced transformants and because a reasonable number of fragments were generated. *Bgl*II recognizes 12 sites in HSV-2 DNA (29), which produces 18 fragments because of the heterogeneity resulting from isomerization of the viral genome (30). The only fragment which altered cells in a way that allowed them to form colonies in methylcellulose or to form dense foci was the *Bgl*II N fragment. These experiments confirm the observations of Reyes et.al. (31).

The *Bgl*II N fragment was cloned into pBR322 (32) and the recombinant plasmid, and subclones derived from it by cleavage with *Pst*I (Fig. 1), were used to transform NIH 3T3 cells and primary rat embryo cells. Examination of the transformed cell DNA by Southern blot hybridizations indicated that some cell lines retained plasmid DNA in less than one copy per cell and that the patterns of viral sequences changed with passage (22). Using a different assay for transformation - continuous passage of hamster embryo cells - Jarriwalla et.al. (33) have described transformation by a different fragment of HSV-2 DNA, *Bgl*II C, mapping to the left of *Bgl*II N at positions 0.43-0.58 (Fig. 2).

From all these results it is at least clear that sequences of HSV-2 DNA can initiate transformation of cells *in vitro*; may persist in the transformants but may not be an obligatory component for the maintenance of transformation (34).

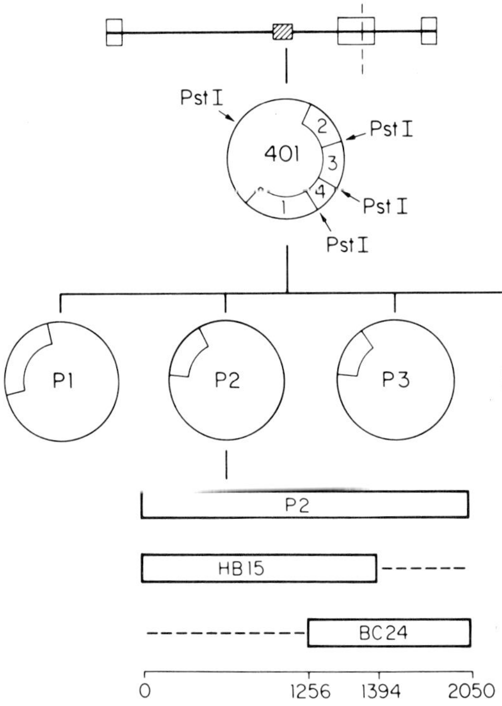

Figure 1. Construction of plasmids containing fragments of the BglII N region of HSV-2 DNA in pDG401. The PstI cleaved viral DNA was re-cloned into the PstI site of pBR322 to produce four new plasmids, P1-P4. The viral sequences in P2 were linearized and deletion mutants constructed by the method of Sakonju et.al. (1980). Two of the deletion fragments HB15 and BC24, were prepared free of plasmid sequences and used as hybridization probes.

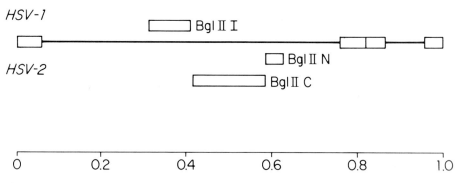

Figure 2. The location of transforming genes along the HSV genome. The blocks indicate the position of restriction enzyme fragments of HSV-1 (above) and HSV-2 (below) transforming DNA along the prototype arrangement of the HSV genome. The numbers are map units.

HERPESVIRUS-SPECIFIC RNA IN CERVICAL CARCINOMA

Recently, several laboratories have reported the detection of HSV-specific RNA in neoplastic cells in the cervix using *in situ* hybridization procedures (12-17). In general, about one-third of all the biopsies examined appear to contain HSV nucleic acid. We used a series of cloned and restriction-endonuclease derived fragments in cytological hybridizations in order to map the RNA transcripts (12). The results indicated that the transcription was limited and restricted to three blocks of sequences located at map positions 0.07-0.40, 0.58-0.63 and 0.82-0.85 (Figs. 3 and 4). Again, less than 100% of the cases were positive suggesting that although HSV-2 might be an etiological factor, other etiologies are probable.

Alternatively, as in the case of *in vitro* transformants described above, viral sequences may be lost after initiation of the neoplastic process. No particular sequence was invariably expressed, making it impossible to define a tumor-specific viral gene.

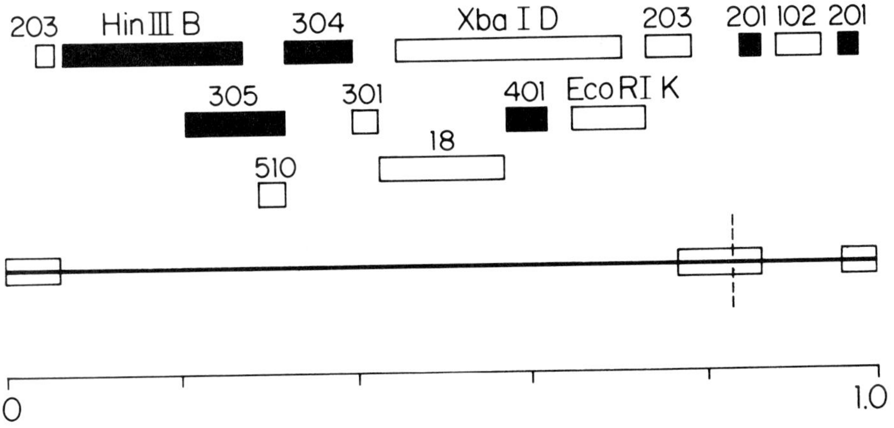

Figure 3. Distribution of HSV-2 DNA probes used for hybridization. The solid boxes indicate positive results for HSV RNA.

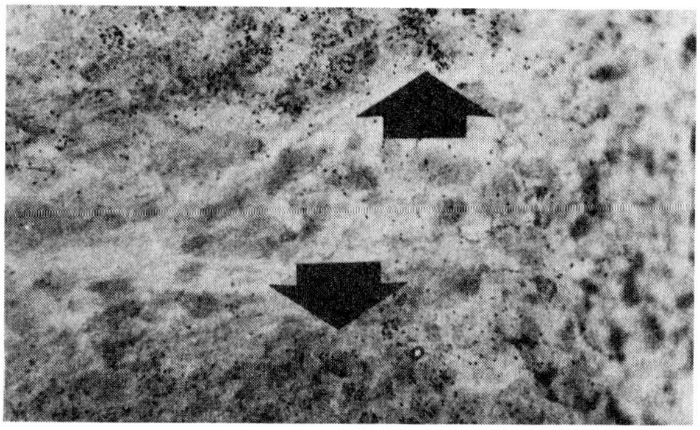

Figure 4. *In situ* hybridization of ^{3}H-labeled, nick translated pDG304 DNA to a frozen section of cervical intraepithelial neoplasia. Silver grains mainly over neoplastic cells.

HERPESVIRUS-SPECIFIC DNA IN CERVICAL CARCINOMA

We have recently examined a series of invasive cervical tumors for evidence of viral DNA by hybridization of cloned viral fragments to tumor DNA by the Southern (40) transfer procedure.

DNA from a series of surgically removed uterine tumors was digested with restriction endonuclease *Eco*Rl, electrophoresed through agarose gels, transferred to nitrocellulose filters, and hybridized to ^{32}P-labelled plasmid DNA containing cloned HSV-2 DNA sequences or to fragments of viral DNA released from the plasmid. The HSV-2 DNA cloned fragments used as probes represented 90% of the genome (39). The results of blot hybridization using purified *Bgl*II J and *Bgl*II N fragments of HSV-2 DNA to probe five of the tumor samples are shown in figure 5. Tumors D, U and W contain sequences which hybridize to HSV-2 DNA but only tumor W hybridizes with both the *Bgl*II J and N fragments. Tumors D and U hybridize with *Bgl*II N or J respectively, and tumors T and H, together with nine other cervical tumors examined but not shown here, are negative for viral sequences (Table 1). Only one of the tumors (C) hybridized to cloned DNA representing all regions of the viral genome, probably indicating virus replication in this tissue. The tumors were also positive when hybridized to HSV-2 *Bgl*II C sequences, which map between 0.43 and 0.58 on the viral genome. It is known that there is homology between the *Bgl*II C fragment and human DNA (41). The probes giving negative results were *Hind*III B, *Bgl*II G, *Bgl*II O, *Eco*RI K, *Hind*III L. *Bam*HI F and *Bam*HI g'h' were also shown to have homology to cell DNA (35) and positive results from these probes were therefore discounted.

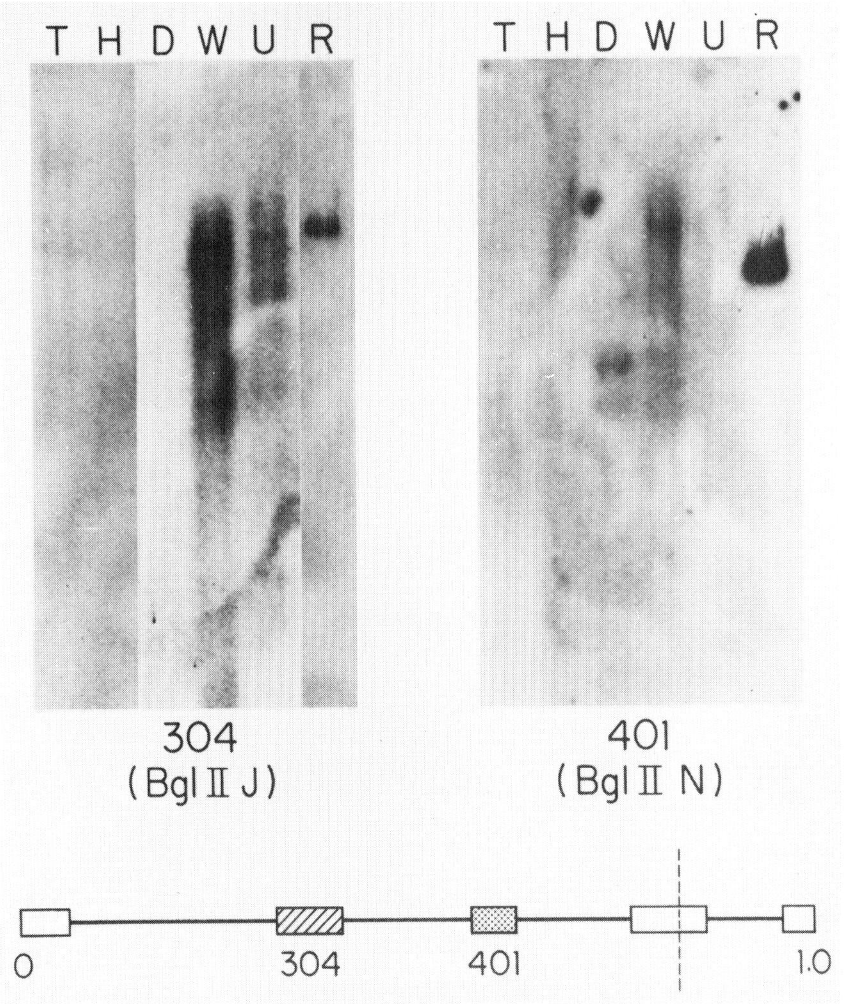

Figure 5. Detection of fragments of tumor DNA that contain sequences hybridizing to HSV-2 DNA cloned fragments after hydrolysis of tumor DNA with endonuclease EcoRI. High molecular weight DNA was extracted from the tumors and 10ug of DNA was digested with EcoRI, fractionated by electrophoresis through 0.7% agarose in a slab gel, hybridized to ^{32}P-labelled viral DNA and exposed for autoradiography. The HSV-2 DNA fragments had been released from the plasmids pDG304 and pDG401 sequences by cleavage with XhoI and BstEII respectively, followed by agarose gel purification. Lanes T,H,D,W, and U each contain 10ug DNA and lane R contains a reconstruction of either pDG304 or pDG401 DNA representing 5 copies of plasmid DNA per diploid genome of human placental DNA. The positions of the viral sequences in pDG304 and pDG401 are shown on the representation of the HSV-2 genome at the bottom of the figure.

TABLE I - HYBRIDIZATION OF CERVICAL CARCINOMA DNA WITH CLONED FRAGMENTS OF HSV-2 DNA

Tumor	Diagnosis	BglII J	BglII N	P2	BC24	HSV-2[1]
U	Adenocarcinoma	+	-	+	NT(3)	-
W	Adenocarcinoma	+	+	+	+	-
H2	Adenosquamous ca.	-	+	+	+	-
D	Adenosquamous ca.	-	+	NT	NT	-
C	Squamous cell ca.	+	+	+	+	+
X[2]	Squamous cell ca.	-	-	-	-	-

[1] Probes representing other regions of the HSV-2 genome as described in text.

[2] Represents eleven other tumor samples

[3] Not tested

For further analysis, the "W" tumor DNA was cleaved with PstI and hybridized with the four PstI fragments of pDG401, constructed as described in Figure 1. Only the left most fragment-P2 hybridized. Two of an extensive collection of deletion mutants of the P2 plasmid (D.A. Galloway et.al. in preparation) were prepared free of pBR322 sequences and hybridized with the PstI cleaved tumor DNA. Since hybridization was achieved with the right-hand BC24 fragment but not with the overlapping HB15 sequences, this result localized the virus-specific DNA to a 656 bp sequence in BC 24 (Fig. 6). Transfection of NIH 3T3 cells and primary rat embryo cells with BC 24 DNA has produced transformed foci (D.A. Galloway et.al. in preparation).

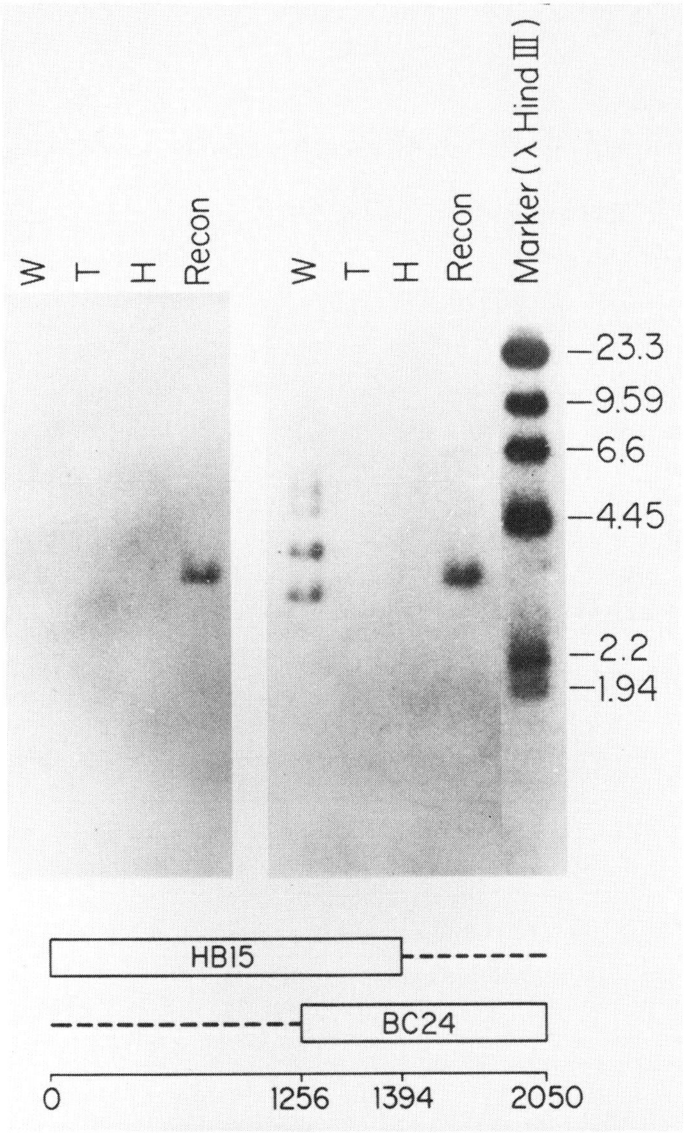

Figure 6. Detection of sequences homologous to HSV-2 DNA deletion fragment BC24 in tumor W. Ten ug of DNA from tumors W,T, and H were digested with endonuclease PstI, fractionated by electrophoresis through 0.7% agarose, hybridized to ^{32}P-labelled viral DNA and exposed for autoradiography. Lane 4 contains a reconstruction of P2 DNA representing 5 copies of plasmid DNA per diploid genome of human placental DNA.

DISCUSSION

These results provide another step towards defining a role for HSV in cervical carcinoma. The *in situ* hybridization studies have described limited transcription from various regions of the HSV-2 genome as occuring in cervical carcinoma cells together with the detection of a number of virus-specific antigens, yet, as described above no one transcript or viral protein has been invariably associated with these tumors. The result presented here on tumor "W" defines a small sequence of tumor DNA which hybridized specifically to HSV-2 DNA sequences. Although the most conclusive result would be one in which an internal fragment from the putative viral sequences in the tumor DNA co-migrated with an authentic viral fragment in a Southern blot experiment; we have been unable to achieve such a result. The small size of the detected sequence, 656 bp, contributes to the difficulty of resolution of this problem.

Cloning of these viral DNA sequences together with presumed flanking cellular sequences, will provide more information on the state of the viral DNA. The problems encountered in attempts to define a single gene in HSV responsible for transformation and which encodes a product universally found in transformed and tumor cells, as is the case for other DNA tumor viruses (36), has engendered speculation about synergistic or mutational events involving HSV infection in the genesis of uterine cancer (34,37). The retention of sequences represented in BC24 may provide support for such hypotheses since we know that this DNA (i) can transform cells *in vitro*: (ii) does not code for a viral "transforming" protein (35) and (iii) contains an IS-like sequence arrangement (D.A. Galloway et.al. in preparation), similar to the structure described by Rechavi et.al. (38) which has been shown to have close homology with the long terminal repeat (LTR) of an intracisternal A-particle (39). Temporary or permanent insertion of this structure into the genome of a cell surviving but exposed to HSV-2 during genital infection could provide initiation and/or promotion functions for neoplastic growth. Our studies on rodent cells transformed with the HSV-2 *Bgl*II N fragment (22), or subsets of those viral sequences, show that permanent retention of the transforming DNA is not a *sine qua non* of HSV transformation but that in some, as yet undefined, circumstances viral sequences may persist.

The link between HSV and cervical cancer remains tantalizing but lacking in definition. Viral nucleic acids, as described above, are found in about one-third of the tissues examined and there is also some evidence that sera from cervical carcinoma patients contain antibodies to HSV proteins and that viral antigens can be detected in cervical squamous cell carcinomas. One of the most extensively described of these is AG4 (or ICP10), reportedly a HSV-2 immediate early protein with a molecular weight of between 140K and 160K, which is expressed on 80% of tumor cells and to which patients have complement-fixing antibody (40). This protein is also reported to be expressed in cells transformed by the HSV-2 BglII C fragment (33). It has been shown however that ICP10 (also known as ICP6) is also partially encoded by sequences within the adjacent BglII N fragment so that the 5' end of the transcript is within BglII C and the 3' end within BglII N. In fact the last 163 amino acids of the protein are encoded within BglII N (D.A. Galloway and M.A. Swain, in press). Recent studies have identified this protein as the virus-specified ribonucleotide reductase (41).

Other antigens which have been demonstrated frequently on cervical tumor cells (10,11,12) react with antisera directed against the major HSV-2 DNA-binding proteins, ICSP 11/12 and ICSP 34/35. The former has also been demonstrated in cells transformed by UV-inactivated HSV-2 (42,43), and is encoded within sequences which have transforming activity from the HSV-1 genome.

All of this leaves us unable to reach any firm conclusion about a possible role for HSV in human tumors. Does the virus act as a mutagen? Recent results from Zur Hausen's laboratory have demonstrated mutagenic activity (44). Definition of the genes involved in mutagenesis will also determine whether they are the same as those involved in transformation.

Is transformation - or mutagenesis - the result of a "hit-and-run" mechanism (34,45)? The failure to detect a consistent set of viral nucleic acid sequences or proteins in tumors or transformed cells at least leaves the question open. Either mutagenesis or a "hit-and-run" mechanism could result in the activation of cellular oncogenes, but we do not so far have any firm evidence in support of that hypothesis.

Finally, what effect do the recent findings of papillomavirus (HPV) DNA sequences in cervical tumors (46,47) have on the HSV results. Could there be an interaction between these viruses, as has been proposed (34,37,48): We have recently detected both HSV and HPV DNA sequences in the same tumor, making it even more tempting to search for an initiation/promotion relationship involving these common human viruses.

ACKNOWLEDGEMENTS

We thank Ann Kritzberger for preparation of the manuscript. These studies are supported by NCI grants CA26001 and CA 29350.

REFERENCES

1. Gardner, H.L.: Am. J. Obstet. Gyn., 135:553, 1979.
2. Boyes. D.A., Worth, A.J. and Anderson, G.H.: Gyn. Oncology, 12: 143, 1981.
3. Nahmias, A.J., Josey, W., Naib, Z.M., Luce, C.F. and Guest, B.: Am. J. Epidemiol., 91:547, 1970.
4. Nahmias, A.J., Naib, Z.M. and Josey, W.: Cancer Res., 34:111, 1974.
5. Rawls, W.E., Tompkins, W.A.F., Figuero, M.E. and Melnick, J.L.: Science, 161:1255, 1968.
6. Rawls, W.E., Tompkins, W.A.F. and Melnick, J.L.. Am. J. Epidemiol., 89:147, 1969.
7. Thomas, D.B. and Rawls, W.E.: Cancer, 42:2716, 1978.
8. Aurelian, L., Strnad, B. and Smith, M.F.: Cancer, 39:1834, 1977.
9. Hollinshead, A. and Tarro, G.: Science, 179:698, 1973.
10. Dreesman, G.R., Burek, J., Adam, E., Kaufman, R.H., Melnick, J.L., Powell, K.L. and Purifoy, D.J.M.: Nature, 283:591, 1980.
11. McDougall, J.K., Galloway, D.A., Purifoy, D.J.M., Powell, K.L., Richart, R.M. and Fenoglio, C.M. in: Esssex, M., Todaro, G., Zur Hausen, H. (Eds.) Viruses in Naturally Occuring Cancers, Cold Spring Harbor Conf. Cell Proliferation, 7:101, 1980.
12. McDougall, J.K., Crum, C.P., Fenoglio, C.M., Goldstein, L.C. and Galloway, D.A.: Proc. Natl. Acad. Sci. USA, 79:3853, 1982.
13. McDougall, J.K., Galloway, D.A. and Fenoglio, C.M.: Int. J. Cancer, 25:1, 1980.
14. ibid, in: Chandra, P. (Ed.) Antiviral Mechanisms in the Control of Neoplasia, Plenum Press, New York, pp. 233-240, 1979.
15. Jones, K.W., Fenoglio, C.M., Schevchuk-Chaban, M., Maitland, N.J. and McDougall, J.K.: in: de The, G., Henle, W. and Rapp, F. (Eds.) Oncogenesis and Herpesviruses III, IARC, Lyon, pp. 917-925, 1978.
16. Eglin, RP., Sharp, F., MacLean, A.B., MacNab, J.C.M., Clements, J.B. and Wilkie, N.M.: Cancer Res., 41:3597, 1981.
17. Maitland, N.J., Kinross, J.H., Busuttil, A., Ludgate, S.M., Smart, G.E. and Jones, K.W.: J. Gen. Virol. 55:123, 1980.
18. Frenkel, N., Roizman, B., Cassai, E. and Nahmias, A.: Proc. Natl. Acad. Sci. USA, 69:3784, 1972.
19. Zur Hausen, H., Schulte-Holthausen, H., Wolf, H., Dorries, K. and Egger H.: Int. J. Cancer 13:657, 1974.
20. Pagano, J.S.: J. Infect. Dis., 132:209, 1975.
21. Copple, C.D. and McDougall, J.K.: Int. J. Cancer, 17:501, 1976.

22. Galloway, D.A. and McDougall, J.K.: J. Virol., 38:749, 1981.

23. Duff, R. and Rapp, F.: Nature (New Biol.), 233:48, 1971.

24. Duff, R. and Rapp, F.: J. Virol., 12:209, 1973.

25. Galloway, D.A., Copple, C.D. and McDougall, J.K.: Proc. Natl. Acad. Sci. USA, 77:880, 1980.

26. Frenkel, N. and Leiden, J. in: Hofschneider, P.H. and Starlinger, P. (Eds.) Integration and Excision of DNA molecules, Springer-Verlag, Heidelberg, pp. 71-77, 1978.

27. Frenkel, N., Locker, H., Cox, B., Roizman, B. and Rapp, F.: J. Virol., 18:885, 1976.

28. Graham, F.L. and van der Eb, A.J.: Virology, 52:456, 1973.

29. Roizman, B.: Cell, 16:2721, 1979.

30. Hayward, G.S., Jacob, R.J., Wadsworth, S.C. and Roizman, B.: Proc. Natl. Acad. Sci. USA., 72:4243, 1975.

31. Reyes, G.R., Lafemina, R., Hayward, S.D. and Hayward, G.S.: Cold Spring Harbor Symp. Quant. Biol., 44:629, 1979.

32. Galloway, D.A. and Swain, M.: Gene, 11:253, 1980.

33. Jariwalla, R.J., Aurelian, L. and T'so, P.O.P.: Proc. Natl. Acad. Sci. USA., 77:2279, 1980.

34. Galloway, D.A. and McDougall, J.K.: Nature, 302:21, 1983.

35. Peden, K., Mounts, P. and Hayward, G.S.: Cell, 31:71, 1982.

36. Tooze, J.: The molecular biology of tumor viruses, Part 2, Cold Spring Harbor, N.Y., 1980.

37. Zur Hausen, J.: Lancet, ii, 1370, 1982.

38. Rechavi, G., Givol, D. and Cannaani, E.: Nature, 300:607, 1982.

39. Kuff, E.L., Feenstra, A., Lueders, K., Rechavi, G., Givol, D. and Canaani, E.: Nature, 302, 547, 1983.

40. Strnad, B.C. and Aurelian, L.: Virology, 73:244, 1976.

41. Huszar, D. and Bachetti, S.: Nature, 302:76, 1983.

42. Lewis, J.G., Kucera, L.S., Eberle, R. and Courtney, R.J.: J. Virol., 42:275, 1982.

43. Flannery, V.L., Courtney, R.J. and Schaffer, P.A.: J. Virol., 21:284, 1977.

44. Schlehofer, J.R. and Zur Hausen, H.: Virology, 122:471, 1982.

45. Skinner, G.: Br. J. exp. Path., 57:361, 1976.

46. Green, M., Brackmann, K.H., Sanders, P.R., Loewenstein, P.M., Freel, J.H., Eisinger, M. and Switlyk, S.A.: Proc. Natl. Acad. Sci. USA, 79:4437, 1982.

47. Durst, M., Gissman, L., Ikenberg, H. and Zur Hausen, H.: Proc. Natl. Acad. Sci. USA, 80:3812, 1983.

48. Fenoglio, C.M., Galloway, D.A., Crum, C.P., Levine, R.U., Richart R.M. and McDougall, J.K. in: Fenoglio, C.M. and Wolff, M. (Eds.) Progress in Surgical Pathology IV, Masson, New York. pp. 45-82, 1982.

The Role of Viruses In Human Cancer. Volume II
G. Giraldo and E. Beth editors

HSV-2 AND CERVICAL CANCER: THE TRANSFORMATION LESSON

LAURE AURELIAN[1,2], MARK M. MANAK[2], AND P.O.P. TS'O[2]
Department of Pharmacology and Experimental Therapeutics, University of Maryland, School of Medicine[1] and Division of Biophysics, The Johns Hopkins Medical Institutions[2], Baltimore, Maryland (U.S.A.)

INTRODUCTION

The suggestion that herpes simplex virus type 2 (HSV-2), plays an etiological role in cervical cancer was originally based on the finding that women with cytologically (37), or serologically (7)proven HSV-2 infections are at a 2-20 fold higher risk of developing cervical dysplasia and/or cancer than women without such infections. It was supported by the detection of viral antigens (3,6), followed by viral RNA (33) and DNA (15), in some tumors. More recently, Wentz *et al*. (48) showed that repeated exposure of the mouse cervix to inactivated HSV-2 causes cervical dysplasia and invasive cancer, and that this response can be prevented by previous immunization with the virus.

Several laboratories have focused on the mechanism of HSV-2 induced malignant transformation, as a means of providing a better understanding of the role played by the virus in cervical cancer. In this report we decribe the results of these studies, and discuss their implications.

TRANSFORMATION BY INACTIVATED HSV 2

Inactivated virus transforms rodent cells from a normal to a neoplastic phenotype. Despite rare claims to the contrary (20,22), there is general agreement that the transformed cells maintain viral DNA (16,17,31,39). Furthermore, using a large number of independently established HSV-2 transformed hamster lines and their clonal derivatives, we have recently shown that maintenance of viral DNA sequences correlates with anchorage independent growth and tumorigenic potential. On the other hand, cellular immortalization and reduced serum requirement for growth are evidenced by HSV-DNA negative lines (32).

Three different approaches were taken by our laboratory in order to determine whether a viral protein is responsible for transformation. The first sought to correlate between the expression of a specific viral protein and the acquisition of a transformed phenotype. We found that all 15 HSV-DNA positive tumorigenic lines, independently established in our laboratory, expressed a 144K protein designated ICP 10 (46), as determined by immunofluorescence (FA), and immunoprecipitation with monospecific antiserum (32). Another viral protein, designated ICP 12/13 (43) was also detected, but only in six of the 11 ICP 10 positive lines studied in this series. Quantitative analysis revealed a strong correlation between the cloning efficiency (CE) of the transformed lines in 0.3% agarose, and the levels of ICP 10 (but not ICP 12/13) that they express (5). Three HSV DNA negative nontumorigenic lines that failed to clone in 0.3% agarose were negative for both ICP 10 and ICP 12/13 (32). Although peptide mapping remains to be performed in order to establish that the proteins in the tranformed and productively infected cells are structurally identical, the data demonstrate that a 144K protein that shares antigenic determinants with ICP 10, is expressed in the HSV-DNA positive lines in relationship to anchorage independent growth and tumorigenicity.

The second approach taken in order to identify a specific viral protein responsible for transformation, was based on previous findings indicating that only 4% of independently isolated, morphologically altered foci grow into immortal lines (29,31). Briefly, 11 foci were independently replica-plated onto coverslip slides using wet discs of Whatman No. 50 filter paper (13). The original foci were isolated and independently subcultured, and the replicas were FA stained with monospecific antisera to ICP 10 and ICP 12/13. Cells from 3 of the 11 foci stained with antiserum to ICP 10, and two of these three foci (RC92F3 and RA81F1) survived subculture and grew into anchorage independent, tumorigenic lines (Table 1). All the other foci, including 2 that stained with anti-ICP 12/13 serum, senesced within 2-5 further subculture passages, indicating that only those foci that express ICP 10 can grow into anchorage independent lines.

In the third series of experiments, designed to identify a viral protein involved in transformation, Syrian hamster embryo (SHE) cells were transformed with HSV-2 temperature sensitive (ts) mutants tsA8 and tsH9, at the permissive temperature (34°C). At various post-infection passages (pip), they were assayed for CE in 0.3% agarose at both the permissive (34°C), and nonpermissive (39°C) temperatures (Table 2). At pip 14-31, the line established with tsA8 (A8-21), but neither one of the two lines established with tsH9 (H9-12), or with the wild type virus (data not shown), evidenced a significantly reduced CE in 0.3% agarose at 39° as compared to 34°C. Beyond this passage level, all lines cloned equally well at 34°C and 39°C (Table 2).

Table 1

ICP 10 AND ANCHORAGE INDEPENDENT GROWTH OF MORPHOLOGICALLY ALTERED FOCI

Focus	Staining with antiserum to[a] ICP 10	ICP 12/13	Growth[b]
RC92F3	+++	++	Line
RC91F1	0	+	Senesced
RC92F2	0	++	Senesced
RC92F4	0	0	Senesced
RC21F1	0	ND	Senesced
RC22F4	0	ND	Senesced
RA1F2	0	ND	Senesced
RA1F3	0	ND	Senesced
RA1F4	+++	0	Lost
RA1F5	0	0	Senesced
RA81F1	+++	0	Line
SHE Control	0	0	Senesced

a 0 = no visible FA positive cells; + = 1-5%, ++ = 5-10% and +++ = 0-25% of the cells are FA positive.

b Lines were anchorage independent and tumorigenic for newborn hamsters (31). Senescence occurred at passage 3-6. RA1F4 survived subculture but was lost at passage 10.

Table 2

ANCHORAGE INDEPENDENT GROWTH OF SHE CELLS TRANSFORMED WITH HSV-2 tsMUTANTS

Cell line[a]	pip	CE at 34°C Exp. 1	34°C Exp. 2	39°C Exp. 1	39°C Exp. 2
A8-21	14	ND	36.0	ND	0.1
	27	14.8	14.4	0.3	0.8
	31	24.0	10.4	1.0	0.0
	41	20.0	12.0	6.0	23.0
H9-12	14	ND	11.6	ND	12.1
	24	19.0	36.0	25.0	31.2
	31	31.0	40.0	35.0	39.2
	35	23.0	12.0	26.0	32.4

a Lines were established at 34°C with tsA8 (A8-21) or tsH9 (H9-12). At increasing pip, cells (10^3) were seeded in 35mm dishes and incubated at 34°C or 39°C. Colonies of >20-30 cells were counted at 2-3 weeks. CE = the number of colonies x $100/10^3$ (31).

TsA8 and tsH9 were mapped in a region of the HSV-2 genome that lies within or near the coding sequences for the 130K major DNA binding protein (10,45), but they failed to complement each other. Therefore, our findings are amenable to the two following interpretations: (i) tsA8 has a secondary, not conditionally lethal, mutation that is involved in anchorage independence, or (ii) the 130K associated mutant phenotype in tsA8 (but not in tsH9), is related to anchorage independence. Presently available data do not differentiate between these two interpretations. However, it should be noted that at 39°C, cells infected with tsA8, but not tsH9, failed to stain with anti-ICP 10 serum (Fig. 1), suggesting that ICP 10 is associated with anchorage independent growth.

Taken in toto, our studies indicate that cellular immortalization and reduced serum requirement for growth, precede the acquisition of anchorage independence and tumorigenicity. The latter two properties are acquired by most, but not all lines, and CE in 0.3% agarose correlates with the maintenance of viral DNA (32). A viral protein expressed in the transformed cells (ICP 10), appears to play a role in the transformation process since: (i) its expression correlates with the ability of a focus of morphologically altered cells to survive subculture and grow into an immortal line, (ii) there is a strong quantitative correlation between CE in 0.3% agarose and the expression of ICP 10, and (iii) cells transformed by a ts mutant, that is ICP 10 negative at the nonpermissive temperature, fail to clone in 0.3% agarose at the nonpermissive temperature. Two other viral proteins are expressed in the transformed cells, but their relationship to transformation is unclear. These are 35K (47) and VP143 (14), also designated ICSP 11/12 (41) or ICP 12/13 (Smith and Aurelian, in preparation). The finding that tsA8 transformed cells clone equally well at 37° and 39°C beyond pip 31 (Table 2) is amenable to three interpretations: (i) viral DNA sequences are randomly retained in the transformed cells, (ii) viral DNA sequences other than those encoding for ICP 10 correlate with the maintenance of anchorage independence, or (iii) the maintenance of anchorage independent growth is due to rearrangement of integrated viral DNA sequences that encode ICP 10, giving rise to anchorage independent lines that are not dependent on ICP 10 for maintenance of this phenotype. Presently available data do not differentiate between these interpretations. However, it should be pointed out that rearrangement of viral DNA sequences has been shown to acompany the appearance of anchorage independent SV40 transformed lines that no longer require viral gene A product for expression of this phenotype (24).

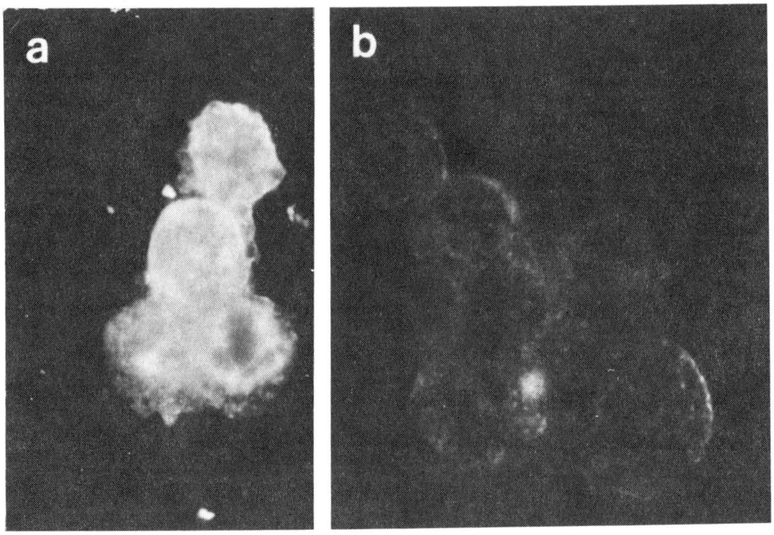

Fig. 1. Immunofluorescent staining with anti-ICP 10 serum of cells infected with tsH9 (a) and tsA8 (b) at 39°C.

CHARACTERIZATION OF VIRAL PROTEINS IN TRANSFORMED CELLS

As discussed earlier, ICP 10, ICP 12/13 and 35K are the only proteins reproducibly identified in HSV-2 transformed cells. ICP 12/13 has the highest affinity for DNA and it is precipitated by monospecific antibody to the major DNA binding protein ICSP 11/12 (41). ICP 10 (the homologue of the HSV-1 encoded ICP 6) is an early 144K protein that is also synthesized in the absence of viral DNA replication (1,46). It is phosphorylated, but the phosphate cycles on and off. It localizes both in the cytoplasm and the nucleus of productively infected cells, but its nuclear proportion decreases from 56% at 4.5 hrs. to 17% at 8.5 hrs. p.i. ICP 10 binds to DNA. However, its binding affinity appears to be reduced by phosphorylation, a finding that might explain the lower proportion of nuclear ICP 10 late in infection (Smith and Aurelian, in preparation). Recent studies indicate that it has ribonucleotide reductase activity, suggesting that it is involved in DNA biosynthesis. Thus, monoclonal antibody 48S (42), precipitates a protein from HSV-2 infected cells that has a proteolytic peptide map identical to that of ICP 10, and inactivates the ribonucleotide reductase activity of HSV-1 and HSV-2 infected cells (8). These findings are in agreement with data indicating that a ts lesion in HSV-2 that affects viral reductase activity (12), co-maps with ICP 10 (25).

The 35K protein detected in transformed cells is structurally identical to the 38K protein in productively infected cells (47). The mRNA for this protein has a common 3' terminus with the mRNA that encodes the 144K (ICP 10) protein, but the reading frame encoding the 38K protein is downstream from the C terminus of the 144K protein (34,35) and monoclonal antibody to ICP 10 does not precipitate the 38K protein (25). Its function remains to be established.

TRANSFORMATION BY VIRAL DNA FRAGMENTS

Another approach used to identify viral genes that are responsible for transformation, has been the use of defined viral DNA fragments. These studies have shown that exposure of established heteroploid cells, such as NIH 3T3, to the Bgl II N fragment of HSV-2 DNA, which maps between positions 0.58 and 0.63, causes them to become morphologically altered and to evidence a reduced serum requirement for growth (18). Further experiments indicated that the transforming region was located within a 2.1 Kb fragment bounded on the left by a BamHI site (18). Cells transformed by Bgl II N did not retain the 2.1 Kb sequences but instead retained sequences from the right hand end of the fragment. Furthermore, not one of the proteins encoded by Bgl II N is fully encoded within the 2.1 Kb transforming sequences (19), and viral proteins were not identified in the cells transformed by Bgl II N (20). These data indicate that viral DNA sequences are not maintained in cells transformed by Bgl II N, and that transformation mediated by this fragment may be explained by the "hit and run" hypothesis (20).

On the other hand, studies of normal diploid SHE cells have unequivocally demonstrated that, in these cells, neoplastic transformation is mediated by the Bgl II C fragment that is positioned between 0.419 and 0.582 map units on the HSV-2 genome (26). Further studies indicated that cells transfected with sequences spanning the left 64% end of this fragment (0.419 - 0.525 map units; plasmid pGR75) (Fig. 2), escaped senescence and grew into immortal lines. However, these lines remained nontumorigenic as late as post-transfection passage (ptp) 60. The failure of the pGR75-induced lines to acquire tumorigenic potential is not due to the amount of pGR75 used in the transfection experiments, since under identical conditions, cells transfected with the cloned Bgl II C fragment (plasmid pGR140) acquired neoplastic potential. Diploid SHE cells transfected with the cloned Bgl II N fragment (plasmid pGR60) at concentrations (0.1 - 0.5 g) at which pGR140 and pGR75 were able to respectively mediate neoplastic transformation or cellular immortalization, senesced at ptp 8-9, together with the normal uninfected SHE cells. Cells transfected with 10-fold higher concentrations of pGR60 continued to grow beyond this passage, but their growth rate decreased and by ptp 16, they finally senesced (27).

These findings are consistent with those obtained with inactivated HSV-2, in that they indicate that normal diploid cells acquire a neoplastic phenotype by passage through at least two stages. The first is the acquisition of continuous cell proliferation (immortalization), which, in the case of HSV-2, is mediated by DNA sequences that map between 0.419 - 0.525 units on the Bgl II C fragment ("immortalizing sequences"). On the other hand, the entire Bgl II C fragment is needed for the acquisition of neoplastic potential. Since Hind III cleavage of Bgl II C (at 0.525 map units) did not inactivate its neoplastic potential, we interpret the data to suggest that the right hand 34% of the Bgl II C fragment extending from 0.525 to 0.582 map units, is required for the conversion of preneoplastic immortal lines to a tumorigenic state ("presumptive neoplastic sequences") (Fig. 2).

Cell-free translation of mRNA homologous to specific fragments of HSV-2 DNA was used in order to map proteins encoded by these sequences (25). Three proteins, including 144K (ICP 10) and two other proteins, with respective mol. weights of 52K and 27K were identified when mRNA was selected with the cloned Bgl II C fragment (pGR140). Cell free translation of mRNA selected by pGR75, and by its BamHI subfragments, indicates that the 27K protein is encoded by sequences bounded on the left by a BamHI site at 0.46 map units. Its exact right hand boundary remains to be established, but it is not beyond 0.52 map units. The order of the other two proteins [144K (ICP 10) and 52K] was deduced from a series of experiments in which mRNA was selected with subfragments of a recombinant clone that contains the Bam HI e fragment of HSV-2 DNA (plasmid pGH17a; 0.525 - 0.585 map units), and of pGR60 a clone that contains the Bgl II N fragment of HSV-2 DNA. The data unequivocally demonstrated that sequences that encode for the 144K (ICP 10) protein are located in the right hand end of pGH17a and extend into the left hand end of pGR60 (Bgl II N) with maximal boundaries of 0.556 - 0.60 map units. The right hand maximal boundary encoding for the 52K protein is 0.548 map units. The 38K protein is encoded by sequences within 0.582 - 0.60 map units on the HSV-2 genome (Fig. 3).

These data are consistent with those of McLaughlan and Clements (34), who found two mRNA species (5.0 Kb and 1.2 Kb, respectively), within map coordi-

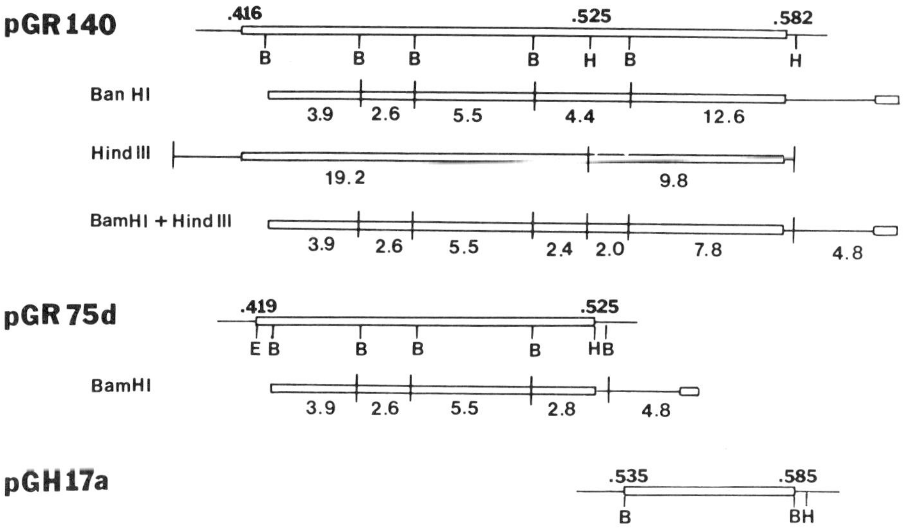

Fig. 2. Sites of Bam HI and Hind III cleavage on plasmids of pGR140, pGR75d and pGR17a respectively containing HSV-2 DNA inserts Bgl II C, EcoRI/Hind III AE and BamHI e. Double digestion of plasmid pGR140 with BamHI and Hind III generates a 7.8 Kb fragment that overlaps with pGH17a.

nates 0.56 and 0.60 on the HSV-1 genome. These mRNAs are unspliced and share a 3' terminus. When translated in vitro, the 5.0 Kb mRNA specified a 140K protein and the 1.2 Kb mRNA specified a 40K protein. With few exceptions, the gene orders of HSV-1 and HSV-2 are thought to be identical, and it is tempting to conclude that these 140K and 40K proteins are homologs of our 144K (ICP 10) and 38K proteins. Indeed, the HSV-1 encoded 140K (ICP 6) protein is precipitated by the same monoclonal antibody that precipitates ICP 10 (42), and it was independently mapped in this region using intertypic recombinants of HSV-1 and HSV-2 (30).

If a viral protein is involved in immortalization, the only candidate identified by our studies is the 27K protein. Similarly, if viral proteins mediate the acquisition of tumorigenic potential by immortalized cells, the 144K (ICP 10) and 52K proteins are the only possible candidates. The 38K protein is excluded from consideration, since with the exception of the initiation codon, it is encoded within Bgl II N (35). Based on our findings, the 144K (ICP 10) protein is a particularly attractive candidate. However, since the mRNA that specifies for this protein is encoded by sequences that overlap with Bgl II N, our data suggest that the reading frame for the 144K (ICP 10) protein is located within the Bgl II C fragment (35) or, alternatively, that a truncated 144K (ICP 10) protein is sufficient in order to establish anchorage independent lines. Indeed, truncated forms of the large T-antigen have been identified in SV40 transformed cells (9).

DNA SEQUENCES HOMOLOGOUS TO THE Bgl II C FRAGMENT ARE PRESENT IN NORMAL AND TRANSFORMED CELLS

Several sites of major sequence homology between the HSV genome and repetitive sequences in mouse and human cell DNA were described by Peden _et al_. (40). The Bam HI Y fragment that maps within coordinates 0.44 - 0.46 on the HSV-2 genome, was shown to represent the most intense hybridization between HSV-2 and human DNA, while the major site of homology with mouse DNA resided in the Bam HI P fragment that maps within the L segment inverted repeats. The hybrids persisted in relatively stringent conditions, implying a relatively high degree of homology between viral and cellular elements (40).

Our studies demonstrated hybridization between hamster cell DNA and the Bgl II C fragment (27). The results of further experiments (49) may be summarized as follows. When plasmids pGR140, pGR75d and pGH17A were probed with ^{32}P-repetitive hamster DNA under relatively stringent conditions (65°C; 0.9M Na^+), we observed a 2.6 Kb band, shared by pGR140 and pGR75d that is presumably the Bam HI Y fragment. When ^{32}P-pGR140 was used as probe on Southern blots of BamHI and EcoRI digested hamster and human DNA, a light background of hybridization was found under slightly relaxed conditions (60°, 0.9M Na^+), suggesting that cellular DNA sequences that are homologous to sequences within pGR140, are highly interspersed in the mammalian cell genome. The viral and cellular elements have a relatively high degree of homology since this background was reduced but not eliminated under more stringent conditions (65°C; 0.9 M Na^+). Bands of different intensities that could result from differences in the degree of homology and copy number of the genomic sequences, were clearly identified over this background. Thus, when ^{32}P-pGR140 or ^{32}P-pGR75d were used as probes, bands of 4.3,2.2, and 2.1 Kb were observed in BamHI digested hamster DNA, and bands of 4.3, 3.8 and 1.7 Kb were observed in blots of EcoRI digested DNA. Based on the cell-free translation data (25), the homologous sequences in HSV represent noncoding sequences. They might have been acquired by recombination with the cell genome or they might have arisen by convergent evolutionary processes. Their role, if any, in cellular immortalization remains to be established.

When ^{32}P-pGH17a (that encompasses the "presumptive neoplastic" sequences) was used as probe, a 7.8 Kb BamHI band was detected in the DNA from tumors established with SHE cells transformed by Hind III cleaved pGR140k. This 7.8 Kb band was present only in the tumor cell DNA and was not observed in normal hamster or human cell DNA probed with pGR75d or pGH17a. The 7.8 Kb band was also detected in tumor cells probed with a 2.2 Kb Sac I subfragment of pGH17a that maps at 0.535 - 0.548 map units on the HSV-2 genome (Fig. 3). However, the data do not exclude the possibility that the 7.8 Kb fragment in the tumor cells, consists of both viral and homologous cellular elements.

HSV AND CERVICAL CANCER

In their totality, the transformation studies suggest that normal diploid cells are transformed by the Bgl II C fragment of HSV-2 DNA while heteroploid cells rendered preneoplastic by factors other than HSV-2, can be transformed by the Bgl II N fragment. Transformation by Bgl II C appears to involve the retention of DNA sequences that are homologous to the left hand end of the fragment (0.525 - 0.582 map units) and that encode for 144K (ICP 10) and a 52K protein. Bgl II N appears to transform according to the "hit and run" hypothesis (20).

Assuming that extrapolation is reasonable, the transformation data predict that cervical cancer patients fall into two distinct groups with respect to the maintenance and expression of HSV-2 DNA sequences. Thus, if the cervical cells are diploid at the time of exposure to HSV-2, transformation might be mediated by Bgl II C, resulting in the maintenance and expression of the "presumptive neoplastic" sequences (0.525 - 0.582 map units). On the other hand, tranformation of cervical cells rendered preneoplastic by factor(s) other than HSV-2 would be mediated by Bgl II N and this may be a "hit and run" phenomenon. In both cases, viral DNA sequences that are unrelated to transformation could be retained and expressed.

With rare exceptions (15), studies designed to identify HSV-2 DNA in cervical cancer tissue have met with difficulties, giving rise to the suggestion that HSV-2 initiates the transformation of susceptible cervical cells,

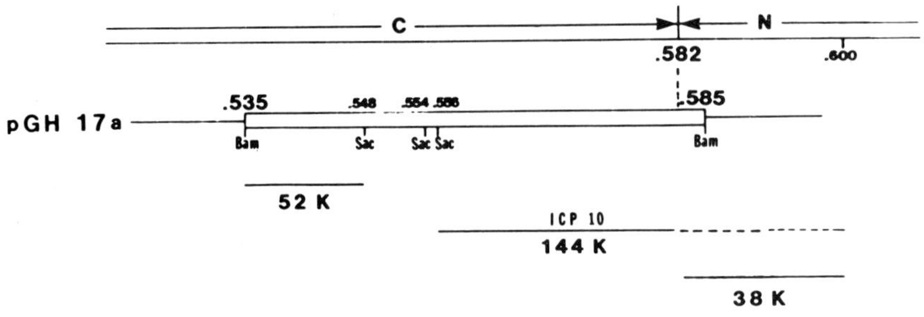

Fig. 3. The location of genes encoded by the right hand end of the HSV-2 Bgl II C and the left hand end Bgl II N fragments. The mRNA that specifies the 144K (ICP 10) protein extends to 0.60 map units. However, the termination codon for this protein may precede the initiation codon for the 38K protein (dotted line) (34,35).

but other factors commit them to neoplastic conversion ("hit and run" hypothesis) (20,50). However, since numerous studies have consistently indicated that viral transcripts and proteins are expressed in cervical tumor cells (4, 6,33,11), the "hit and run" interpretation suggests that they reflect the expression of randomly retained viral DNA sequences. Implicit in these interpretations is the prediction that whatever their relationship to carcinogenesis, viral DNA sequences must be detected in cervical cancer tissue. The failure to detect such sequences can best be explained by technical difficulties, thereby precluding any conclusions pertaining to the role of HSV-2 in cervical carcinogenesis. Indeed, the Southern blot hybridization has failed to detect HSV-2 DNA sequences in transformed rat cells and the tumors that they induce, even though the entire viral genome is maintained in these cells (39). Furthermore, there is major homology between viral DNA sequences within the immortalizing portion of the Bgl II C fragment and human cell DNA sequences, including the cervix (27,40,49).

Studies from our laboratory using upwards of 2000 patients have shown that ICP 10 is expressed in cervical tissue from 65-85% of patients with dysplasia - carcinoma (3,4,,6,7), and similar results were independently obtained in different geographic areas (2,23,28,38). Most significantly, mimicking its apparent role in the acquisition of anchorage independent growth, the expression of ICP 10 reflects the growth of the cervical tumor (4).

Another viral protein originally described by us (21,43,44) in a large scale study of 393 patients and later by other investigators (11), is the major DNA binding protein, observed in 31% of patients with dysplasia and 50% of those with carcinoma. It is also identified in HSV-2 transformed cells. However, its expression in these cells does not correlate with the acquisition of anchorage independent growth (Table 2), and it is only detected in 6 of 11 ICP 10 positive cultures (5). In cervical cancer patients, it is expressed in 54% of patients with cervical tumor cells that are positive for ICP 10 (6).

The argument that the expression of these viral proteins is irrelevant (50), or inconsistent (20), is based on the assumption that immunological and seroepidemiological studies are of an inherently lower logical and experimental validity than molecular hybridization data. Not only is this a very naive concept, but, in fact, conclusions pertaining to the consistent or relevant aspect of any finding, must perforce depend on the study of numerous subjects, such as can only be done by the epidemiologic approach. The hybridization studies that still suffer from technical and interpretative difficulties, have not yet met this criterion. However, the sheer number of cervical cancer patients that are positive for ICP 10 indicates that viral DNA sequences that encode them are invariably expressed in cervical tumor tissue. Together with the findings that: (i) ICP 10 is involved in the acquisition of anchorage independent growth by cells transformed *in vitro* with HSV-2, (ii) is encoded by the "presumptive neoplastic" sequences that are maintained in tumor cells induced by the Bgl II C fragment of HSV 2 DNA, and (iii) is expressed in a proportion of cervical cancer patients in a fashion that reflects tumor growth (3), the data provide strong evidence that the expression of ICP 10 encoding viral DNA sequences is involved in carcinogenesis.

REFERENCES

1. Anderson, K.P., Frink, R.J., Devi, G.B., Gaylord, B.H. and Wagner, E.K. J. Virol. 37:1011, 1981.
2. Arsenakis, M., Georgiou, G.M., Welsh, J.K., Cauchi, M.N., and May, J.T. Int. J. Cancer 25:67, 1980.
3. Aurelian, L. Fed. Proc. 31:1651, 1972.
4. Aurelian, L., Kessler, I.I., Rosenshein, N.B., and Barbour, G. Cancer 48:455, 1981.
5. Aurelian, L., Manak, M., and Ts'o. P.O.P. Proc. International Workshop on Herpesviruses, Bologna, Italy, p 216, 1981.
6. Aurelian, L., Smith, C.C., Klacsmann, K.T., Gupta, P.K. and Frost, J.K. Cancer Investigation, in press, 1983.

7. Aurelian, L. In: Gynecologic Oncology. (Eds. Griffiths, C.T. and Fuller, A.F.). Martinus Nijhoff, Boston, Mass. pp 1-62, 1983.

8. Bacchetti, S., Huszar, D. and Muirhead, B. Proc. Int. Herpesvirus Workshop. p 297. 1983.

9. Chaudry, F., Harvey, R. and Smith, A.E. J. Virol. 44:54, 1982.

10. Dixon, R.A.F., Sabourin, D.J. and Schaffer, P.A. J. Virol. 45:343, 1983.

11. Dreesman, G.R., Burck, J., Ada, E., Kaufman, R.H., Melnik, J.L., Powell, K.L. and Purifoy, D.J. Nature 283:591, 1980.

12. Dutia, B.M. J. Gen. Virol. 64:513, 1983.

13. Esko, J. and Raetz, C. Proc. Nat. Acad. Sci. 75:1190, 1978.

14. Flannery, V., Courtney, H. and Schafer, P. J. Virol. 21:284,, 1977.

15. Frenkel, N., Roizman, B., Cassai, E. and Nahmias, A.J. Proc. Nat. Acad. Sci. 69:3784, 1972.

16. Frenkel, N., Locker, H., Cox, B., Roizman, B. and Rapp, F. J. Virol. 18: 885, 1976.

17. Galloway, D.A., Copple, C.D. and McDougall, J.K. Proc. Nat. Acad. Sci. 77:880, 1980.

18. Galloway, D.A. and McDougall, J.K. J. Virol. 38:749, 1981.

19. Galloway, D.A., Goldstein, L.C. and Lewis, J.B. J. Virol. 42:530, 1982.

20. Galloway, D.A. and McDougall, J.K. Nature 302:21, 1983.

21. Gupta, P.K., Aurelian, L., Carpenter, J.M., Klacsmann, K.T., Rosenshein, N.B. and Frost, J.K. Gynec. Onc. 12:S232, 1981.

22. Hampar, B., and Boyd, A. In: Oncogenesis and Herpesviruses III, p 583, G. de The, W. Henle, F. Rapp (eds.), IARC Publications, 1978.

23. Heise, E.R., Kucera, L.S., Raben, M. and Homesley, H. Cancer Res. 39: 4022, 1979.

24. Hiscott, J., Murphy, D. and Defendi, V. Cell 22:535, 1980.

25. Iwasaka, T., Smith, C.C., Aurelian, L. and Ts'o, P.O.P. Proc. Int. Herpesvirus Workshop. p 75, 1983.

26. Jariwalla, R.J., Aurelian, L. and Ts'o, P.O.P. Proc. Nat. Acad. Sci. 77:2279, 1980.

27. Jariwalla, R.J., Aurelian, L. and Ts'o, P.O.P. Proc. Nat. Acad. Sci. In press. 1983.

28. Kawana, T., Sakamoto, S., Kasamatsu, T. and Aurelian, L. Gann 69: 589, 1978.

29. Kimura, S., Flannery, V.B., Levy, B. and Schafer, P.A. Int. J. Cancer 15:786, 1975.

30. Knipe, D., Batterson, W., Noral, C., Roizman, B. and Buchan, A. J. Virol. 38:539, 1981.

31. Manak, M.M., Aurelian, L. and Ts'o, P.O.P. J. Virol. 40:289, 1981.

32. Manak, M.M., Smith, C.C., Aurelian, L. and Ts'o, P.O.P. Submitted. 1983.

33. McDougall, J.K, Crum, C.P., Fenoglio, C.M., Goldstein, L.C. and Galloway, G.A. Proc. Nat. Acad. Sci. 79:3853, 1982.

34. McLaughlan, J. and Clements, J.B. J. Gen. Virol. 64:997, 1983.

35. McLaughlan, J. and Clements, J.B. Proc. Int. Herpesvirus Workshop, p 78, 1983.

36. Nahmias, A.J., Naib, Z.M., Josey, W.E., Franklin, E. and Jenkins, R. Cancer Res. 33:1491, 1973.

37. Naib, Z.M., Nahimas, A.J., Josey, W.E. and Zolis, S.A. Cancer Res. 33: 1452, 1973.

38. Notter, M.F.D. and Docherty, J.J. J. Nat. Cancer Res. 57:483, 1976.

39. Park, M. and Macnab, J.C. J. Gen. Virol. 64:755, 1983.

40. Peden, K., Mounts, P. and Harward, G.S. Cell 31:71, 1982.

41. Powell, K., Sittler, E. and Purifoy, D. J. Virol. 39:894, 1981.

42. Showalter, S.D., Zweig, M. and Hampar, B. Inf. Imm. 34:684, 1981.

43. Smith, C.C. and Aurelian, L. Virol. 98:255, 1979.

44. Smith, C.C. and Aurelian, L. Nature 292:388, 1981.

45. Spang, A.E., Godowski, P.J. and Knipe, D. J. Virol. 45:332, 1983.

46. Strnad, B.C. and Aurelian, L. Virol. 87:401, 1978.

47. Suh, M. J. Virol. 41:1095, 1982.

48. Wentz, W.B., Haggie, A.D., Anthony, D.D. and Reagan, J.W. Proc. Int. Herpesvirus Workshop. p 196, 1983.

49. Wu, J.R., Dieffenbach, C.W., Torres, D.M., Ts'o, P.O.P. and Aurelian, L. Am. Soc. Cell. Biol. Meeting, submitted, 1983.

50. Zurhausen, H. Lancet 2:1370, 1982.

ACKNOWLEDGEMENTS

We thank Dr. G. Hayward for the kind gift of cloned HSV-2 DNA fragments, Dr. M. Zweig for monoclonal antibody 48S, Dr. K. Powell for antiserum to ICSP 11/12 and Dr. P. Schafer for the HSV-2 ts mutants. The studies done in our laboratory were supported by Grant CA 16043 from the National Cancer Institute.

The Role of Viruses in Human Cancer. Volume II
G. Giraldo and E. Beth editors

ANTIVIRAL CHEMOTHERAPY

HERBERT E. KAUFMAN AND EMILY D. VARNELL

Lions Eye Research Laboratories, LSU Eye Center, Louisiana State University Medical Center School of Medicine, 136 S. Roman St., New Orleans, LA 70112 (U.S.A.)

INTRODUCTION

Dendritic keratitis is one manifestation of ocular herpetic infection, and thus far, the only one for which we have reliable therapy. Deeper infections in the eye, such as stromal necrosis, iritis, and retinitis, are still unsolved therapeutic problems. Effective antiviral chemotherapy against ocular herpetic infection originated with the discovery of idoxuridine (IDU) in 1962,[1] followed by the development of vidarabine (Ara-A) a few years later.[2] Disadvantages of these compounds, however, include toxicity and insolubility, i.e., topical application is by means of an ointment that blurs vision. Several years ago, trifluorothymidine (trifluridine) was shown to be superior in a number of ways in the treatment of dendritic keratitis. First, trifluorothymidine heals 97% of dendritic ulcers within two weeks, compared to no more than 80% with previous drugs.[3] Second, trifluorothymidine is relatively soluble, and therefore can be applied topically as a clear drop, which increases patient compliance.

The newest antiviral agent, acycloguanosine (acyclovir), is one of a class of drugs that has a much reduced potential for toxicity as a result of its mechanism of action.[4] Acyclovir, which has received FDA approval for cutaneous use in the United States, and a similar compound, BVDU, which is still in the investigational stages, are activated by the viral thymidine kinase, and therefore, do not become involved in the replication processes of normal cells. However, this mechanism of action may also carry an increased risk of development of drug-resistant virus strains, either by deletion of the viral enzyme or by modification of the viral DNA polymerase.

Recently, there has been considerable interest in evaluating combinations of antiviral agents, as well as combinations of various antiviral agents with interferon, with the hope of finding additive or even synergistic therapeutic effects. In such a case, smaller doses of each drug could be used, thereby minimizing toxicity and the development of drug-resistant strains. In 1982,

Sundmacher et al.[5] showed that human leukocyte interferon combined with trifluorothymidine drops is more effective in treating dendritic keratitis in monkey and human eyes, compared to any single therapeutic modality tested. In the past year, also, Colin et al.[6] treated human patients with acyclovir plus human leukocyte interferon, and found that the mean healing time for established dendritic keratitis was significantly reduced, compared to healing times with acyclovir plus placebo.

We used the rabbit model of herpetic keratitis to test the effect of three therapeutic combinations: trifluorothymidine plus acyclovir; vidarabine plus acyclovir; and vidarabine plus human leukocyte interferon. The two drug combinations showed an enhanced effect, compared to results obtained with either drug alone. Vidarabine plus interferon, however, produced an antagonistic effect, with the resulting ocular disease as severe as that seen in untreated controls.

MATERIALS AND METHODS

Both eyes of 58 New Zealand white rabbits (2-3 kg; both sexes) were infected with McKrae strain herpesvirus type 1 (HSV-1). The corneas were minimally scarified, and 2 drops of virus suspension (10^5 PFU/ml; twice the minimal infective dose) were placed into the cul-de-sac. The lids were closed and rubbed gently over the corneas for 30 seconds.

Beginning three days after infection, 9 rabbits were treated twice a day for five days with trifluorothymidine and acyclovir. Each treatment consisted of the application of 2 drops of 1% trifluorothymidine followed 30 minutes later by a 2 mm strip of 3% acyclovir ointment. Ten rabbits were treated on the same schedule with a 2 mm strip of 3% vidarabine ointment combined with the same amount of 3% acyclovir ointment. Ten rabbits received 2 drops of human leukocyte interferon (5×10^6 IU/ml), followed 30 minutes later by a 2 mm strip of 3% vidarabine ointment. Twenty-nine rabbits were treated with a single drug: ten with 3% vidarabine ointment; nine with 3% acyclovir ointment; and ten with 2 drops of 1% trifluorothymidine. Ten rabbits were untreated controls.

Treatment and evaluation of resulting disease were performed on a masked basis. Fluorescein was instilled into the eyes to delineate the extent of corneal epithelial ulceration, and the severity of the keratitis was graded as follows: clear cornea, 0; 1/4 area of cornea ulcerated, 1; 1/2 corneal surface ulcerated, 2; 3/4 ulceration, 3; and total corneal ulceration, 4. Examinations were performed by means of slit lamp biomicroscopy, with cobalt blue illumination.

Vidarabine (Vira-A®) was purchased from Parke, Davis Co., Morris Plains, New Jersey, and trifluorothymidine (Viroptic®) from Burroughs Wellcome Co., Research Triangle Park, North Carolina. Acyclovir (Zovirax®) was donated by Burroughs Wellcome Co. Human leukocyte interferon was obtained from the Revlon Health Care Group, Meloy Laboratories, Springfield, Virginia.

RESULTS

The two drug combinations, trifluorothymidine plus acyclovir and vidarabine plus acyclovir, appeared to have an enhanced therapeutic effect, compared to the effect of each drug alone (TABLE). However, vidarabine plus interferon produced an antagonistic effect, whereby the disease seen in the eyes treated with this combination was as severe as that seen in the untreated control eyes (TABLE). Thus, the two agents together appeared to negate the therapeutic effect normally obtained with the antiviral drug alone.

TABLE

SEVERITY OF KERATITIS IN RABBIT EYES TREATED WITH CHEMOTHERAPEUTIC AGENTS

Severity of keratitis graded on a scale of 0-4, from clear cornea to completely ulcerated cornea. Values below are averages of all eyes graded in a given category on a given day. Treatment was begun three days after infection with McKrae strain herpesvirus.

	Treatment Day				
Agent(s)	1	2	3	4	5
Control	0.65	1.78	1.78	2.00	2.11
TFT	0.63	1.13	0.50	0.41	0.15
AraA	0.60	1.13	0.91	1.05	0.98
ACV	0.65	0.78	0.38	0.29	0.31
TFT+ACV	0.67	0.88	0.35	0.05	0.09
AraA+ACV	0.60	0.60	0.24	0.14	0.18
AraA+If	0.65	1.03	1.43	1.56	1.63

TFT, trifluorothymidine; AraA, vidarabine; ACV, acyclovir; If, human leukocyte interferon; Control, untreated.

DISCUSSION

With the recent upsurge in interest in herpetic infections and antiviral drugs, and with the development of new methods of producing large quantities of relatively inexpensive interferon, considerable effort is being made to tie all of this current technology together to provide a way of preventing and/or curing this virus disease.

In this work, we have shown that the combination of vidarabine and acyclovir yields an enhanced therapeutic effect in the rabbit eye, and in other studies done in our laboratories, a similar result was obtained in tissue culture.[7] In tissue culture, vidarabine plus human leukocyte interferon showed a slightly antagonistic effect, and an even more pronounced antagonism was seen in rabbit eyes.

However, other work in our laboratories indicates that there may be some difficulty in extrapolating the results of interferon studies from one species to another, and finally, from animal models to humans. Sanitato et al.[8] showed that human leukocyte interferon reduced the severity of ocular herpetic infection in rabbit eyes, but recombinant interferon had no effect. In contrast, both types of interferon were effective in monkey eyes. In these two experiments, the effective interferons were applied prophylactically and therapeutically, i.e., before and after infection, but therapeutic interferon alone, i.e. applied only after infection, had no effect on the disease. So it seems that results in the rabbit model may not be predictive of primate results, and monkey results may, then, also not be predictive of human results.

There are other problems with the use of interferon. We found, in the course of the study reported here, that standard human leukocyte interferon titers against VSV (vesicular stomatitis virus) did not correlate with titer against HSV. Several lots of human leukocyte interferon were tested for efficacy against herpes keratitis, and two lots had no effect on the course of the disease in _in vivo_ testing and in tissue culture assay. The titers against VSV, however, remained the same before and after the HSV testing.

It may be that the results of antiviral therapy are more dependent on the genetics of the virus strain than we had thought previously. In human patients, it has been noted that some herpetic ocular infections are worsened by corticosteroid drugs, while others are not. It also appears that, experimentally, there is some connection between the genome of the infecting virus strain and the response to corticosteroids seen in rabbit eyes infected with recombinant HSV strains.[9] It may be that the explanation for the clinical observations is similarly tied to the genetic character of the infecting virus.

It has been suggested that a major factor in the induction of stromal disease is the production of large amounts of particular, highly antigenic glycoproteins characteristic of particular herpesvirus strains.[10] If this is so, then glycoprotein inhibitors, such as tunicamycin and 2-deoxy-D-glucose, might be capable of mitigating stromal disease; however, rabbit eyes infected with RE strain herpesvirus showed no diminution of disease with the application of these compounds.[11] It may be that topically applied drugs are not effective, or that lower molecular weight proteins or heterosaccharides also contribute to the antigenic stimulus.

Chronic oral administration of low toxicity antiviral drugs, such as acyclovir, has been considered as a means of preventing recurrent herpetic disease. However, we have shown that chronic oral administration of acyclovir or BVDU in rabbits infected with the McKrae strain herpesvirus did not prevent shedding in the the tears or the re-appearance of active ocular disease, nor did such chronic drug therapy promote the emergence of resistant virus in the tears.[12] In a similar study, in which massive oral, topical, and intramuscular doses of acyclovir were given beginning 36 days after infection in combination with artifical herpes reactivation by epinephrine iontophoresis, no virus was found in the tears during the 60 days of the study.[13] It may be that virus shedding can be prevented by sufficient doses, but it is not clear whether such large doses can be safely given over long periods of time, and whether this approach can ever be used in a practical manner in humans.

In summary, much work is being done on antiviral agents and interferon for the treatment of herpetic disease, and some of the newer antiviral agents hold considerable promise for improved therapeutic benefits for patients with ocular herpetic disease, and possibly for other types of systemic viral infections. However, a clearer understanding of the role and usefulness of animal models in these studies is needed, in order to provide information that can be extended to the treatment of human disease.

ACKNOWLEDGEMENTS

This work was supported in part by PHS grants EY02672 and EY02377 from the National Eye Institute, National Institutes of Health, Bethesda, MD.

REFERENCES

1. Kaufman, H.E., Nesburn, A.B., and Maloney, E.D.: Arch. Ophthalmol. 67:396, 1962.

2. Pavan-Langston, D., and Dohlman, C.H.: Am. J. Ophthalmol. 74:81, 1972.

3. Wellings, P.C., Awdry, P.N., Bors, P.H., Jones, B.R., Brown, D.C., and Kaufman, H.E.: Am. J. Ophthalmol. 73:932, 1972.

4. Park, N.H., Pavan-Langston, D., and McLean, S.L.: J. Infect. Dis. 140: 802, 1979.

5. Sundmacher, R., Neumann-Haefelin, D., and Cantell, K.: In: Herpetic Eye Diseases. Vol 12, Sundmacher, R. (ed.), J.F. Bergmann Verlag, Munich, 1981, p. 401.

6. Colin, J., Chastel, C., Renard, G., and Cantell, K.: Am. J. Ophthalmol. 95:346, 1983.

7. Hubbard, A.E., and Centifanto-Fitzgerald, Y.M.: In vitro effects of combined drug therapy on herpes simplex virus type 1. Proc. Soc. Exp. Biol. Med., Submitted.

8. Sanitato, J.J., Varnell, E.D., Kaufman, H.E., and Raju, V.K.: Differences in native and recombinant interferon for herpes keratitis in two animal models. Invest. Ophthalmol. Vis. Sci., Submitted.

9. Kaufman, H.E., Varnell, E.D., and Centifanto-Fitzgerald, Y.M.: Virus genome regulates response of herpetic infection to corticosteroids. In preparation.

10. Smeraglia, R., Hochadel, J., Varnell, E.D., Kaufman, H.E., and Centifanto-Fitzgerald, Y.M.: Exp. Eye Res. 35:443, 1982.

11. Raju, V.K., Varnell, E.D., and Kaufman, H.E.: The lack of effect of tunicamycin and 2-deoxy-D-glucose on corneal stromal herpes in rabbits. Invest. Ophthalmol. Vis. Sci., In Press.

12. Kaufman, H.E., Varnell, E.D., Centifanto-Fitzgerald, Y.M., De Clercq, E., and Kissling, G.E.: Oral antiviral drugs in experimental herpes simplex keratitis. Antimicrob. Agents Chemother., Submitted.

13. Nesburn, A.B., Willey, D.E., and Trousdale, M.D.: Proc. Soc. Exp. Biol. Med. 172:316, 1983.

The Role of Viruses in Human Cancer. Volume II
G. Giraldo and E. Beth editors

EPSTEIN-BARR VIRUS AND BURKITT'S LYMPHOMA

M.A. EPSTEIN
Department of Pathology, University of Bristol Medical School,
University Walk, Bristol BS8 1TD (England)

INTRODUCTION

The realisation by Burkitt that a variety of lymphoid tumours in children in Africa were all manifestations of a hitherto unrecognised, single, malignant lymphoma syndrome (1) was an important discovery. Even more significant was the further finding that this tumour, now universally known as Burkitt's lymphoma (BL), had a geographical distribution determined by climatic factors (2, 3). For, if temperature and rainfall influenced the incidence of BL, some biological agent must clearly be involved in causation and at the outset an oncogenic virus carried by an arthropod vector seemed the most likely. It was this possibility which was directly responsible for the search for unusual viruses in BL material, and for the discovery of Epstein-Barr (EB) virus (4).

Although epidemiological observations soon made it plain that the tumour could not arise by case-to-case infection mediated by a climate-dependent vector, the idea of an infectious cause has been continually strengthened by evidence accumulated over many years and other explanations for the climate-dependence of BL have emerged in connection with a co-factor, hyperendemic malaria (5).

The extroardinary close association of EB virus with BL in endemic areas has been fully documented and reviewed (6), and quite recent work on cellular oncogene activation in BL suggests possible explanations for this. Thus, as discussed elsewhere (7), the virus and hyperendemic malaria seem to interact to facilitate characteristic chromosomal translocations which in turn activate the cellular *myc* oncogene (8, 9, 10); even newer findings also implicate the HuBlym-1 oncogene (11).

Apart from BL, EB virus has a remarkable association with one other human cancer, undifferentiated nasopharyngeal carcinoma (NPC) (12). The form of this second association (13) shows close similarities to that of the virus and endemic BL, but although less is known in the case of NPC, many experts now believe that the virus is likely to have an even more directly causative role in this malignancy.

Although the accumulation of information both on the general biological behaviour of EB virus and on its role in human malignancies is scientifically interesting, it has seemed for some years that such activities would be considerably more worthwhile if they could lead to intervention against the virus infection which might in consequence reduce the incidence of the associated cancers. It was in this context that proposals were first put forward for a vaccine against EB virus (14) and the present paper reviews recent progress which has been made in this direction.

RATIONALE FOR A VACCINE AGAINST EB VIRUS

Reference has already been made to the striking evidence implicating EB virus, together with co-factors, in the causation of endemic BL and undifferentiated NPC. Although the former tumour is not of great numerical importance, undifferentiated NPC is of considerable significance in world cancer terms since it has a very high incidence amongst Southern Chinese (12) and Eskimos (15), and areas with moderately high incidence levels have been recognised in North Africa (16), East Africa (17), and through most of South East Asia (12). The control of a naturally occurring herpesvirus-induced lymphoma of chickens, Marek's disease (18, 19), by inoculation with apathogenic virus (20, 21) provided the first example of anti-viral vaccination affecting the frequency of a cancer, whilst other work with the malignant lymphoma which can be induced experimentally by inoculation of *Herpesvirus saimiri* in certain subhuman primates (22) has shown that animals given killed vaccine were protected against infection and did not therefore get tumours (23). By analogy, prevention of EB virus infection might be expected greatly to reduce subsequent development of the associated human tumours.

EB VIRUS MEMBRANE ANTIGEN AS THE BASIS FOR A VACCINE

It has long been known that antibodies from EB virus-infected individuals which neutralize the virus are those directed against the virus-determined cell surface membrane antigen (MA) (24, 25, 26, 27) and this information prompted the suggestion that MA be used as an anti-viral vaccine (14). Antigen-containing membranes from cells infected with Marek's disease herpesvirus have markedly reduced lymphoma incidence when used as an experimental vaccine in chickens (28) and cell membrane preparations from *Herpesvirus saimiri*-infected cells have induced neutralizing antibodies when inoculated into marmosets (29).

Investigations into the molecular structure of MA have identified two high molecular weight components (340,000 and 270,000 daltons - gp340 and gp270) (30, 31, 32, 33, 34) and the concordance between human antibodies to MA and EB virus neutralization has been formally explained by the demonstration of these same glycoproteins in both the viral envelope and cell membrane MA (33). gp340/270 can elicit virus- neutralizing antibodies (35) and monoclonal antibodies which react with both components are likewise virus-neutralizing (36, 37).

As regards sources of gp340 and gp270, most EB virus-carrying, virion-producing lymphoid cell lines synthesize roughly equal amounts of the two molecules, but in this respect the B95-8 marmoset lymphoblastoid line (38) is anomalous in expressing almost exclusively gp340, thus providing an important advantage for molecular weight-based purification procedures.

A radioimmunoassay for MA gp340

For the elaboration of an efficient purification procedure a sensitive assay is essential to quantify the product and permit modifications designed to maximise yields. A highly sensitive, quantitative radioimmunoassay (RIA) for gp340 was therefore developed based on very small amounts of extremely pure radioiodinated glycoprotein. The antigenicity of this ^{125}I-gp340 was established by the finding that up to 75% was precipitated specifically by naturally occurring human antisera to MA and the radiolabelled material was thereafter used in a conventional competition RIA to quantify unlabelled gp340 in test samples. One unit of antiserum was arbitrarily defined as that amount which caused a 50% reduction in ^{125}I-gp340 binding to a standard reference serum. A full account of this work has been given elsewhere (39).

Isolation of MA gp340

With the crucial help of the RIA a preparative sodium dodecyl sulphate polyacrylamide gel electrophoresis (SDS-PAGE) procedure was worked out for the isolation of gp340 from B95-8 cell membranes. A most important step involved renaturation to ensure that the final product was obtained in an antigenic form; this was achieved by removing the SDS under conditions where protein refolding was prevented by urea followed by removal of urea during dialysis against buffer containing non-ionic detergent (either 0.1% Triton X100 or 0.87% octylglucoside). When monitored by the RIA the procedure gave a fifty fold increase in recovery as compared to conventional techniques and SDS-PAGE analysis of the product demonstrated its homogeneity. Details of this preparation method have recently been described (40).

Immunogenicity of gp340

Preliminary experiments with mice and rabbits made it clear that gp340 in

Freund's adjuvant was only weakly immunogenic since microgram amounts only gave rise to virus-neutralising antibodies after multiple injections. In order to overcome the disadvantages of repeated immunization with scarce purified glycoprotein and the need for Freind's adjuvant which could never be suitable for administration to man, alternative immunization methods were sought. Reports of the successful use of artificial liposomes containing Semliki Forest virus spike protein (41) or hepatitis B virus surface antigen (42) suggested an alternative and it was found that mice given a single small dose of gp340 incorporated into liposomes and injected *i.p.* together with *B. pertussis* organisms responded with high titre antibodies (35). Immunogenicity was thus enhanced and Freund's adjuvant avoided.

Further experiments were therefore undertaken to compare systematically the efficiency of this procedure with both conventional and other novel immunization techniques using various routes of administration in two species of banal laboratory animal, the mouse and the rabbit, and in the cotton-top tamarin *(Saguinus oedipus oedipus)*, a subhuman primate known to be susceptible to EB virus infection (43). Any method using liposomes gave better results as assessed by antibody titres than antigen presented with Freund's adjuvant. Furthermore, where mice were immunized by the *i.p.* route, antigen in liposomes administered with *B. pertussis* was more effective than without, although antigen in liposomes alone was highly immunogenic when administered *i.v.* More important, irrespective of the route of injection and whether or not doses were split, presentation methods involving liposomes with the lipid A component of *E. coli* lipopolysaccharide (44) always gave superior results; if threshold amounts of antigen were given to mice and rabbits antibodies were only seen where the gp340 was incorporated in liposomes with lipid A. The response of the cotton-top tamarins followed a similar pattern; antibodies were seen earlier after immunization with liposomes incorporating lipid A and it is of importance that these antibodies were virus-neutralizing.

All the antisera were specific and reacted only with the gp340 component of MA from B95-8 cells and failed to recognise other molecules either from the surface or the interior of such cells. This specificity held good when the sera were tested against other lymphoid cells expressing gp340, and in addition, where the cells also expressed the structurally related gp270 MA component this antigen was recognised as well, indicating once again that antigenic determinants are shared by the two molecules. These experiments have been reported in full (45).

Structure of gp340

The preparation in the long term of an EB virus MA subunit vaccine for administration to humans is likely to make use either of recombinant DNA technology or of chemical synthesis of peptides. An understanding of the structure of gp340 and the contribution, if any, of the sugar moiety to antigenicity is an obvious prerequisite to such procedures, and experiments were therefore undertaken in which gp340 was analysed after treatment with a battery of glycosidases and V8 protease. Following treatment with tunicamycin during synthesis or directly with glycosidases, the carbohydrate component has been found to consist of both O-linked and N-linked types and to constitute about 50% of the molecular mass. Digestion studies with neuraminidase and oligosaccharidase indicated that gp340 is heavily sialated with most of the sialic acid located on O-linked sugars. The high carbohydrate content of the molecule appears to confer resistance to proteolysis since V8 protease was only effective at high concentrations (above lmgm/ml); removal of sialic acid before V8 protease digestion did not alter this pattern nor affect the antigenicity of the digestion fragments. The antigenicity of the intact molecule was likewise unaffected by removal of sialic acid nor were the O-linked and N-linked carbohydrate moieties essential for this property. Finally, the binding of virus-neutralizing human sera and monoclonal antibody to gp340 from which either O-linked or N-linked sugars had been removed indicates that the sites on the molecule which generate virus-neutralizing antibodies are present in the protein component. These observations have been described in full in a separate report (46).

DISCUSSION AND CONCLUSIONS

The advances considered above clearly demonstrate that the high molecular weight component of EB virus MA glycoprotein can be prepared in tractable quantities by a simple molecular weight-based procedure which leaves the molecule active antigenically. In addition, incorporation of gp340 in artificial liposomes confers significantly increased immunogenicity and this is enhanced further by the use of lipid A.

As already described, the immunogenicity studies were pursued in the first place in mice and rabbits in order to clarify the effect of route and mode of administration of gp340 in liposomes with various adjuvants, and work out the most effective protocol. Thereafter, the work was extended to the cotton-top tamarins.

Only two kinds of animal are known to be fully susceptible to experimental infection with EB virus, the owl monkey *(Aotus trivirgatus)* (47, 48, 49) and the cotton-top tamarin (50; see 43 for review). However, the former "species" has recently been found to be very heterogeneous with at least nine different karyotypes (51, 52, 53), and shows considerable variation in susceptibility to certain infections. The cotton-top tamarin is therefore preferable for experimental *in vivo* studies with EB virus despite having been on the endangered species list for some years. For, although rather little was hitherto known about this animal and the possibility of its successful propagation in captivity, recent efforts have resulted in the definition of the necessary management conditions and flourishing breeding colonies have been established (54, 55, 56).

Having demonstrated the capacity of gp340 in liposomes to induce virus-neutralizing antibodies in cotton-top tamarins the way is now clear for key experiments on the protective effects of such immunization against challenge with disease-inducing doses of virulent virus, as a preliminary to the evaluation of liposomal gp340 as a vaccine for man. The efficacy of subunit antiviral vaccines has been demonstrated in both human and veterinary diseases (57, 58), and the development of an EB virus vaccine will likewise have to proceed in this direction. If such a vaccine can indeed be shown to protect cotton-top tamarins, the most satisfactory test for it in an EB virus-related disease would be in the context of young adults known to be at risk for infectious mononucleosis. Information on such protection could readily be obtained amongst young people in Western countries and thereafter the effect of vaccination and consequential prevention of disease could be assessed in the context of endemic BL. The peak incidence of endemic BL is around the age of seven (59) and the influence of vaccination on tumour incidence would therefore perhaps be apparent within a decade. If this were successful there would then be good reason for attacking the far more difficult problem of intervention against undifferentiated NPC, a disease of middle and later life in high incidence areas (12) calling therefore for the maintenance of immunity over many years.

ACKNOWLEDGEMENTS

This work was supported by the Medical Research Council, London (Special Project Grant SPG/978/32/5), together with the Cancer Research Campaign, London (out of funds donated by the Bradbury Investment Company of Hong Kong).

REFERENCES

1. Burkitt, D. (1958) A sarcoma involving the jaws in African children, Brit. J. Surg., 46, 218.

2. Burkitt, D. (1962) Determining the climatic limitations of a children's cancer common in Africa, Brit. Med. J., 2, 1019.

3. Burkitt, D. (1962) A children's cancer dependent on climatic factors, Nature, 194, 232.

4. Epstein, M.A., Achong, B.G. and Barr, Y.M. (1964) Virus particles in cultured lymphoblasts from Burkitt's lymphoma, Lancet i, 702.

5. Burkitt, D.P. (1969) Burkitt's lymphoma - an alternative hypothesis to a vectored virus, J. Natl. Cancer Inst., 42, 19.

6. Epstein, M.A. and Achong, B.G. (1979) The relationship of the virus to Burkitt's lymphoma in: The Epstein-Barr Virus. Epstein, M.A., and Achong, B.G. (Eds.), Springer, Berlin, Heidelberg and New York, pp. 321-337.

7. Epstein, M.A. and Morgan, A.J. (1983) Clinical consequences of Epstein-Barr virus infection and possible control by an antiviral vaccine, Clin. Exp. Immunol., 53, 257.

8. Lenoir, G.M., Preud'homme, J.L., Bernheim, A. and Berger, R. (1982) Correlation between immunoglobulin light chain expression and variant translocation in Burkitt's lymphoma, Nature, 298, 474.

9. Dalla-Favera, R., Bregni, M., Erikson, J., Patterson, D., Gallo, R.C. and Croce, C.M. (1982) Human c-*myc onc* gene is located on the region of chromosome 8 that is translocated in Burkitt lymphoma cells, Proc. Nat. Acad. Sci., 79, 7824.

10. Taub, R., Kirsch, I., Morton, C., Lenoir, G., Swan, D., Tronick, S., Aaronson, S. and Leder, P. (1982) Translocation of the c-*myc* gene into the immunoglobulin heavy chain locus in human Burkitt lymphoma and murine plasmacytoma cells, Proc. Nat. Acad. Sci., 79, 7837.

11. Diamond, A., Cooper, G.M., Ritz, J. and Lane, M.A. (1983) Identification and molecular cloning of the human Blym transforming gene activated in Burkitt's lymphomas, Nature, in press.

12. Shanmugaratnum, K. (1971) Studies on the etiology of nasopharyngeal carcinoma in: International Review of Experimental Pathology. Richter, G.W. and Epstein, M.A. (Eds.), Academic Press Inc., New York and London, 10, pp 361-413.

13. Klein, G. (1979) The relationship of the virus to nasopharyngeal carcinoma in: The Epstein-Barr Virus. Epstein, M.A. and Achong, B.G. (Eds.), Springer, Berlin Heidelberg and New York, pp. 339-350.

14. Epstein, M.A. (1976) Epstein-Barr virus - is it time to develop a vaccine program?, J. Natl. Cancer Inst., 56, 697.

15. Lanier, A., Bender, T., Talbot, M., Wilmeth, S., Tschopp, C., Henle, W., Henle, G., Ritter, D. and Terasaki, P. (1980) Nasopharyngeal carcinoma in Alaskan Eskimos, Indians and Aleuts: a review of cases and study of Epstein-Barr virus, HLA and environmental risk factors, Cancer, 46, 2100.

16. Cammoun, M., Hoerner, G.V. and Mourali, N. (1974) Tumors of the nasopharynx in Tunisia: an anatomic and clinical study based on 143 cases, Cancer, 33, 184.

17. Clifford, P. (1970) A Review: on the epidemiology of nasopharyngeal carcinoma, Int. J. Cancer, 5, 287.

18. Marek, J. (1907) Multiple Nervenentzündung (Polyneuritis) bei Hühnern, Deutsch. Tierärztl. Wschr., 15, 417.

19. Payne, L.N., Frazier, J.A. and Powell, P.C. (1976) Pathogenesis of Marek's disease in: International Review of Experimental Pathology. Richter, G.W. and Epstein, M.A. (Eds.), Academic Press Inc., New York, San Francisco, London, 16, pp. 59-154.

20. Churchill, A.E., Payne, L.N. and Chubb, R.C. (1969) Immunization against Marek's disease using a live attenuated virus, Nature, 221, 744.

21. Okazaki, W., Purchase, H.G. and Burmester, B.R. (1970) Protection against Marek's disease by vaccination with a herpesvirus of turkeys, Avian Dis., 14, 413.

22. Meléndez, L.V., Hunt, R.D., Daniel, M.D., Garcia, F.G. and Fraser, C.E.O. (1969) Herpesvirus saimiri.II. An experimentally induced primate disease resembling reticulum cell sarcoma, Lab. Animal Care, 19, 378.

23. Laufs, R. and Steinke, H. (1975) Vaccination of non-human primates against malignant lymphoma, Nature, 253, 71.

24. Pearson, G., Dewey, F., Klein, G., Henle, G. and Henle, W. (1970) Relation between neutralization of Epstein-Barr virus and antibodies to cell-membrane antigens induced by the virus, J. Natl. Cancer Inst., 45, 989.

25. Pearson, G., Henle, G. and Henle, W. (1971) Production of antigens associated with Epstein-Barr virus in experimentally infected lymphoblastoid cell lines, J. Natl. Cancer Inst., 46, 1243.

26. Gergely, L., Klein, G. and Ernberg, I. (1971) Appearance of Epstein-Barr virus-associated antigens in infected Raji cells, Virology, 45, 10.

27. De Schryver, A., Klein, G., Hewetson, J., Rocchi, G., Henle, W., Henle, G., Moss, D.J. and Pope, J.H. (1974) Comparison of EBV neutralization tests based on abortive infection or transformation of lymphoid cells and their relation to membrane reactive antibodies (anti MA), Int. J. Cancer, 13, 353.

28. Kaaden, O.R. and Dietzschold, B. (1974) Alterations of the immunological specificity of plasma membranes of cells infected with Marek's disease and turkey herpesviruses, J. Gen. Virol., 25, 1.

29. Pearson, G.R. and Scott, R.E. (1977) Isolation of virus-free herpesvirus saimiri antigen-positive plasma membrane vesicles, Proc. Nat. Acad. Sci., 74, 2546.

30. Qualtière, L.F. and Pearson, G.R. (1979) Epstein-Barr virus-induced membrane antigens: immunochemical characterisation of Triton X100 solubilized viral membrane antigens from EBV-superinfected Raji cells, Int. J. Cancer, 23, 808.

31. Strnad, B.C., Neubauer, R.H., Rabin, H. and Mazur, R.A. (1979) Correlation between Epstein-Barr virus membrane antigen and three large cell surface glycoproteins, J. Virol., 32, 885.

32. Thorley-Lawson, D.A. and Edson, C.M. (1979) The polypeptides of the Epstein-Barr virus membrane antigen complex, J. Virol., 32, 458.

33. North, J.R., Morgan, A.J. and Epstein, M.A. (1980) Observations on the EB virus envelope and virus-determined membrane antigen (MA) polypeptides, Int. J. Cancer, 26, 231.

34. Qualtiere, L.F. and Pearson, G.R. (1980) Radioimmune precipitation study comparing the Epstein-Barr virus membrane antigens expressed on P_3HR-1 virus-superinfected Raji cells to those expressed on cells in a B95-8 virus-transformed producer culture activated with tumor-promoting agent (TPA), Virology, 102, 360.

35. North, J.R., Morgan, A.J., Thompson, J.L. and Epstein, M.A. (1982) Purified EB virus gp340 induces potent virus-neutralising antibodies when incorporated in liposomes, Proc. Nat. Acad. Sci., 79, 7504.

36. Hoffman, G.J., Lazarowitz, S.G. and Hayward, S.D. (1980) Monoclonal antibody against a 250,000-dalton glycoprotein of Epstein-Barr virus identifies a membrane antigen and a neutralising antigen, Proc. Nat. Acad. Sci., 77, 2979.

37. Thorley-Lawson, D.A. and Geilinger, K. (1980) Monoclonal antibodies against the major glycoprotein (gp350/220) of Epstein-Barr virus neutralise infectivity, Proc. Nat. Acad. Sci., 77, 5307.

38. Miller, G., Shope, T., Lisco, H., Stitt, D. and Lipman, M. (1972) Epstein-Barr virus: transformation, cytopathic changes, and viral antigens in squirrel monkey and marmoset leukocytes, Proc. Nat. Acad. Sci., 69, 383.

39. North, J.R., Morgan, A.J., Thompson, J.L. and Epstein, M.A. (1982) Quantification of an EB virus-associated membrane antigen (MA) component, J. Virol. Methods, 5, 55.

40. Morgan, A.J., North, J.R. and Epstein, M.A. (1983) Purification and properties of the gp340 component of Epstein-Barr (EB) virus membrane antigen (MA) in an immunogenic form, J. Gen. Virol., 64, 455.

41. Morein, B., Helenius, A., Simons, K., Pettersson, R., Käriäinen, L. and Schirrmacher, V. (1978) Effective subunit vaccines against an enveloped animal virus, Nature, 276, 715.

42. Manesis, E.K., Cameron, C.H. and Gregoriadis, G. (1979) Hepatitis B surface antigen-containing liposomes enhance humoral and cell-mediated immunity to the antigen, FEBS Lett., 102, 107.

43. Miller, G. (1979) Experimental carcinogenicity by the virus *in vivo* in: The Epstein-Barr Virus. Epstein, M.A. and Achong, B.G. (Eds.), Springer, Berlin, Heidelberg and New York, pp. 351-372.

44. Naylor, P.T., Larsen, H.L., Huang, L. and Rouse, B.T. (1982) *In vivo* induction of anti-Herpes simplex virus immune response by Type 1 antigens and Lipid A incorporated into liposomes, Infection and Immunity, 36, 1209.

45. Morgan, A.J. Epstein, M.A. and North, J.R. (1983) Comparative immunogenicity studies on Epstein-Barr (EB) virus membrane antigen (MA) with novel adjuvants in mice, rabbits and cotton-top tamarins, J. Med. Virol., in press.

46. Morgan, A.J., Smith, A.R., Barker, R.N. and Epstein, M.A. (1983) A structural investigation of the Epstein-Barr (EB) virus membrane antigen glycoprotein gp340, J. Gen. Virol., in press.

47. Epstein, M.A., Hunt, R.D. and Rabin, H. (1973) Pilot experiments with EB virus in owl monkeys *(Aotus trivirgatus) I.* Reticuloproliferative disease in an inoculated animal, Int. J. Cancer, 12, 309.

48. Epstein, M.A., Rabin, H., Ball, G., Rickinson, A.B., Jarvis, J. and Meléndez, L.V. (1973) Pilot experiments with EB virus in owl monkeys *(Aotus Trivirgatus) II.* EB virus in a cell line from an animal with reticuloproliferative disease, Int. J. Cancer, 12, 319.

49. Epstein, M.A., Zur Hausen, H., Ball, G. and Rabin, H. (1975) Pilot experiments with EB virus in owl monkeys *(Aotus trivirgatus) III.* Serological and biochemical findings in an animal with reticuloproliferative disease, Int. J. Cancer, 15, 17.

50. Shope, T., Dechairo, D. and Miller, G. (1973) Malignant lymphoma in cotton-top marmosets after inoculation with Epstein-Barr virus, Proc. Nat. Acad. Sci., 70, 2487.

51. Ma, N.S.F., Jones, T.C., Miller, A.C., Morgan, L.M. and Adams, E.A. (1976) Chromosome polymorphism and banding patterns in the owl monkey *(Aotus)*, Lab. Animal Sci., 26, 1022.

52. Ma, N.S.F., Rossan, R.N., Kelley, S.T., Harper, J.S., Bedard, M.T. and Jones, T.C. (1978) Banding patterns of the chromosomes of two new karyotypes of the owl monkey, *Aotus*, captured in Panama, J. Med. Primatol., 7, 146.

53. Ma, N.S.F. (1981) Chromosome evolution in the owl monkey, *Aotus*, Amer. J. Phys. Anthropol., 54, 293.

54. Brand, H.M. (1981) Husbandry and breeding of a newly established colony of cotton-topped tamarins *(Saguinus oedipus)*, Lab. Animals, 15, 7.

55. Kirkwood, J.K., Epstein, M.A. and Terlecki, A.J. (1983) Factors influencing population growth of a colony of cotton-top tamarins, Laboratory Animals, 17, 35.

56. Kirkwood, J.K. (1983) Effects of diet on health, weight and litter size in captive cotton-top tamarins *Saguinus oedipus oedipus*, Primates, in press.

57. Szmuness, W., Stevens, C.E., Zang, E.A., Harley, E.J. and Kellner, A. (1981) A controlled clinical trial of the efficacy of the Hepatitis B vaccine (Heptavax B): a final report, Hepatology, 1, 377.

58. Bittle, J.L., Houghten, R.A., Alexander, H., Shinnick, T.M., Sutcliffe, J.G., Lerner, R.A., Rowlands, D.J. and Brown, F. (1982) Protection against foot and mouth disease by immunization with a chemically synthesized peptide predicted from the viral nucleotide sequence, Nature, 298, 30.

59. Burkitt, D. (1963) A lymphoma syndrome in tropical Africa in: International Review of Experimental Pathology. Richter, G.W. and Epstein, M.A. (Eds.), Academic Press Inc., New York and London, 2, pp. 67-138.

The Role of Viruses in Human Cancer. Volume II
G. Giraldo and E. Beth editors

LATENT INFECTION AND GROWTH TRANSFORMATION BY EPSTEIN-BARR VIRUS

ELLIOTT KIEFF, TIMOTHY DAMBAUGH, KEVIN HENNESSY, SUSAN FENNEWALD, MARK HELLER, TAKUMI MATSUO AND MARY HUMMEL
Marjorie B. Kovler Laboratories, The University of Chicago, 910 East 58th Street, Chicago, Illinois 60637

INTRODUCTION

Epstein-Barr Virus is singularly efficient in its ability to induce proliferation of human B lymphocytes (1,2,3,4). Most or all B lymphocytes infected with this virus proliferate and maintain the virus genome in a state characterized by the expression of three latency transforming (LT) virus genes and no expression virus genes associated with early or late stages of virus replication (for review see ref. 5). Because of the efficiency and rapidity by which this virus transforms lymphocytes the virus is likely to include among the genes expressed in transformed cells, one or more genes which are capable of stimulating lymphocyte proliferation. These virus genes are likely to be of fundamental biologic interest because of the association of latent virus infection with human malignancies, because regulation of lymphocyte growth may be important to human immunology and because knowledge of lymphocyte growth regulation may be of broad importance to studies of regulation of cell growth. Since this virus usually does not go through an acute infection-virus replication cycle in cells in culture, genetic studies with the virus have been limited. Further, attempts in several laboratories to accomplish growth transformation with whole viral DNA have been unsuccessful thereby clouding a direct approach at defining transforming functions. Our approach is to define the viral genome and its transcripts, messenger RNAs and proteins within growth transformed cells. This should make it possible to determine the role of the virus DNA and of each of the virus gene products in the transformation process.

The viral genome. The EBV genome (Figure 1) is approximately 170 Kbp (6-10); 57% guanine and cytosine (6). Virion DNA is nicked, linear and double stranded. There are five classes of repeat sequence elements: a 538 bp terminal direct repeat, TR (10,11); a 3071 bp internal direct repeat, IR1 (12-14); a 124 bp direct repeat, IR2 (15); a simple triplet nucleotide repeat array, IR3; and a 103 bp direct repeat, IR4 (15). The genome is arranged as 8 TR-U1-8 IR1-U2-11 IR2-U3-300 IR3-U4-25 IR4-U5-8 TR (18,25); where U1-U2-U3-U4 and U5 are largely unique sequence domains of 11, 3, 40, 57 and 18 x 10^3 bp and the numbers preceeding the repeats indicate the average repeat number among different virus isolates. The number of copies of each repeat varies among each virus isolate (18,25).

Virus DNA in Transformed cells. After infection of lymphocytes and concomitant growth transformation, the entire virus DNA persists as a multicopy episome (20,21) and as an integrated viral DNA molecule (22). The multicopy episome is formed by covalent linkage between the TR sequences at both ends of linear DNA (8). With continued passage of the latently infected, growth transformed cells, defective viral DNA molecules are found in addition to the prototype forms (18). Recent experiments summarized in Figure 2 (Heller, Matsuo, OShiro and Kieff, in preparation) and previously published chromosome in situ hybridizations (22) with the Namalwa African Burkitt tumor cell line and the IB4 in vitro EBV transformed cell line demonstrate that (i) the entire EBV genome is integrated as a single site on chromosome 1 in the Namalwa Burkitt tumor cell line, (ii) there is no episomal DNA in Namalwa cells, (iii) integrated EBV DNA in Namalwa cells has the same organization as linear virion DNA except that the terminal repeat at each end of EBV DNA is covalently attached to cell DNA, (iv) cell DNA has been deleted from Namalwa cells at the site of EBV integration, (v) IB4 cells have several episomal copies of EBV DNA and one integrated copy of the EBV genome associated with IB4 chromosome 4 and (vi) the rightward end of one copy of the EBV genome in IB4 is not linked to the left end of the genome indicating that the right (and probably also the left) has recombined with cell DNA. In Namalwa, viral DNA is integrated near the locus of the human B lym gene (23).

Viral RNA in Transformed cells. At least three different messenger RNAs are regularly detected within latently infected transformed cells (16,25-27, Figure 1). These are encoded by DNA domains designated LT1, LT2 and LT3 which are shown in Figure 1 panel a-c, respectively. The LT1 region begins within IR1 and ends in U2; the LT2 region begins within U3 and ends in U4; and

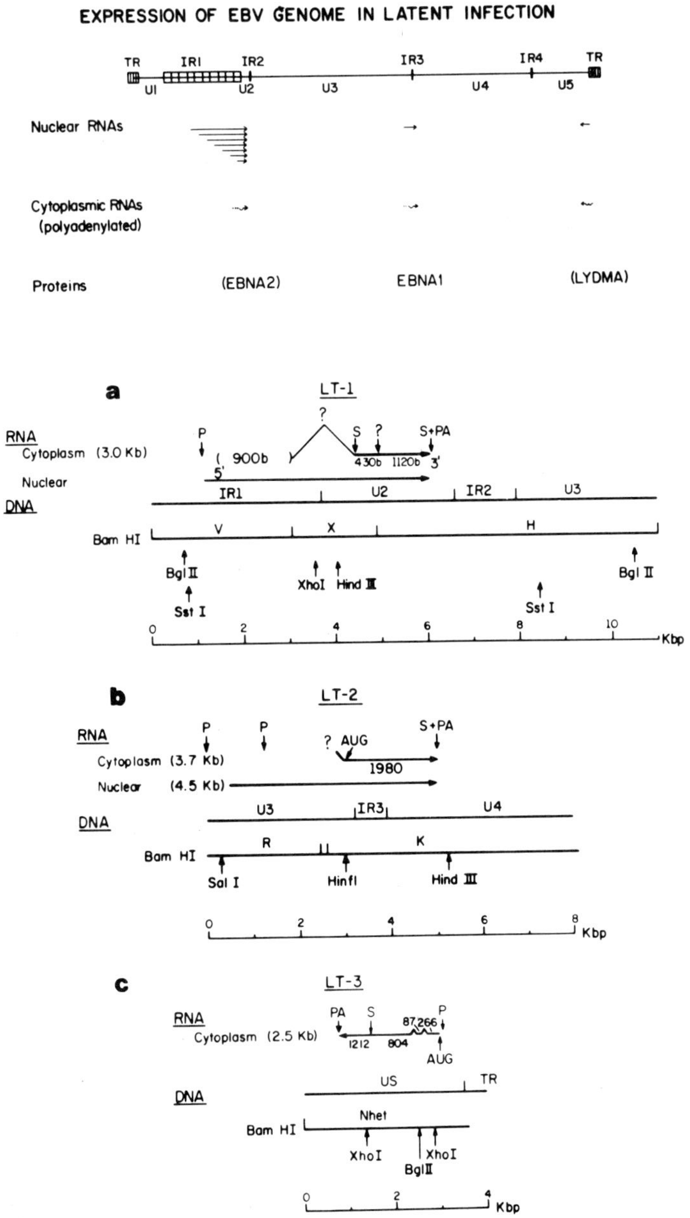

Fig. 1: Schematic structure of the EBV genome indicating largely unique (U) and tandom direct repetitive (R) elements. Also indicated are the nuclear and cytoplasmic polyadenylated RNAs and proteins encoded by EBV in latently infected, growth transformed IB4 cells. The IR1-U2, U3-IR3-U4 and U5 domains which encode RNA in latent transforming infection (LT-1, LT-2 and LT-3, respectively) are shown in detail in panel a, b and c.

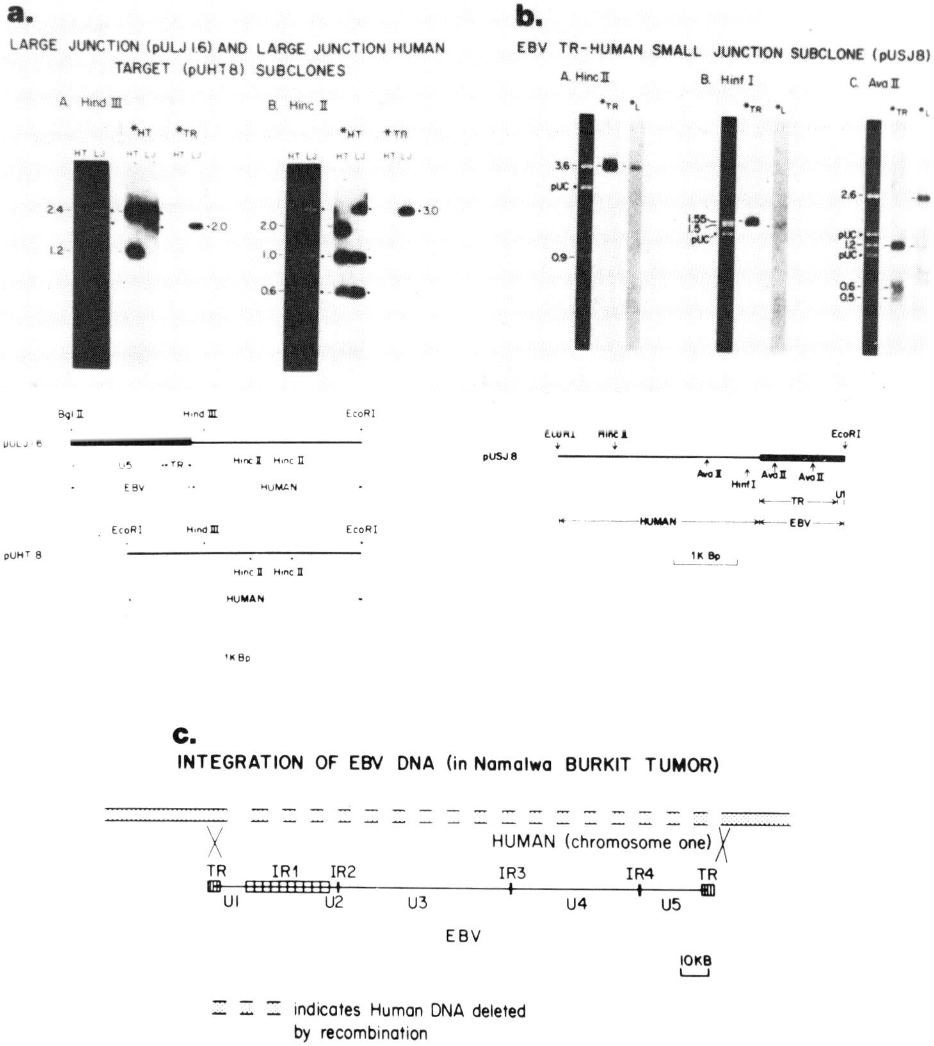

Fig. 2: Integration of EBV DNA in the Namalwa cell line. The junction fragments between EBV, U5-TR and cell DNA (large junction fragment, LJ, panel a) or between EBV U1-TR and cell DNA (small junction fragment, SJ, panel b) were cloned from Namalwa cells and used to identify clones of unrecombined cell DNA from a library of human cell DNA (24). Clones of the unrecombined cell DNA identified by Namalwa LJ (designated HT in panel A) are colinear with LJ cell DNA to the point of linkage of LJ to viral DNA. Similar results are found with SJ. However, LJ and SJ recombination sites are not in the same fragment of cell DNA indicating that there has been a deletion of cell DNA in Namalwa at the point of integration of EBV (summarized in panel c).

the LT3 region is entirely within U5. The LT1 and LT2 regions are transcribed left to right (16,26,27); while LT3 is transcribed right to left (26,Fennewald, van Santen and Kieff, in preparation).

The LT1 RNA is transcribed beginning to the right of a promotor in any repeat of IR1 past a polyadenylation signal in U2 (14,27, Dambaugh and Kieff, manuscript in preparation). Polyadenylated LT1 RNAs accumulate in the nucleus (27). These nuclear RNAs extend from just to the right of the promotor to the polyadenylation site. Non polyadenylated RNAs also accumulate in the nucleus. These RNAs map 3' to the polyadenylation site and may represent the end of the primary transcript. There are also numerous partially spliced intermediates in the nucleus (27). In the cytoplasm there are approximately three copies of a 3 kb polyadenylated RNA which is believed to be the LT1 mRNA. S1 experiments using contiguous clones define possibly contiguous 500 b and 1300 b exons at the right end of the BamHI X and the left end of the BamHI H (Figure 1a, 27). The 1300 b exon terminates at a polyadenylation site and it and the 500 b exon are entirely open in one continuous reading frame. The protein encoded by this open reading frame includes proline, glycine-arginine and lysine rich domains (Figure 4, Dambaugh and Kieff, manuscript in preparation).

The LT2 region encodes a 3.7 kb cytoplasmic polyadenylated mRNA (16,26, Figure 1b). There are approximately 3 copies of this RNA per transformed cell. S1 experiments define a single 2 kb exon which includes the right end of U3, all of IR3 and the left end of U4 (16,26,28,29). Other exon(s) of the mRNA have not been mapped. A long open reading frame within the 2 kb exon encodes most or all of the EBNA protein (16,26,28,29, Figure 3). A 4.5 kb polyadenylated RNA accumulates in the nucleus and may be the unspliced precursor to the 3.7 kb mRNA (16). If so, the promotor for the 4.5 and 3.7 kb RNAs could be either of the promotor like sequences indicated in Figure 1b (Barrell et al., personal communication).

The LT3 region encodes a 2.7 kb cytoplasmic polyadenylated RNA (26, Figure 1c). No other RNAs from this region have been detected in the nucleus. There are approximately 50 copies of this RNA per cell. It has therefore been possible to prepare cDNA clones to this RNA, although none of the clones thus far derived is a complete copy of the mRNA (Fennewald, van Santen and Kieff, manuscript in preparation). S1 experiments define three exons to the left of a promotor like sequence. The exons are 320, 90 and 2,000 b in length. At the 3 prime end of the 2 kb exon is a polyadenylation site. The first two exons and half of the third exon are open to translation in one reading frame. After this point, there are multiple stops in all three reading frames (Fennewald, van Santen and Kieff, manuscript in preparation). The anticipated protein size is approximately 42 kd. This is smaller than the estimated size of the protein translated in vitro from hybrid selected LT3 RNA (54 kd on SDS acrylamide gels) (Hummel and Kieff, unpublished observations).

Other RNAs have been detected in latently infected growth transformed cells including 2.3 and 2.0 kb polyadenylated RNAs encoded by the EcoRI J fragment of U1 (26). It is not known if these RNAs are regularly expressed in latent transforming infection and they have not been further characterized. Two polymerase III transcribed small RNAs similar in nucleotide sequence to adeno VA RNAs are found in latent transformed cells and in tumor tissue (26,30-32). Unlike the LT RNAs, whose abundance decreases when cells are induced to permit virus replication, transcription of the EBV small RNAs is markedly increased early in virus replication (30,33). The EBV small RNAs can substitute for adeno VA RNA in mutants deleted for both adeno VA RNAs (Thimappaya, personal communication). Thus, these RNAs are not as specifically associated with the latent transformed state as are the RNAs described in the preceeding sections. The small RNAs could be important in processing or translation of both latently and productively infected cell viral RNAs; or, their presence in latent transforming infection could result from leaky regulation of transcription of these RNAs whose biologic role may be solely in virus replication. Other polyadenylated RNAs have been detected in the cytoplasm of some latently infected growth transformed cells (30,34) and not in others and in some Burkitt or Nasopharyngeal cancers and not in others (30,35-38). These RNAs are regularly detected in large amounts early in the virus replicative cycle (33) and encode early antigens (39,40). The presence of these RNAs in some latently infected growth transformed cells and tumor tissue presumably reflects a very small fraction of cells which are expressing early antigens.

Viral Proteins in Transformed cells. Complement enhanced immunofluorescence with some EBV immune human sera can detect a new antigen in nuclei of latently infected, growth transformed cells (41). This new antigen is designated EBNA for Epstein-Barr Virus Nuclear Antigen. EBV related viruses induce a related but not identical antigen in human cells providing indirect evidence that EBNA is virus encoded (42-45). Immunoblots using immune human serum identify EBNA as a polymorphic protein product whose size varies among cell lines transformed with different EBV isolates (46). The size of the IR3 repeat element varies among the DNAs of EBV isolates and the variation quantitatively correlates with the difference among isolates in EBNA size (16,28,29, 46). Further, transfection with a DNA fragment including IR3 induces an EBNA like reactivity in mouse cells (47). The reading frame for translation of this region into EBNA and the definitive proof that LT2 encodes EBNA was obtained by synthesis of an IR3-Beta galactosidase fusion protein (29). Human sera with EBNA reactivity specifically react with this fusion protein; and, rabbit antisera against the fusion protein react specifically with EBNA on immunoblots of latently infected cell proteins (29). The open reading frame of the 2 kb exon indicated in figure 1b begins with an AUG and correlates with the open reading frame of the antigenic fusion protein. The translated protein from the open reading frame would be 641 AA or 56 kd (Figure 3). The EBNA actually encoded by this virus has an apparent size of 75kd on gels. The size discrepancy could be due to anomalous migration which might be expected because of low SDS binding and less than average mass per amino acid of the glycine alanine copolymer domain encoded by IR3 or it could be due to a short amino terminal polypeptide sequence provided by another exon. Translation from another exon is unlikely since the BamHI-Hind III fragment which includes little more than the 2 kb exon, induces a full EBNA size protein when transfected into cells (Summers, W. et al., and Levine, A. et al., 1983, Herpes Virus Workshop, Oxford, England); and, although there is a splice acceptor site a few nucleotides '5 to the AUG which presumably is the 5' terminus of

10 20 30

MET SER ASP GLU GLY PRO GLY THR GLY PRO GLY ASN GLY LEU GLY GLU LYS GLY ASP THR SER GLY PRO GLU GLY SER GLY GLY SER

40 50 60

GLY PRO GLN ARG ARG GLY GLY ASP ASN HIS GLY ARG GLY ARG GLY ARG GLY ARG GLY ARG GLY GLY GLY ARG PRO GLY ALA PRO GLY GLY

70 80 90

SER GLY SER GLY PRO ARG HIS ARG ASP GLY VAL ARG ARG PRO GLN LYS ARG PRO SER CYS ILE GLY CYS LYS GLY THR HIS GLY GLY THR

100 110 120

GLY ALA GLY ALA GLY ALA GLY GLY ALA GLY ALA GLY GLY ALA GLY ALA GLY GLY GLY ALA GLY ALA GLY GLY GLY ALA GLY GLY ALA GLY

130 140 150

GLY ALA GLY GLY ALA GLY ALA GLY GLY GLY ALA GLY ALA GLY GLY GLY ALA GLY GLY ALA GLY GLY ALA GLY ALA GLY GLY GLY ALA GLY

160 170 180

ALA GLY GLY GLY ALA GLY GLY ALA GLY ALA GLY GLY GLY ALA GLY GLY ALA GLY GLY ALA GLY ALA GLY GLY GLY ALA GLY ALA GLY GLY

190 200 210

GLY ALA GLY GLY ALA GLY ALA GLY GLY GLY ALA GLY GLY ALA GLY GLY ALA GLY ALA GLY GLY GLY ALA GLY ALA GLY GLY ALA GLY GLY

220 230 240

ALA GLY GLY ALA GLY ALA GLY GLY ALA GLY ALA GLY GLY GLY ALA GLY GLY ALA GLY GLY ALA GLY ALA GLY GLY ALA GLY ALA GLY GLY

250 260 270

ALA GLY ALA GLY GLY ALA GLY ALA GLY GLY ALA GLY GLY ALA GLY ALA GLY GLY ALA GLY GLY ALA GLY ALA GLY GLY ALA GLY GLY ALA

280 290 300

GLY ALA GLY GLY GLY ALA GLY GLY ALA GLY ALA GLY GLY GLY ALA GLY GLY ALA GLY ALA GLY GLY ALA GLY GLY ALA GLY ALA GLY GLY

310 320 330

ALA GLY GLY ALA GLY ALA GLY GLY ALA GLY GLY ALA GLY ALA GLY GLY GLY ALA GLY ALA GLY GLY ALA GLY ALA GLY GLY GLY GLY ARG

340 350 360

GLY ARG GLY GLY SER GLY GLY ARG GLY ARG GLY GLY SER GLY GLY ARG GLY ARG GLY GLY SER GLY GLY ARG ARG GLY ARG GLY ARG GLU

370 380 390

ARG ALA ARG GLY GLY SER ARG GLU ARG ALA ARG GLY ARG GLY ARG GLY ARG GLY GLU LYS ARG PRO ARG SER PRO SER SER GLN SER SER

400 410 420

SER SER GLY SER PRO PRO ARG ARG PRO PRO PRO GLY ARG ARG PRO PHE PHE HIS PRO VAL GLY GLU ALA ASP TYR PHE GLU TYR HIS GLN

430 440 450

GLU GLY GLY PRO ASP GLY GLU PRO ASP VAL PRO PRO GLY ALA ILE GLU GLN GLY PRO ALA ASP ASP PRO GLY GLU GLY PRO SER THR GLY

460 470 480

PRO ARG GLY GLN GLY ASP GLY GLY ARG ARG LYS LYS GLY GLY TRP PHE GLY LYS HIS ARG GLY GLN GLY GLY SER ASN PRO LYS PHE GLU

490 500 510

ASN ILE ALA GLU GLY LEU ARG ALA LEU LEU ALA ARG SER HIS VAL GLU ARG THR THR ASP GLU GLY THR TRP VAL ALA GLY VAL PHE VAL

520 530 540

TYR GLY GLY SER LYS THR SER LEU TYR ASN LEU ARG ARG GLY THR ALA LEU ALA ILE PRO GLN CYS ARG LEU THR PRO LEU SER ARG LEU

550 560 570

PRO PHE GLY MET ALA PRO GLY PRO GLY PRO GLN PRO GLY PRO LEU ARG GLU SER ILE VAL CYS TYR PHE MET VAL PHE LEU GLN THR HIS

580 590 600

ILE PHE ALA GLU VAL LEU LYS ASP ALA ILE LYS ASP LEU VAL MET THR LYS PRO ALA PRO THR CYS ASN ILE ARG VAL THR VAL CYS SER

610 620 630

PHE ASP ASP GLY VAL ASP LEU PRO PRO TRP PHE PRO PRO MET VAL GLU GLY ALA ALA ALA GLU GLY ASP ASP GLY ASP ASP GLY ASP GLU

640

GLY GLY ASP GLY ASP GLU GLY GLU GLU GLY GLN GLU

Fig. 3. Amino acid sequence of EBNA1 projected from the nucleotide sequence and reading frame as described in text. Nucleotides encoding amino acids 384-642 were sequenced by Barrell, B., Bankier, A., Baer, R., Biggin, M., Deininger, P., Farrell, P., Gibson, T., Hatfull, G., Hudson, G., Satchwell, S. and Sequin, C. at MRC, Cambridge, England. The glycine alanine copolymer domain encoded by IR3, the basic glycine arginine domain and the acidic aspartic acid glycine domains are underlined.

the 2 kb exon (16), there is no likely splice donor between the acceptor and the BamHI site (16). Thus, since the protein in cells transfected with the BamHI-HindIII fragment is "full size", it is likely that translation of EBNA begins at the methionine which begins the exon indicated in figure 1b. The nascent protein is rich in arginine and therefore would be basic (Figure 3). The multiple isoelectric forms which have been identified to range from basic to acidic presumably represent steps in post translational processing (48,49). Since the more acidic forms are not larger than the more basic forms, phosphorylation is a likely basis for the multiple forms differing in isoelectric property. The more acidic forms bind DNA less avidly (48,49). Early data suggest that EBNA may be a multimer and that it may be specifically complexed with cellular protein(s) (50,51).

Antisera raised in rabbits against the glycine alanine copolymer domain of EBNA react principally with EBNA, but also react faintly with a smaller protein in immunoblots of EBV infected and uninfected lymphoblastoid cells (29, Hennessy and Kieff, unpublished observations). These sera do not react with any protein larger than EBNA. Analysis of human cellular DNAs homologous to the EBV IR3 (17, Heller, Deininger and Kieff, manuscript in preparation) which encodes the glycine-alanine copolymer domain of EBNA and studies of expression of the human IR3 homologues in cells are being pursued.

Some human sera with EBNA reactivity identify EBNA and another protein of 82 kd in latently infected growth transformed cells (29). Some of these immune human sera react more strongly with the 82 kd protein than with EBNA. The 82 kd protein does not vary in size among cell lines which are infected with EBV isolates whose EBNA protein varies from 65-80 kd (29). Anti EBNA rabbit sera have no reactivity with the 82 kd protein. These data clearly indicate that the 82 kd protein does not include the glycine alanine copolymer domain of EBNA. Lymphoblastoid cells not infected with EBV do not have the 82 kd protein. The 82 kd protein is a nuclear protein (29). Sera which preferentially react with the 82 kd protein identify an intranuclear antigen in EBV infected cells in indirect immunofluorescence. The staining pattern is similar to that observed for EBNA using complement enhanced fluorescence (29). Preliminary data suggest that the 82 kd protein is encoded by LT1. First, P3HR-1 cells are deleted for the U2 part of LT1 (18,52-54) and are unique among EBV infected cell lines in that the virus produced is unable to growth transform lymphocytes (55,56). P3HR-1 cells are also unique in the absence of the 82 kd protein (Hennessy and Kieff, manuscript in preparation). Second, the amino acid sequence of the protein translated from the open reading frame at the 3' end of the LT1 RNA is arginine and lysine rich and highly basic as might be expected for a nuclear DNA binding protein (Figure 4).

Y

TRANSLATION OF : XHCS112

```
        1429        1439        1449        1459        1469        1479        1489        1499        1509
TGA ACC TGT GGT TGG GCA GGT ACA TGC CAA CAA CCT TCT AAG CAC CCG CGC TTG TGT TTT GCT TTA TCT GCC GCC ATC ATG CCT ACA TTC
*** Thr Cys Gly Trp Ala Gly Thr Cys Gln Gln Pro Ser Lys His Pro Arg Leu Cys Phe Ala Leu Ser Ala Ala Ile Met Pro Thr Phe

        1519        1529        1539        1549        1559        1569        1579        1589        1599
TAT CTT GCG TTA CAT GGG GGA CAA ACA TAT CAT CTA ATT GTT GAC ACG GAT AGT CTT GGA AAC CCG TCA CTC TCA GTA ATT CCC TCG AAT
Tyr Leu Ala Leu His Gly Gly Gln Thr Tyr His Leu Ile Val Asp Thr Asp Ser Leu Gly Asn Pro Ser Leu Ser Val Ile Pro Ser Asn

        1609        1619        1629        1639        1649        1659        1669        1679        1689
CCC TAC CAG GAA CAA CTG TCA GAC ACT CCA TTA ATT CCA CTA ACA ATC TTT GTT GGG GAA AAC ACG GGG GTG CCC CCA CCA CTC CCA CCA
Pro Tyr Gln Glu Gln Leu Ser Asp Thr Pro Leu Ile Pro Leu Thr Ile Phe Val Gly Glu Asn Thr Gly Val Pro Pro Pro Leu Pro Pro

        1699        1709        1719        1729        1739        1749        1759        1769        1779
CCC CCC CCA CCA CCA CCC CCA CCA CCC CCA CCA CCC CCA CCA CCC CCA CCA CCC CCA CCA CCC CCA CCA CCT CCA CCA CCT TCA CCA CCA
Pro Pro Pro Pro Pro Pro Pro Pro Pro Pro Pro Pro Pro Pro Pro Pro Pro Pro Pro Pro Pro Pro Pro Pro Pro Pro Pro Ser Pro Pro

        1789        1799        1809        1819        1829        1839        1849        1859        1869
CCC CCG CCC CCA CCA CCC CCA CCA CCT CAG CGC AGG GAT GCC TGG ACA CAA GAG CCA TCA CCT CTT GAT AGG GAT CCG CTA GGA TAT GAC
Pro Pro Pro Pro Pro Pro Pro Pro Pro Gln Arg Arg Asp Ala Trp Thr Gln Glu Pro Ser Pro Leu Asp Arg Asp Pro Leu Gly Tyr Asp

        1879        1889        1899        1909        1919        1929        1939        1949        1959
GTC GGG CAT GGA CCT CTA GCA TCT GCT ATG CGA ATG CTT TGG ATG GCT AAT TAT ATT GTA AGA CAA TCA CGG GGT GAC CGG GGC CTT ATT
Val Gly His Gly Pro Leu Ala Ser Ala Met Arg Met Leu Trp Met Ala Asn Tyr Ile Val Arg Gln Ser Arg Gly Asp Arg Gly Leu Ile

        1969        1979        1989        1999        2009        2019        2029        2039        2049
TTG CCA CAA GGC CCA CAA ACA GCC CCT CAG GCC AGG TTC GTC CAG CCA CAT GTC CCC CCT CTA CGC CCG ACA GCA CCC ACC ATT TTG TCA
Leu Pro Gln Gly Pro Gln Thr Ala Pro Gln Ala Arg Leu Val Gln Pro His Val Pro Pro Leu Arg Pro Thr Ala Pro Thr Ile Leu Ser

        2059        2069        2079        2089        2099        2109        2119        2129        2139
CCT CTG TCA CAA CCG AGG CTT ACC CCT CCA CAA CCA CTC ATG ATG CCA CCA AGG CCT ACC CCT CCT ACC CCT CTG CCA CCT GCA ACA CTA
Pro Leu Ser Gln Pro Arg Leu Thr Pro Pro Gln Pro Leu Met Met Pro Pro Arg Pro Thr Pro Pro Thr Pro Leu Pro Pro Ala Thr Leu

        2149        2159        2169        2179        2189        2199        2209        2219        2229
ACG GTG CCA CCA AGG CCT ACC CGT CCT ACC ACT CTG CCA CCC ACA CCA CTA CTC ACG GTA CTA CAA AGG CCT ACC GAA CTT CAA CCC ACA
Thr Val Pro Pro Arg Pro Thr Arg Pro Thr Thr Leu Pro Pro Thr Pro Leu Leu Thr Val Leu Gln Arg Pro Thr Glu Leu Gln Pro Thr

        2239        2249        2259        2269        2279        2289        2299        2309        2319
CCA TCA CCA CCA CGC ATG CAT CTC CCT GTC TTG CAT GTG CCA GAC CAA TCA ATG CAC CCT CTT ACT CAT CAA AGC ACC CCA AAT GAT CCA
Pro Ser Pro Pro Arg Met His Leu Pro Val Leu His Val Pro Asp Gln Ser Met His Pro Leu Thr His Gln Ser Thr Pro Asn Asp Pro

        2329        2339        2349        2359        2369        2379        2389        2399        2409
GAT AGT CCA GAA CCA CGG TCC CCG ACT GTA TTT TAT AAC ATT CCA CCT ATG CCA TTA CCC CCC TCA CAA TTG CCA CCA CCA GCA GCA CCA
Asp Ser Pro Glu Pro Arg Ser Pro Thr Val Phe Tyr Asn Ile Pro Pro Met Pro Leu Pro Pro Ser Gln Leu Pro Pro Pro Ala Ala Pro

        2419        2429        2439        2449        2459        2469        2479        2489        2499
GCA CAG CCA CCT CCA GGG GTC ATC AAC GAC CAA CAA TTA CAT CAT CTA CCC TCG GGG CCA CCA TGG TGG CCA CCC ATC TGC GAC CCC CCG
Ala Gln Pro Pro Pro Gly Val Ile Asn Asp Gln Gln Leu His His Leu Pro Ser Gly Pro Pro Trp Trp Pro Pro Ile Cys Asp Pro Pro

        2509        2519        2529        2539        2549        2559        2569        2579        2589
CAA CCC TCT AAG ACT CAA GGC CAG AGC CGG GGA CAG AGC AGG GGG AGG GGC AGG GGC AGG GGC AGG GGC AGG GGC AAG GGC AAG TCC AGG
Gln Pro Ser Lys Thr Gln Gly Gln Ser Arg Gly Gln Ser Arg Gly Arg Gly Arg Gly Arg Gly Arg Gly Arg Gly Lys Gly Lys Ser Arg

        2599        2609        2619        2629        2639        2649        2659        2669        2679
GAC AAG CAA CGC AAG CCC GGT GGA CCT TGG AGA CCA GAG CCA AAC ACC TCC AGT CCT AGC ATG CCT GAA CTA AGT CCA GTC CTC GGT CTT
Asp Lys Gln Arg Lys Pro Gly Gly Pro Trp Arg Pro Glu Pro Asn Thr Ser Ser Pro Ser Met Pro Glu Leu Ser Pro Val Leu Gly Leu

        2689        2699        2709        2719        2729        2739        2749        2759        2769
CAT CAG GGA CAA GGG GCT GGG GAC TCA CCA ACT CCT GGC CCA TCC AAT GCC GCC CCC GTT TGT AGA AAT TCA CAC ACG GCA ACC CCT AAC
His Gln Gly Gln Gly Ala Gly Asp Ser Pro Thr Pro Gly Pro Ser Asn Ala Ala Pro Val Cys Arg Asn Ser His Thr Ala Thr Pro Asn

        2779        2789        2799        2809        2819        2829        2839        2849        2859
GTT TCA CCA ATA CAT GAA CCG GAG TCC CAT AAT AGC CCA GAG GCT CCC ATT CTC TTC CCC GAT GAT TGG TAT CCT CCA TCT ATA GAC CCC
Val Ser Pro Ile His Glu Pro Glu Ser His Asn Ser Pro Glu Ala Pro Ile Leu Phe Pro Asp Asp Trp Tyr Pro Pro Ser Ile Asp Pro

        2869        2879        2889        2899        2909        2919        2929        2939        2949
GCA GAC TTA GAC GAA AGT TGG GAT TAC ATT TTT GAG ACA ACA GAA TCT CCT AGC TCA GAT GAA GAT TAT GTG GAG GGA CCC AGT AAA AGA
Ala Asp Leu Asp Glu Ser Trp Asp Tyr Ile Phe Glu Thr Thr Glu Ser Pro Ser Ser Asp Glu Asp Tyr Val Glu Gly Pro Ser Lys Arg

        2959        2969        2979        2989        2999        3009        3019        3029        3039
CCT CGC CCC TCC ATC CAG TAA AAA CCC TTG CCC TCT CCA GCA ACC AAT GTA TCC CAA ATA AAT GTT ACT TCT TTT GCT CTT AAC CAT TGA
Pro Arg Pro Ser Ile Gln *** Lys Pro Leu Pro Ser Pro Ala Thr Asn Val Ser Gln Ile Asn Val Thr Ser Phe Ala Leu Asn His ***
```

Fig. 4. Nucleotide sequence and translated protein through the exon at the 3' end of the LT1 RNA shown in Figure 1a. The BamHI site between BamHI H and X is indicated by a box around nucleotides 1851-1856. The BamHl X portion of the protein is proline rich. The BamHI H portion includes a glycine arginine rich region at nucleotides 2539-2579 (Dambaugh and Kieff, manuscript in preparation).

```
             640             650             660             670             680             690             700             710             720
ACA CGC GTT ACT CTG ACG TAG CCG CCC TAC ATA AGC CTC TCA CAC TGC TCT GCC CCC TTC TTT CCT CAA CTG CCT TGC TCC TGA CAC ACT

             730             740             750             760             770             780             790             800             810
GCC CTG AGG ATG GAA CAC GAC CTT GAG AGG GGC CCA CCG GGC CCG CGA CGG CCC CCT CGA GGA CCC CCC CTC TCC TCT TCC CTA GGC CTT
            Met Glu His Asp Leu Glu Arg Gly Pro Pro Gly Pro Arg Arg Pro Pro Arg Gly Pro Pro Leu Ser Ser Ser Leu Gly Leu
             820             830             840             850             860             870             880             890             900
GCT CTC CTT CTC CTC CTC TTG GCG CTA CTG TTT TGG CTG TAC ATC GTT ATG AGT GAC TGG ACT GGA GGA GCC CTC CTT GTC CTC TAT TCC
Ala Leu Leu Leu Leu Leu Leu Ala Leu Leu Phe Trp Leu Tyr Ile Val Met Ser Asp Trp Thr Gly Gly Ala Leu Leu Val Leu Tyr Ser
             910             920             930             940             950             960             970             980             990
TTT GCT CTC ATG CTT ATA ATT ATA ATT TTG ATC ATC TTT ATC TTC AGA AGA GAC CTT CTC TGT CCA CTT GGA GCC CTT TGT ATA CTC CTA
Phe Ala Leu Met Leu Ile Ile Ile Ile Leu Ile Ile Phe Ile Phe Arg Arg Asp Leu Leu Cys Pro Leu Gly Ala Leu Cys Ile Leu Leu
             1000            1010            1020            1030            1040            1050            1060            1070            1080
CTG ATG AGT AAG TAT TAC ACC CTT TGC CCC ACA CCC CCT TTC CCT TAC TCT TCC TTC TCT AAC GCA CTT TCT CCT CTT TCC CCA GTC ACC
Leu Met                                                                                                             Ile Thr
             1090            1100            1110            1120            1130            1140            1150            1160            1170
CTC CTG CTC ATC GCT CTC TGG AAT TTG CAC GGA CAG GCA TTG TTC CTT GGA ATT GTG CTG TTC ATC TTC GGG TGC TTA CTT GGT AAG ATC
Leu Leu Leu Ile Ala Leu Trp Asn Leu His Gly Gln Ala Leu Phe Leu Gly Ile Val Leu Phe Ile Phe Gly Cys Leu Leu
             1180            1190            1200            1210            1220            1230            1240            1250            1260
TAAC ATT CCC TAG GAA TTA TTT ACC ACA CCC CCA CTT TTC CAA CCC TAA CAC TCT TTT TTC AAC GCA GTC TTA GGT ATC TGG ATC TAC TTA
                                                                                          Val Leu Gly Ile Trp Ile Tyr Leu
             1270            1280            1290            1300            1310            1320            1330            1340            1350
TTG GAG ATG CTC TGG CGA CTT GGT GCC ACC ATC TGG CAG CTT TTG GCC TTC TTC CTA GCC TTC TTC CTA GAC CTC ATC CTG CTC ATT ATT
Leu Glu Met Leu Trp Arg Leu Gly Ala Thr Ile Trp Gln Leu Leu Ala Phe Phe Leu Ala Phe Phe Leu Asp Leu Ile Leu Leu Ile Ile
             1360            1370            1380            1390            1400            1410            1420            1430            1440
GCT CTC TAT CTA CAA CAA AAC TGG TGG ACT CTA TTG GTT GAT CTC CTT TGG CTC CTC CTG TTT CTG GCG ATT TTA ATC TGG ATG TAT TAC
Ala Leu Tyr Leu Gln Gln Asn Trp Trp Thr Leu Leu Val Asp Leu Leu Trp Leu Leu Leu Phe Leu Ala Ile Leu Ile Trp Met Tyr Tyr
             1450            1460            1470            1480            1490            1500            1510            1520            1530
CAT GGA CAA CGA CAC AGT GAT GAA CAC CAC CAC GAT GAC TCC CTC CCG CAC CCT CAA CAA GCT ACC GAT GAT TCT GGC CAT GAA TCT GAC
His Gly Gln Arg His Ser Asp Glu His His His Asp Asp Ser Leu Pro His Pro Gln Gln Ala Thr Asp Asp Ser Gly His Glu Ser Asp
             1540            1550            1560            1570            1580            1590            1600            1610            1620
TCT AAC TCC AAC GAG GGC AGA CAC CAC CTG CTC GTG AGT GGA GCC GGC GAC GGA CCC CCA CTC TGC TCT CAA AAC CTA GGC GCA CCT GGA
Ser Asn Ser Asn Glu Gly Arg His His Leu Leu Val Ser Gly Ala Gly Asp Gly Pro Pro Leu Cys Ser Gln Asn Leu Gly Ala Pro Gly
             1630            1640            1650            1660            1670            1680            1690            1700            1710
GGT GGT CCT GAC AAT GGC CCA CAG GAC CCT GAC AAC ACT GAT GAC AAT GGC CCA CAG GAC CCT GAC AAC ACT GAT GAC AAT GGC CCA CAT
Gly Gly Pro Asp Asn Gly Pro Gln Asp Pro Asp Asn Thr Asp Asp Asn Gly Pro Gln Asp Pro Asp Asn Thr Asp Asp Asn Gly Pro His
             1720            1730            1740            1750            1760            1770            1780            1790            1800
GAC CCG CTG CCT CAG GAC CCT GAC AAC ACT GAT GAC AAT GGC CCA CAG GAC CCT GAC AAC ACT GAT GAC AAT GGC CCA CAT GAC CCG CTG
Asp Pro Leu Pro Gln Asp Pro Asp Asn Thr Asp Asp Asn Gly Pro Gln Asp Pro Asp Asn Thr Asp Asp Asn Gly Pro His Asp Pro Leu
             1810            1820            1830            1840            1850            1860            1870            1880            1890
CCT CAT AGC CCT AGC GAC TCT GCT GGA AAT GAT GGA GGC CCT CCA CAA TTG ACG GAA GAG GTT GAA AAC AAA GGA GGT GAC CAG GGC CCG
Pro His Ser Pro Ser Asp Ser Ala Gly Asn Asp Gly Gly Pro Pro Gln Leu Thr Glu Glu Val Glu Asn Lys Gly Gly Asp Gln Gly Pro
             1900            1910            1920            1930            1940            1950            1960            1970            1980
CCT TTG ATG ACA GAC GGA GGC GGC GGT CAT AGT CAT GAT TCC GGC CAT GGC GGC GGT GAT CCA CAC CTT CCT ACG CTG CTT TTG GGT TCT
Pro Leu Met Thr Asp Gly Gly Gly Gly His Ser His Asp Ser Gly His Gly Gly Gly Asp Pro His Leu Pro Thr Leu Leu Leu Gly Ser
             1990            2000            2010            2020            2030            2040            2050            2060            2070
TCT GGT TCC GGT GGA GAT GAT GAC GAC CCC CAC GGC CCA GTT CAG CTA AGC TAC TAT GAC TAA CCT TTC TTT ACT TCT AGG CAT TAC CAT
Ser Gly Ser Gly Gly Asp Asp Asp Asp Pro His Gly Pro Val Gln Leu Ser Tyr Tyr Asp ***
```

Fig. 5. Nucleotide and projected amino acid sequence of the exons of the LT3 domain. (Fennewald, van Santen and Kieff, manuscript in preparation)

Latently infected, growth transformed cells are killed by EBV immune T lymphocytes (57). The killing is HLA restricted (58). A virus induced or virus encoded antigen is therefore presumed to be present on the surface of latently infected cells. Attempts to raise monoclonal antibodies to the new surface antigen have yielded antibodies to antigens which also appear on the surface of stimulated, but not necessarily virus infected cells (59,60). Such cells are presumably not killed by immune T cells. EBNA1 and EBNA2 are nuclear proteins. Either could be a cell surface protein as well. However, most abundant RNA in latently infected cells is encoded by the LT3 region; and, the amino acid sequence of the protein projected from the nucleotide sequence of the LT3 message includes several hydrophobic domains which could be transmembrane portions of a membrane protein. Furthermore, more than half of the protein is hydrophobic as might be compatible with a membrane protein but not with a soluble cytoplasmic or nuclear protein (Figure 5).

To summarize briefly: We have identified three genes each of which appears to encode a single RNA and a single protein in latently infected growth transformed cells. Identification of these genes is a first step to discerning their function in cells. We are making parts of each of these genes in bacteria. The bacterial products enable us to make antisera to identify the viral proteins within cells. They also can be used to directly study mechanism of association of EBNA1 and 2 with DNA and of Lydma association with the cell membrane. Some of these biosynthetic peptides such as EBNA1 are useful as new diagnostic reagents to detect antibody to EBV. It is clear that we are only on the first or second leg of a sgnificant and lengthy journey.

ACKNOWLEDGEMENTS

This research was supported by grants from the U. S. Public Health Service, CA 17281 and CA 19264, and a grant from the American Cancer Society, ACS MV32I. K.H. and M.H. are recipients of National Research Service Awards from the National Institutes of Health, GM 07183 and AI 07099, respectively. T.M. is the recipient of a Cancer Research Campaign International Fellowship from the International Union Against Cancer. E.K. is the recipient of a Faculty Research Award from the American Cancer Society. We thank Terri Cole, Larvan Chamnankit and Patricia Morrison for excellent assistance.

REFERENCES

1. Pope, J., Horne, M. and Scott, W. (1968) Int. J. Cancer 3:857-866.

2. Henle, W., Diehl, V., Kohn, G., zur Hausen, H. and Henle, G. (1967) Science 157:1064-1065.

3. Sugden, B. and Mark, W. (1977) J. Virol. 23:503-508.

4. Henderson, E., Miller, G., Robinson, J. and Heston, L. (1977) Virology 76:152-163.

5. Kieff, E., Dambaugh, T., Hummel, M. and Heller, M. (1983) Advances in Viral Oncology. Vol. 3:133-182, Raven Press.

6. Pritchett, R., Hayward, S. and Kieff, E. (1975) J. Virol. 15:556-569.

7. Hayward, S. and Kieff, E. (1977) J. Virol. 23:421-429.

8. Dambaugh, T., Beisel, C., Hummel, M., King, W., Fennewald, S., Cheung, A., Heller, M., Raab-Traub, N. and Kieff, E. (1980) PNAS 77:299-3003.

9. Raab-Traub, N., Dambaugh, T. and Kieff, E. (1980) Cell 22:257-267.

10. Given, D. and Kieff, E. (1978) J. Virol. 28:524-542.

11. Given, D., Yee, D., Griem, K. and Kieff, E. (1979) J. Virol. 30:852-862.

12. Given, G. and Kieff, E. (1979) J. Virol. 31:315-324.

13. Cheung, A. and Kieff, E. (1981) J. Virol. 40:501-507.

14. Cheung, A. and Kieff, E. (1982) J. Virol. 44:286-294.

15. Dambaugh, T. and Kieff, E. (1982) J. Virol. 44:823-833.

16. Heller, M., van Santen, V. and Kieff, E. (1982) J. Virol. 44:311-320.

17. Heller, M., Henderson, A. and Kieff, E. (1982) PNAS 79:5916-5920.

18. Heller, M., Dambaugh, T. and Kieff, E. (1981) J. Virol. 38:632-648.

19. Dambaugh, T., Raab-Traub, N., Heller, M., Beisel, C., Hummel,

M., Cheung, A., Fennewald, S., King, W. and Kieff, E. (1980) Annals N.Y. Acad. Sci. 602:711-719.

20. Nonoyama, M. and Pagano, J. (1972) Nature New Biol. 238:169-171.

21. Lindahl, T., Adams, A. Bjursell, G., Bornkamm, G., Kaschka-Dierich, G. and Jehn, U. (1967) J. Mol. Biol. 102:511-530.

22. Henderson, A., Ripley, S. Heller, M. and Kieff, E. (1983) PNAS 80:1987-1991.

23. Morton, C., Taub, R., Diamond, A., Lane M., Cooper, G. and Leder, P. (1984) Science 223:173-175.

24. Lawn, R., Fritsch, E., Parker, R., Blake, G. and Maniatis, T. (1978) Cell 15:1157-1174.

25. King, W., Thomas-Powell, A., Raab-Traub, N., Hawke, M. and Kieff, E. (1980) J. Virol. 36:506-518.

26. van Santen, V., Cheung, A. and Kieff, E. (1981) PNAS 78:1930-1934.

27. van Santen, V., Cheung, A., Hummel, M. and Kieff, E. (1983) J. Virol. 46:424-433.

28. Hennessy, K., Heller, M., van Santen, V. and Kieff, E. (1983) Science 220:1396-1398.

29. Hennessy, K. and Kieff, E. (1983). PNAS 80:5665-5669.

30. King, W., van Santen, V. and Kieff, E. (1981) J. Virol. 38:649-660.

31. Lerner, M., Andrews, N., Miller, G. and Steitz, J. (1981) PNAS 78:805-809.

32. Rosa, M., Sattieb, C., Lerner, M. and Steitz, J. (1981) Mol. Cell Biol. 1:785-796.

33. Hummel, M. and Kieff, E. (1982) J. Virol. 43:262-272.

34. Weigel, R. and Miller, G. (1983) Virology 125:287-298.

35. Powell, A., King, W. and Kieff, E. (1979) J. Virol. 29:261-274.

36. Dambaugh, T., Nkrumah, F., Biggar, R. and Kieff, E. (1979) Cell 16:313-322.

37. Arrand, J., Walsh-Arrand, J. and Rymo, L. (1983) EMBO J. 2:1673-1683.

38. Raab-Traub, N., Hood, R., Yang, C., Henry, B. and Pagano, J. (1983) J. Virol. 48:580-590.

39. Hummel, M. and Kieff, E. (1982) PNAS 79:5698-5702.

40. Pearson, G., Vroman, B., Chase, B., Sculley, T., Hummel, M. and Kieff, E. (1983) J. Virol. 47:193-201.

41. Reedman, B.M. and Klein, G. (1973) Int. J. Cancer 11:499-520.

42. Falk, L., Deinhardt, F., Nonoyama, M., Wolfe, L., Bergholz, C., Lapin, B. Yakoleva, L., Agrba, V., Henle, G., and Henle, W., (1976) Int. J. Cancer 18:798-807.

43. Gerber, P. and Birch, S. (1967) PNAS 58:478-484.

44. Gerber, P., Pritchett, R. and Kieff, E. (1976) J. Virol. 19:1090-1100.

45. Gerber, P., Kalter, S., Schidlovsky, G., Peterson, W. and Daniel, M. (1977) Int. J. Cancer 20:448-459.

46. Strnad, B., Schuster, T. Hopkins, R., Neubauer, R. and Rabin, H. (1981) J. Virol. 38:996-1004.

47. Summers, W., Grogan, E., Sheed, D., Robert, M., Lei, C. and Miller, G. (1982) PNAS 79:5688-5692.

48. Spelsberg, T., Sculley, T., Pikler, G., Gilbert, J. and Pearson, G. (1982) J. Virol. 43:555-565.

49. Scully, T., Kreofsky, T., Pearson, G. and Spelsberg, T. (1983) J. Biol. Chem. 258:3974-3982.

50. Lenoir, G., Bethelon, M., Favre, M. and de The, G. (1976) J. Virol. 17:672-674.

51. Luka, J., Lindahl, T. and Klein, G. (1978) J. Virol. 27:604-611.

52. King, W., Dambaugh, T., Heller, M., Dowling, J. and Kieff, E. (1982) J. Virol. 43:979-986.

53. Bornkamm, G., Hudewentz, J., Freese, U. and Zimber, V. (1982) J. Virol. 43:952-968.

54. Hayward, D., Nazarowitz, S., Hayward, S. (1982) J. Virol. 43:201-212.

55. Miller, G., Robinson, J., Heston, N. and Lipman, M. (1974) PNAS 71:4006-4010.

56. Ragona, G., Ernberg, J. and Klein, G. (1980) Virology 101:553-557.

57. Svedmyr, E., Jondal, M., Henle, W., Weiland, O., Rombo, L. and Klein, G. (1978) J. Clin. Lab. Immunol. 1(3):225-232.

58. Rickinson, A., Wallace, L. and Epstein, M. (1980) Nature 283:865-867.

59. Sugden, B. and Metzenberg, S. (1983) J. Virol. 46:800-807.

60. Thorley-Lawson, D., Schooley, R., Bhan, A. and Nadler, L. (1982) Cell 30:415-425.

© 1984 Elsevier Science Publishers B.V.
The Role of Viruses in Human Cancer. Volume II
G. Giraldo and E. Beth editors

EPSTEIN-BARR VIRUS - ONCOGENESIS IN IMMUNE DEFICIENT INDIVIDUALS

David T. Purtilo, M.D.
Departments of Pathology and Laboratory Medicine, and Pediatrics, and the Eppley Institute for Research in Cancer and Allied Diseases, University of Nebraska Medical Center, 42nd & Dewey Avenue, Omaha, NE 68105 (U.S.A.)

INTRODUCTION

A broad spectrum of diseases has been associated with Epstein-Barr virus (EBV) since its discovery two decades ago. Traditionally, Burkitt's lymphoma (BL), nasopharyngeal carcinoma (NPC), and infectious mononucleosis (IM) have been linked etiologically with the virus (1). Research during the first decade following the discovery of EBV focused on the role of the virus in the induction of BL, NPC, and IM. Paralleling the development of immunology in the 1970's was the accrual of knowledge regarding immunological events occurring during IM. During the subsequent decade, new diseases induced by the virus in immune deficient individuals have continued to be elaborated (2).

Summarized here is the immunobiology of EBV, particularly the clinical consequences resulting from immune deficient responses to the virus. Prior to discussing malignant lymphoma (ML), EBV, and immune deficiency, the normal immune defenses against the virus will be summarized. Acquired progressive immune deficiency disorders triggered in part by EBV will be discussed as it concerns destruction of thymus gland and impairment of immune regulation in individuals with preexisting primary or acquired immune deficiency. On primary infection, EBV induces a polyclonal proliferation of B cells which are normally dealt with efficiently by watertight immune surveillance mechanisms. However, in immune deficient individuals, we have proposed that a specific cytogenetic error (i.e., chromosomal 8;14 translocation) occurs in a B cell. This event likely endows the altered cell with the capacity to escape immune surveillance mechanisms. This process will be discussed and illustrated. Examples of primary immune deficiency syndromes and the development of a spectrum of diseases (Figure 1), especially ML will be elaborated. Similarly, individuals

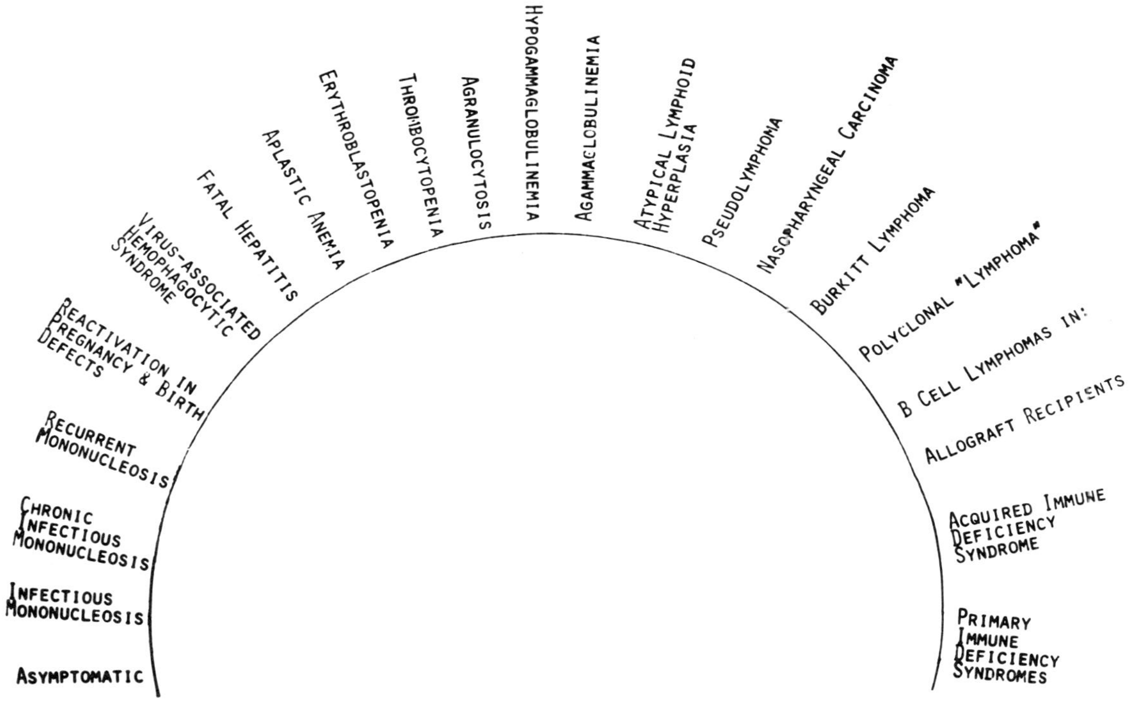

SPECTRUM OF EBV-ASSOCIATED DISEASES

Fig. 1. Sectrum of EBV-associated diseases. Many of the manifestations of EBV infections which cause abnormalities in the blood are shown in the diagram. The underlying basis for the majority of these diseases is abnormal immune regulation. Cytogenetic alterations probably account for the evolution to monoclonal malignancies. Nasopharyngeal carcinoma stands apart from the other disorders. Thus far, immune deficiency has not been studied in detail as a factor in the development of this tumor.

with acquired immune deficiency disorders including renal transplant recipients and male homosexuals with acquired immune deficiency syndrome (AIDS) can develop lymphoproliferative diseases induced by the virus. This discussion concludes with a guide to recognition of EBV in immune deficient individuals and potential therapeutic interventions.

Immune defenses against EBV

This herpesvirus has evolved from the herpesvirus group approximately 1 million years ago and adapted to Homo sapiens. This ubiquitous virus has established a very successful host-parasite relationship. This biological harmony may explain, in part, the genetic stability of the virus. Virus isolated from IM, NPC, and BL appear to be one strain. Hence, the immune system, cofactors (i.e., chemicals), genetic, and cytogenetic events probably govern the outcome of primary and secondary EBV infection. Perhaps of greatest importance in determining the outcome of infection is the immune response, or lack of it, in individuals.

EBV has adapted itself to B cells which alone contain receptors. Hence, the target cell is restricted. However, the virus undoubtedly enters epithelial cells despite their lacking receptors. The virus likely enters the oropharynx from another person shedding virus from oral secretions. Following a 4-6 week incubation period, IM occurs in approximately one-half to two-thirds of individuals who are infected during adolescence or early adulthood. Seldom does IM occur in third world countries. There and elsewhere the infection usually occurs during early childhood. Multiple failsafe-type mechanisms prevent the virus from destroying the host. Among the first lines of defense are natural killer cells (NK). These cells are likely boosted in their activity by interferon liberated by the infected B cell. Rapidly, powerful activation of suppressor/lytic thymus-derived lymphocytes occurs. These cells can be marked and enumerated in the laboratory by monoclonal antibodies. For example, the usual ratio of subpopulations of peripheral blood T cells is 2:1 for T4 (helper/

inducer):T8 (suppressor/lytic) populations. During acute IM the ratio often drops to 0.2 (reviewed in 3). Approximately 1 to 20% of circulating B cells exhibit EB nuclear-associated antigen (EBNA) during the initial week. In contrast, during latency only about 1 per million cells harbor virus. Humoral immune responses include an array of antibodies against viral capsid antigen (VCA) of IgM, IgA, and IgG isotypes, early antigen (EA), and late appearing is the anti-EBNA response. Antibody-dependent cellular cytotoxicity (ADCC) also is evident. Memory T cells which can be assessed by the outgrowth inhibition assay probably play a major role in maintaining latency. Classical textbooks persist today in asserting that IM does not recur nor is chronicity established. However rare these events may be, they do occur. Breakdown in immune surveillance accounts for the reactivation. Even within normal population 5-25% of individuals show reactivation of virus.

Prior to describing and illustrating various types of immune deficiency associated with EBV-induced lymphoproliferative diseases, a summary of the role of destruction of thymus gland and cytogenetic conversion from polyclonal to monoclonal proliferation is provided to indicate influences of the virus on immune competence and the genetic apparatus. These events may be pivotal in progression of virally induced immune deficiency and the final step to lymphomagenesis.

Thymic epithelial destruction associated with progressive immune deficiency

The pathogenesis of EBV-induced lymphomagenesis is probably a multi-step process. In inherited and acquired immune deficiency disorders the pathogenesis of the immune defects may include thymic epithelial destruction as a critical lesion in the pathogenesis. Recently Seemayer (4) has gathered clinical and experimental evidence demonstrating that thymic epithelial destruction is an important event in some cases of severe combined immune deficiency syndrome and AIDS. Seemingly, both lymphoid cells and virus can initiate this lesion. Among the conditions associated with the destruction of thymic epithelial are graft-vs-host response in F_1 mouse hybrids, maternal-fetal transfusion leading to severe combined immune deficiency (4), fatal infectious mononucleosis in X-linked lymphoproliferative syndrome (XLP) (5), Steinbrinck-Chediak-Higashi syndrome (6), fatal IM (5), and acquired immune deficiency syndrome (4). Malignant lymphomas or pseudolymphomas occur in many of these patients associated with EBV. Merino has suggested that the terminal lymphoproliferative disease arising in children with Steinbrinck-Chediak-Higashi syndrome is due to EBV (7). We have documented destruction of thymic epithelium in children dying with the IM phenotype of XLP (5) (Figure 2).

The mechanism for destruction of the thymic epithelium is unknown. Hypotheses to be investigated include the alteration of HLA antigens by EBV in the virally-infected thymus gland with subsequent misdirected cytotoxic T cells which could destroy epithelium. Alternatively, the epithelium could be an innocent bystander to cytotoxic T cells or NK cells attempting to destroy infected B cells. Among other possibilities would be the fusion of infected B cells with thymic epithelium with subsequent unleashing of productive cycle of the virus with lysis of the epithelial cells leading to their death (8). Following infection by EBV, individuals with XLP show defects in NK function and inverted T4/T8 ratios (9). We have attributed the progression of the immune deficiency to destruction of thymic epithelium in the children. Persistent polyclonal B cell lymphoproliferation driven by transforming EBV can set the stage for conversion to monoclonal malignancy. This conversion to monoclonality in XLP and in other immune deficient patients likely occurs owing to a specific cytogenetic alteration (Figure 2).

Cytogenetic conversion of B cells from polyclonality to monoclonality

The polyclonal proliferation of B cells initiated by EBV infection, if permitted to continue unchecked, can lead to fatal IM. ML can occur if the proliferation is sustained sufficiently to allow a cytogenetically altered cell to clone out into a monoclonal malignancy (10). This conceptual model is based

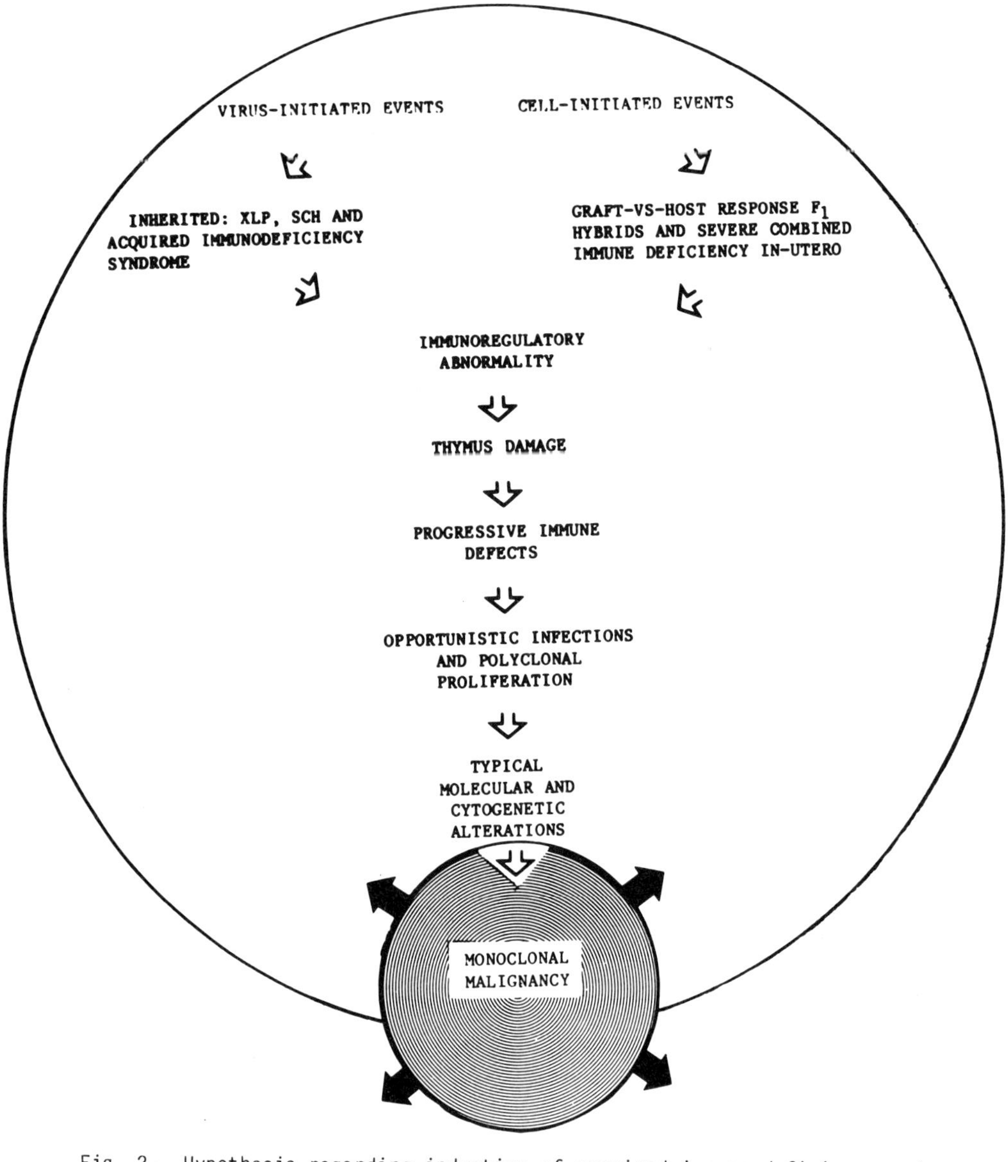

Fig. 2. Hypothesis regarding induction of acquired immune deficiency and malignancy owing to either viral infection or cell initiated events. This process can be initiated by virus or immune competent lymphoid cells, leads to destruction of thymic epithelium and thereafter further decline in immunity is found. Viral burden in this setting can increase and if lytic infection does not occur and the infected target cells proliferate, a specific cytogenetic event may transpire which would convert a polyclonal cellular proliferation to monoclonality. In those with severe immune defects, opportunistic infections would emerge. Abbreviations: XLP, X-linked lymphoproliferative syndrome; and SCH, Steinbrinck-Chediak-Higashi syndrome.

on studies and hypotheses of Klein (11) on BL. This tumor is monoclonal as are other classical MLs. Specific cytogenetic alterations are found in BL. Manolov and Manolova (12) described the 14q+ alteration in BL in 1971. During the past several years exciting molecular cytogenetic studies have demonstrated that the break points they described in the chromosomes involved in the translocations observed in BL cells are precisely at genetic loci coding for immunoglobulin. Chromosome 14 contains the heavy chain genetic loci, #2 kappa and 22 lambda genes. Chromosome 8 is always involved in the reciprocal translocations with 14, 2 or 22. The breakpoint on the long arm of 8 has been identified to contain the c-myc oncogene. Hypothetically, the juxtaposition of an immunoglobulin loci with the c-myc oncogene leads to enhanced lymphoproliferation and a resistance to host immune surveillance mechanisms (reviewed in ref. 8). Cytogenetically altered tumor cells have been identified in many individuals with inherited or acquired immune deficiency. In some of the cases oligoclonal lymphoid populations have been detected suggesting the "tumor" was in progress of converting from polyclonality to monoclonality when examined (13). Children with fatal IM, renal transplant recipients and patients with AIDS have exhibited such karyotypes (reviewed ref. 8).

Chaganti (14) and we (15) have described BL-like monoclonal malignancies in male homosexuals with AIDS. These patients have exhibited specific cytogenetic alterations typical for BL (Figure 3). Often these patients manifest chronic lymphadenomegaly as a prodromal phase prior to the acquisition of AIDS and the ML. Moreover, such lymph nodes often contain EBV genome (16).

During the past five years, the author and his collaborators have continued to test the hypothesis that polyclonal B cell proliferation initiated by EBV can lead to a monoclonal ML owing to the acquisition of a specific cytogenetic error. Increasingly, cases are being reported which support this hypothesis. Opportunities to evaluate the role of immune deficiency in the pathogenesis of ML are available in experiments of nature (i.e., children with primary immune

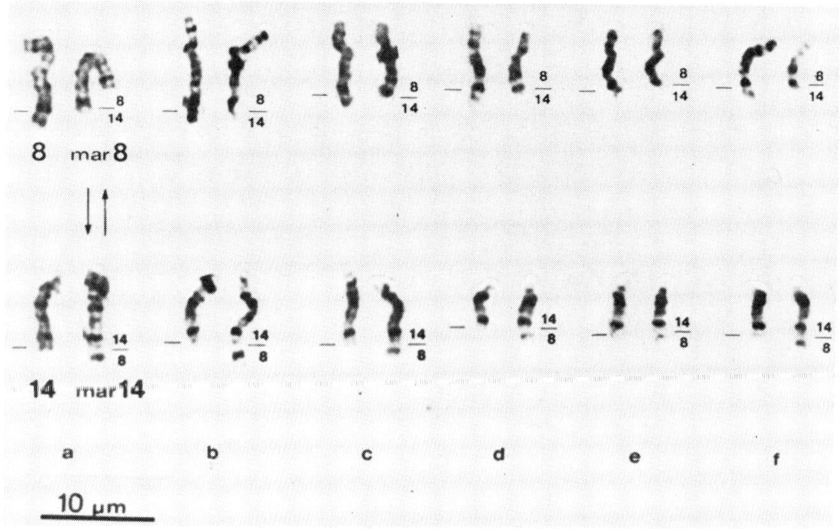

Fig. 3. Karyotype of Burkitt-like lymphoma from male homosexual with acquired immune deficiency syndrome. The tumor was EBV genome positive. (Karyotype performed by George Manolov and Yanka Manolova).

deficiency disorders) and also in individuals who develop acquired immune deficiency which may result from recent adoption of alternate lifestyles or medical interventions (i.e., allograft recipients).

Primary immune deficiency and EBV-induced ML

In the early 1970's, Gatti and Good (17) noted an increased frequency of malignancy in children with primary immune deficiency and Starzl and Penn (18) noted a similar phenomenon in renal transplant allograft recipients. These observations, in part, fueled the imaginations of individuals regarding immune surveillance against malignancy. At that time, investigators thought that immune surveillance against tumor-specific antigens eliminated tumors. Breakdown of immune surveillance was reasoned to be responsible for a development of a wide variety of malignancies. However, after much arduous research in experimental models and patients, it appears that immune surveillance is relevant chiefly to virally-induced malignancies in immune deficient individuals (proposal argued in ref. 19).

Noteworthy are the restricted type of tumors occurring in immune deficient patients. In the child with inherited immune deficiency, the predominant malignancy is ML. The commonly occurring malignancies in adults and children are not common. A noted paucity of carcinomas and Hodgkin's disease is found. The Immune Deficiency-Cancer Registry of Kersey at the University of Minnesota (20) has documented frequencies of malignancy ranging from approximately 0.7% in Bruton's agammaglobulinemia to 15% in Wiskott-Aldrich. More than one-half of the malignancies are ML. Perhaps ML is infrequent in agammaglobulinemia since EBV receptors are lacking on the cells arrested in the pre-B cell state. At the opposite extreme, males with XLP show approximately 35% occurrence of the ML phenotype (reviewed in ref. 5).

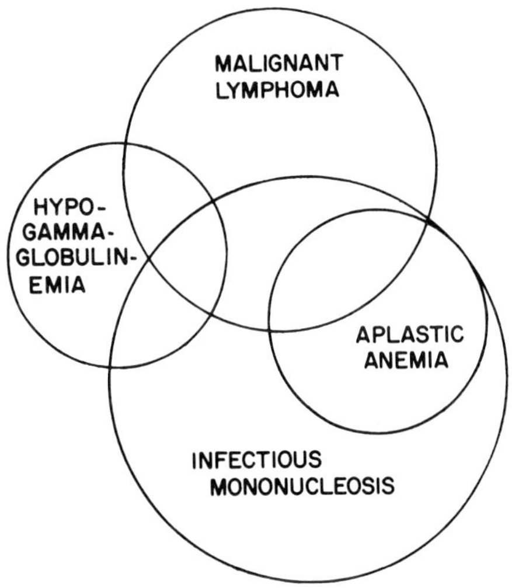

Fig. 4. Venn diagram of the initial 100 cases of X-linked lymphoproliferative syndrome in the XLP Registry. The overlapping expressions of EBV infections show relationships between the various phenotypes. The relative size of each circle indicates the frequency of the disorder. (Published with permission of Yorke Publishers).

X-linked lymphoproliferative syndrome, EBV, and ML

In 1975 we described X-linked recessive progressive combined variable immune deficiency (Duncan's disease) manifested as fatal IM, aplastic anemia, agammaglobulinemia, or MI (21). Subsequently, we have developed a registry and laboratory to identify diagnostic features and to explain mechanisms of EBV-induced diseases (5). To date, more than 30 kindreds, distributed worldwide, with XLP have been identified and studied in our registry.

Diagnosis of XLP requires two or more maternally related males to manifest the phenotypes described above following EBV infection (5). The patients invariably show a lack of anti-EBNA response. This humoral defect reflects T cell deficiency. Also adding to diagnostic certainty is the finding of elevated anti-VCA titers, often of IgM and IgA isotypes, and anti-EA antibodies in the carrier female. Owing to the partial immune deficiency to EBV with the hemizygous state for the XLP locus, the virus becomes chronically active. Importantly, the carrier females have not exhibited the phenotypes of males.

The Venn diagram (Figure 4) demonstrates the overlap of phenotypes and the relative frequency of each in the initial 100 cases studied in our registry (5). The overlapping circles indicate a progression of phenotypes or simultaneous occurrence of phenotypes in individual cases. Bidirectional progression or conversion between the specific disease categories has occurred. We have hypothesized that the phenotype (disease) is largely the result of the variable immune responses to EBV in the patients.

The malignant lymphomas in the males with XLP have all been of the B cell phenotype. BL, histiocytic lymphoma (Rappaport classification), immunoblastic sarcoma B cells (Lukes-Collins classification), and other histological types in the B cell spectrum have been identified. The tumors occur predominantly in extranodal sites in the terminal ileum, liver, kidneys, and central nervous system. The tumor cells show monomorphous features and do not disseminate to organs other than the primary site. These biological features strongly implicate monoclonality for these tumors. Disappointing has been our inability, so far, to obtain adequate fresh tissue for immunophenotyping. Contrasting with the ML is the IM phenotype which widely disseminates. Moreover, a heterogeneous population of lymphoid cells with plasma cell endstage differentiation is found (22). A "gray zone" where it is difficult histologically to distinguish between the IM and ML phenotypes is often encountered. Atypical hyperplasia of Peyer's patches, tonsils, and lymph nodes and sheets of infiltrating cells in other organs is frequently encountered.

In the individuals surviving beyond 4 weeks of IM, there is usually destruction of thymus gland (Figure 2) and depletion of T cell-dependent regions in lymph nodes and spleen. Bone marrow becomes invaded by the infected B cells late in the clinical course (22).

The patients with ML appear to have a superior immune competence compared to those who succumb to IM. We have conjectured that the immune competence of such individuals allows a sustained polyclonal B cell proliferation to continue for time sufficient for a specific cytogenetic event (i.e., t(8;14) to supervene and convert the polyclonal proliferation to monoclonality (23).

The patients with XLP show no or only subtle evidence of immune deficiency prior to infection by EBV (24). Variable phenotypic expressions occur following infection by the virus (21). These clinical outcomes are likely governed by the

TABLE 1

DEFECTIVE IMMUNE FUNCTIONS IN X-LINKED LYMPHOPROLIFERATIVE SYNDROME

Defects
Immunoregulatory function
Suppression of mitogen-induced Ig production by B cells
Deficient secondary response to ØX174, i.e., poor switching from IgM to IgG
Poor development of memory T cells to EBV
Abnormal levels and ratios of OKT4 cells and OKT8 cells
Thymic epithelial destruction and T-cell depletion after EBV infection
Low natural killer cell activity, acquired after EBV infection
EBV-specific antibody production, especially anti-EBNA, is lacking
Normal
Normal white blood differential count when not acutely infected
Normal levels of OKT3-positive, Ia-positive, and surface-Ig-positive cells
Normal proliferative responses to mitogens (PHA, PWM, EBV)

TABLE 2

IMMUNE DEFICIENCY AND EBV-INDUCED LYMPHOPROLIFERATIVE DISORDERS[a]

Factor	SCID	AT	XLP	WAS	XAg	RTR	CTM
Genetics	AR,XR	AR	XR	XR	XR	--	Chimera ±
Lymphoma	1%	5%	35%	14%	0.5%	1%	30%
Fatal IM	?	?	65%	?	0	Rare + ML	?35%
Other	1%	5% ALL+HD	36% -HD	1% -HD	0 -HD	Carcinoma	
NK defect		+	+acquired	+,-	-	+,-	-
T defect	++	+	++	+	-	+	+
B defect	+,-	+IgA	+	+	Pre-B	-,?	-
Secretion EBV		?	+	?	?	+ ATG	+
Anti-EA	?	++	-,+	-,	?	+	+
Anti-VCA	?	++	-,+	-,±	?	+ATG delay	+
Anti-EBNA	?	-	-	+	?	+,-	-
EBV DNA	+	+	+	+	?	+	+
Monoclonal	-	?	+	+	?	+	-
Polyclonal	+	?	+	?	?	+	+
Karyotype	?	14q+	?	+	?	-,+	Diploid
Comments	Post-thymic transplant lymphoma	Breakage and immune defect	Immunoregulatory defect	Immunoregulatory defect	B-cell target defect	ATG-EBV severe	unnatural

[a]SCID = severe combined immunodeficiency, AT = ataxia telangiectasia, XLP = X-linked lymphoproliferative syndrome, WAS = Wiskott-Aldrich syndrome, XAg = X-linked agammaglobulinemia, RTR = renal transplant recipient, CTM = cotton-topped marmoset, AR = autosomal recessive, XR = X-linked recessive, IM = infectious mononucleosis, NK = natural killer cells, + = present, - = absent, ATG = antithymocyte globulin, ALL = acute lymphocytic leukemia, HD = Hodgkin's disease, EA = early antigen, VCA = viral capsid.

individuals immune responses to the virus. Listed in Table 1 are the various immune defects detected in the syndrome. XLP continues to offer a great opportunity to investigate the pathogenesis of EBV-induced ML and many other diseases summarized in figure 1.

The role of EBV in the other primary immune deficiency disorders as a factor in lymphomagenesis is in a beginning stage of exploration. EBV-carrying ML have been identified in patients with ataxia telangiectasia, common variable immune deficiency, and severe combined immune deficiency following thymic epithelial transplant (Table 2).

A growing number of case reports of IM converting to ML have appeared in the literature (reviewed in 3). These individuals range in age from early childhood to late adulthood. Likely, the children have an underlying unidentified primary immune deficiency, whereas the adults have acquired immune defects for a variety of reasons. It appears that IM is a benign counterpart to EBV-induced ML. The cases that are studied virologically, immunologically, and cytogenetically in transit from IM to ML have substantiated our hypothesis. Acquired immune deficiency is increasing in frequency due to cultural changes (i.e., AIDS) and medical intervention.

Acquired immune deficiency disorders and EBV-induced lymphomagenesis

EBV genome has been found in malignant lymphomas occurring in renal transplant recipients (RTR) (25), cardiac transplant recipients (CTR) (26), thymic epithelial transplant recipients (27,28). In contrast, ML is extremely rare in bone marrow transplant recipients (29). Approximately 1 in 100 RTR have exhibited fatal lymphoproliferative diseases (30). Young RTR fail to respond to primary EBV infection and exhibit IM-like fatal lymphoproliferation. These patients exhibit serological responses with a lack of anti-EBNA similar to males with XLP. Adult RTR manifest ML with high titer anti-VCA, EA and EBNA reflective of reactivation of latent EBV. Usually the tumors are polyclonal, however, Hanto et al. (31) have documented the emergence of a monoclonal EBV-carrying ML in a man with sustained immunosuppression. Withdrawal of the immune suppression in the patient permitted regression of the lymphoid hyperplasia. Acyclovir® treatment had caused cessation of shedding of EBV in his oral secretions. Further diagnostic detection of patients who are at increased risk for ML who are immune suppressed for grafting need to be developed.

Cardiac transplant recipients also develop EBV-carrying ML. The initial cases occurred predominantly in males with idiopathic cardiomyopathy (reviewed in ref. 26). Cardiac allograft patients exhibit an alarming frequency of ML especially in patients who have sustained their grafts for several years. Bieber and colleagues have developed an animal model in cynomolgus monkey. Cyclosporin A and allografting can allow an oncogenic herpesvirus to induce ML analogous to those found in CTR. The cotton-topped marmoset has also been used for an animal model for BL. Following infection by EBV, three events usually transpire (reviewed in 23). Approximately one-third of the animals die following a short incubation period of disseminated IM-like disease. Another group of animals develop tumorous masses and a third group remains asymptomatic. Characteristic of all three groups is a lack of anti-EBNA. This is seen in XLP and in patients with ataxia telangiectasia and is a manifestation of T cell deficiency. Johnson et al. (32) has shown that the lymphoproliferative lesions in cotton-topped marmosets induced by the virus are polyclonal. Diploid karyotypes and polyclonal immunotypes were found. Hence, the cotton-topped marmoset is more akin to fatal IM than to BL. Animal models need to be developed to further investigate the mechanisms of EBV-induced lymphomagenesis.

Borzy and colleagues (27) described development of immunoblastic sarcoma in 10% of children receiving thymic epithelial transplant for immunoreconstitution of severe combined immune deficiency. Their initial three patients with the lymphoproliferative disease showed polyclonal elevation of immunoglobulins suggesting that the cases were polyclonal. They did not perform immunotyping or cytogenetic analysis to determine clonality. A group of investigators in

Montreal (28,30) described polyclonal immunoblastic sarcoma in a similar patient. The tumor contained EBV genome. The patients died within a short period of time following transplantation. Likely, time was insufficient for an evolution of a monoclonal malignancy.

Bone marrow transplant recipients seldom show increased frequency of malignancy. Neudorf et al. (29,31) have reviewed bone marrow transplant registries and summarized impressive data displaying an absence of malignancy in children immunologically reconstituted by bone marrow grafting. The lack of tumors in bone marrow transplant recipients could be due to the short duration of immune suppression and hence rapid recovery of immune surveillance mechanisms against EBV occurs. Often the bone marrow transplant recipient displays increased suppressor T cell populations which may affect suppression outgrowth of infected B cells.

Acquired immune deficiency syndrome (AIDS) has emerged as a major public health problem during this decade. Diagnostic criteria include presence of Kaposi's sarcoma in patients under 60 years of age or opportunistic infections with no known basis for immune deficiency (33). The etiological bases for AIDS is not known. The two major hypotheses include a new infectious agent or alternatively, repeated exposure to multiple microbial infections which overload the immune system (34). In addition, to the opportunistic infections, Kaposi's sarcoma, ML and squamous cell carcinomas arise. These tumors are also those found in RTR. This empirical observation suggests immune depression may allow ubiquitous viruses to become oncogenic.

Chronic lymphadenomegaly appears to be a prodromal phase of AIDS in many male homosexuals. We have documented EBV genome in the lymph nodes of such patients. Their antibody responses to do not reflect the activity of virus in the individuals. The immune competence seems to preclude mounting antibodies to EA demonstrating reactivation of virus. Our study of a group of 150 male homosexuals from Greenwich Village, revealed all were seropositive for EBV (16).

Demonstrated earlier, our collaborative studies have revealed EBV genome in a monoclonal Burkitt-like ML. The patient displayed characteristic 8;14 translocation found in BL (15). Quinan and Armstrong (personal communication) have found simultaneous EBV infection and cytomegalovirus in peripheral blood of Kaposi's sarcoma in AIDS patients. These two viruses probably become very active following destruction of thymus epithelium (8). Since EBV and CMV infect the immune system, further decline in immunity occurs leading to the irreversible stage of AIDS. This hypothesis requires substantial investigation.

One can anticipate an ever increasing number of opportunistic viral induced malignancies, including ML, owing to changes in cultural practices (i.e., intravenous drug abuse, and anal intercourse in a subset of male homosexuals) and the increasing use of allografting.

Documentation of EBV is difficult at times in immune deficient patients. Serology alone may not reflect presence of virus. Absence of anti-EBV antibodies may not indicate absence of virus. Hence throat washings to transform cord cells, spontaneous cell line formation of peripheral blood lymphocytes, use of monoclonal antibodies to EBV antigens, EBNA staining of touch imprints and molecular hybridization for EBV genome may be required to document EBV (35).

Reduction of viral load and immune modulation are rationale approaches to therapy. Gammamune® and Acyclovir® have been worthwhile. Other considerations include use of interferon, thymosin, marrow transplantation and monoclonal antibodies against EBV-specific antigens. Treatment may potentially preclude conversion of polyclonal B cell proliferation to monoclonal malignancy.

ACKNOWLEDGMENTS

This research was support in part by PHS grant 30196, awarded by the National Cancer Institute, DHHS, the American Cancer Society RD161, the Nebraska Department of Health LB506, and the Lymphoproliferative Research Fund.

REFERENCES

1. Epstein, M.A. and Achong, B. (1979) Epstein-Barr Virus, Springer-Verlag, Berlin.

2. Klein, G. and Purtilo, D.T. (1981) Cancer Res., 41, 4302.

3. Purtilo, D.T., Tatsumi, E., Manolov, G., Manolova, Y., Harada, S., Lipscomb, H. and Krueger, G. (in press) in: Richter, G.W. and Epstein, M.A. (Eds.), International Review of Experimental Pathology, Academic Press, New York.

4. Seemayer, T.A., Gartner, J.G. and Lapp, W.S. (1983) Human Pathol., 14, 3.

5. Purtilo, D.T., Sakamoto, K., Barnabei, V., Seeley, J., Bechtold, T., Rogers, G., Yetz, J. and Harada, S. (1982) Am. J. Med., 73, 49.

6. Krueger, G., Bedoya, V. and Grimley, P.M. (1971) Virchow's Archiv. path. Anat., 353, 273.

7. Merino, F. (1983) in: Purtilo, D.T. (Ed.), Immune Deficiency and Cancer: Epstein-Barr Virus and Lymphoproliferative Malignancies, Plenum Press, New York.

8. Purtilo, D.T. and Manolov, G. (in press) AIDS Res.

9. Seeley, J.K., Bechtold, T., Lindsten, T. and Purtilo, D.T. (1982) in: Herberman, R.B. (Ed.), NK Cells and Other Natural Effector Cells, Academic Press, New York, pp. 1211-1218.

10. Purtilo, D.T. (1980) Lancet, 1, 300.

11. Klein, G. (1979) Proc. Natl. Acad. Sci. USA, 76, 2442.

12. Manolov, G., Manolova, Y., Levan, A. and Klein, G. (1971) Hereditas, 68, 235.

13. Frizzera, G., Hanto, D.W., Gajl-Peczalska, J., Rosai, J., McKenna, R.W., Sibley, R.K., Holahan, K.P. and Lindquist, L.L. (1981) Cancer Res., 41, 4262.

14. Chaganti, R.S.K. (1983) Blood, 61, 1269.

15. Petersen, J.M., Tubbs, R.R., Savage, R.A., Calabrese, L.C., Proffitt, M.R., Manolova, Y., Manolov, G., Schumaker, A., Tatsumi, E., McClain, K. and Purtilo, D.T. Submitted for publication.

16. Lipscomb, H., Tatsumi, E., Harada, S., Sonnabend, J., Wallace, J., Yetz, J., Davis, J., McClain, K., Metroka, C., Tubbs, R. and Purtilo, D.T. (in press) AIDS Res.

17. Gatti, R.A. and Good, R.A. (1971) Cancer, 28, 89.

18. Starzl, T. and Penn, I. (1971) Transplant. Rev., 7, 112.

19. Purtilo, D.T. and Linder, J. (1983) J. Clin. Immunol., 3, 197.

20. Filipovich, A.H., Spector, B.D. and Kersey, J. (1980) Prevent. Med., 9, 252.

21. Purtilo, D.T., Yang, J.P.S., Cassel, C.K., Harper, P., Stephenson, S.R., Landing, B.H. and Vawter, G.F. (1975) Lancet 1, 935.

22. Purtilo, D.T. (1983) in: Purtilo, D.T. (Ed.), Immune Deficiency and Cancer: Epstein-Barr Virus and Lymphoproliferative Malignancies, Plenum Press, New York.

23. Purtilo, D.T. (1980) in: Klein, G. and Weinhouse, S. (Eds.), Advances in Cancer Research, Academic Press, New York, pp. 279-312.

24. Lindsten, T., Seeley, J.K., Sakamoto, K., Yetz, J., Harada, S., Bechtold, T., Rogers, G. and Purtilo, D.T. (1982) J. Immunology, 129, 2536.

25. Saemundsen, A.K., Purtilo, D.T., Sakamoto, K., Sullivan, J.L., Synnerholm, A.C., Hanto, D., Simmons, R., Anvret, M., Collins, R. and Klein, G. (1981) Cancer Res., 41, 4237.

26. Bierber, C.P., Herberling, R.L., Oyer, P.E., Cleary, M., Warnke, R., Saemundsen, A., Klein, G., Werner, H. and Stinson, E.B. (1983) in: Purtilo, D.T. (Ed.), Immune Deficiency and Cancer: Epstein-Barr Virus and Lymphoproliferative Malignancies, Plenum Press, New York.

27. Borzy, M., Hong, R., Horowitz, S., Gilbert, E., Kaufman, D., DeMendonca, W., Oxelius, V-A., Dictor, M. and Pachman, L. (1979) N. Engl. J. Med., 301, 565.

28. Reece, E.R., Gartner, J.G., Seemayer, T.A., Joncas, J.H. and Pagano, J.S. (1981) Cancer Res., 41, 4243.

29. Neudorf, S.M.L., Filipovich, A.H. and Kersey, J.H. (1983) in: Purtilo, D.T. (Ed.), Immune Deficiency and Cancer: Epstein-Barr Virus and Lymphoproliferative Malignancies, Plenum Press, New York.

30. Hanto, D.W., Frizzera, G., Gajl-Peczalska, K.J., Purtilo, D.T. and Simmons, R.L. (1983) in: Purtilo, D.T. (Ed.), Immune Deficiency and Cancer: Epstein-Barr Virus and Lymphoproliferative Malignancies, Plenum Press, New York.

31. Hanto, D.W., Frizzera, G., Gajl-Peczalska, K.J., Sakamoto, K., Purtilo, D.T., Balfour, H.H., Simmons, R.L. and Najarian, J.S. (1982) N. Engl. J. Med., 306, 913.

32. Johnson, D.R. (1983) in: Purtilo, D.T. (Ed.), Immune Deficiency and Cancer: Epstein-Barr Virus and Lymphoproliferative Malignancies, Plenum Press, New York.

33. Curran, J.W., Evatt, B.L. and Lawrence, D.N. (1983) Ann. Intern. Med., 98, 401.

34. Sonnabend, J., Witkin, S. and Purtilo, D.T. (1983) J. Am. Med. Assoc., 249, 2370.

35. Purtilo, D.T., Sakamoto, K., Saemundsen, A.K., Sullivan, J.L., Synnerholm, A.C., Andersson-Anvret, M., Pritchard, J., Sloper, C., Sieff, C., Pincott, J., Pachman, L., Rich, K., Cruzi, F., Cornet, J., Collins, R., Barnes, N., Knight, J., Sandstedt, B. and Klein, G. (1981) Cancer Res., 41, 4248.

The Role of Viruses in Human Cancer. Volume II
G. Giraldo and E. Beth editors

ASSOCIATION OF EBV WITH SOME CARCINOMAS ORIGINATING IN WALDEYER RING OUTSIDE NASOPHARYNX

B.BŘICHÁČEK[1], I.HIRSCH[1], A.SUCHÁNKOVÁ[1], E.VILIKUSOVÁ[2], O.ŠÍBL[3], H.ZÁVADOVÁ[1] AND V.VONKA[1]

[1]Department of Experimental Virology, Institute of Sera and Vaccines, Prague; [2]Second Institute of Pathology, Medical Faculty, Charles University, Prague; and [3]Department of Otorhinolaryngology, Bulovka Hospital, Prague /Czechoslovakia/

INTRODUCTION

The epidemiological, immunological, and molecular-biological findings have linked Epstein-Barr virus /EBV/, the etiological agent in infectious mononucleosis /1,2,3/, with pathogenesis of Burkitt lymphoma /4,5,6,7,8,9/, nasopharyngeal carcinoma /NPC/ /10,11,12,13/, and some malignant lymphoproliferative diseases in persons suffering from inherited or acquired immunodeficiency /14,15,16/. In a few past years we have investigated the association of EBV with two other carcinomas, namely, carcinoma of palatine tonsils /TC/ and carcinoma of the supraglottic part of larynx /SGLC/. Our interest in these neoplasias was motivated mainly by similar embyogenetic origin of palatine tonsils, supraglottic part of larynx and nasopharynx. They are parts of Waldeyer ring, the region of close contact of all three germ layers in the early stages of embryogenesis. The massive infiltration of epithelial tissues with lymphocytes is a characteristic feature of this area later in ontogenesis. Our interest in SGLC and TC stemmed out also from the fact that the incidence rates of these tumors in

Czechoslovakia and in most other European countries are much higher than that of NPC; these rates are approximately 5 and 1.8 per 100.000 per year respectively. SGLC is the fourth most common carcinoma in the male population of Czechoslovakia. Recently Vonka and coworkers /17/ observed increased titres against EBV antigens in patients suffering from TC. Similar results were also obtained in other laboratories /18,19/.

We have been able to demonstrate the presence of EBV DNA in some biopsy specimens obtained from TC by reassociation kinetics /20/. However, because of the frequent presence of infiltrating B lymphocytes, which can carry EBV genomes, in these tissues of Waldeyer ring /21,22/, only hybridization in situ and anticomplementary immunofluorescence in the parallel thin sections enabled us to prove that EBV specific macromolecules /EBV DNA and EBNA/ reside really in the tumor cells.

MATERIALS AND METHODS

Patients and tumor material. Fifty eight TC and 6 SGLC patients, and 16 patients in whom tonsillectomy had been performed because of recurrent exudative tonsillits /RT/, were selected for this study. All the patients were under treatment at the Department of Otorhinolaryngology, Bulovka Hospital, Prague. The diagnosis was based on prior probatory biopsy examination. For the present investigation we used material removed at surgery. Immediately after removal, each specimen was placed in cold tissue culture medium and transported to the laboratory, where it was frozen in petrol-ether and stored at -70°C, or frozen and stored in liquid nitrogen.

Cell lines. Human lymphoblastoid cell lines Raji, Ramos and P3HR-1 were used as controls in serological and hybridization tests or as a source of EBV DNA and were the same as in the previous experiments /20,23/. Human diploid cells LEP /24/ were used for herpes simplex virus /HSV/ DNA production.

Preparation of labeled EBV and HSV DNAs. Different methods of EBV DNA labeling were used over the period of collecting our results. It led, due to the increasing specific activity, to the increase of sensitivity of the detection of EBV specific DNA sequences. Originally, EBV DNA was prepared from Raji cells superinfected with P3HR-1 EBV /25/ in the presence of ^{3}H-thymidine. The specific activity of this ^{3}H-EBV DNA was approximately $7x10^{5}$ cpm/µg. Later, EBV DNA was extracted from the virions purified from tumor-promoting agent 12-O-tetradecanoylphorbol-13--acetate /TPA/ induced P3HR-1 cells /26/ and labeled by nick translation reaction /27/ to the specific activity of $8x10^{6}$ cpm/µg for ^{3}H-labeled probe EBV DNA and $5x10^{7}$ cpm/µg for ^{32}P-labeled probe EBV DNA /20/. HSV DNA was isolated from LEP cells infected with HSV type 2, strain 646 Savage by procedure described in ref.28. The specific activity of nick translated ^{3}H-labeled HSV DNA was $6.5x10^{6}$ cpm/µg.

Extraction and purification of test DNA and reassociation kinetics experiments. Tumor biopsy specimens and control non-tumor tissues were minced and homogenized. DNAs from the biopsy homogenates and from the tissue cultures were extracted and purified as described in ref.29. Briefly, the hybridization solution contained labeled viral probe DNA and non-labeled test DNA in 0.1 M Tris-HCl pH 8.1, 0.025 M EDTA, 1 M Na Cl, and 20% formamide. The heat denaturated mixtures were allowed to hybridize at 60°C for various time periods. Formation of double stranded DNA was monitored by S1 nuclease / Sigma Chem.Co./ digestion and TCA precipitation in liquid scintillation beta-counter. The reassociation kinetics data were evaluated by regression analysis as described in ref.20.

In situ experiments. Thin sections /about 5 µm thick/ were prepared from the frozen bioptic materials. Parallel sections were examined /i/ histologically, /ii/ for the presence of EBNA and /iii/ for the presence of EBV specific nucleotide sequences. For histological examination the standard procedure with haematoxylin-eosin staining was used. To test the presence of EBNA, sections were fixed in a mixture of methanol and aceton /1:1/ at 4°C for 10 min, dipped without drying into PBS /30/ and stained using the anti-complement immunofluorescence technique /31/. High anti-EBNA antibody titres /1:160 or 1:320/ sera and sera free of anti-EBNA antibodies /from acute-stage infectious mononucleosis patients/ were used at a 1:10 dilution for EBNA detection in parallel thin sections. The nucleic-acid hybridization in situ procedure was described in ref.32. Since under the conditions of the experiment the labeled probe could hybridize with EBV DNA or RNA, the strength of the reaction expressed a high content of EBV genome equivalents or increased amount of viral RNA in the tumor cells. P3HR-1 cells, TPA induced P3HR-1 cells, and Raji cells served as positive controls. Ramos cells, LEP cells, and sections from non-tumor tissue specimens served as negative controls. In these tests a proportion of P3HR-1 cells corresponding to the virus-producing cells was strongly positive and Raji cells were weakly but clearly positive, this indicating that the sensitivity of the test was such as to detect less than 50 EBV genome equivalents per cell.

EBV related serology. Sera of 48 TC patients, the same number of controls matched by age and sex, 6 SGLC patients, and 16 RT patients were examined. As controls healthy persons or persons suffering from various non-malignant diseases were selected. Serological test for the presence of antibodies against EBV capsid antigen /VCA/ were performed as in previous studies /17/. The anti-early antigen /EA/ antibody titres were determined on Raji cells induced by a mixture of n-butyrate and TPA /33/. In calculations of geometric mean titres /GMT/ sera negative at a 1:10 dilution were considered positive at a 1:5 dilution.

RESULTS

Table 1 presents the serological data obtained in 48 TC patients, 48 controls matched to these TC patients, 6 SGLC patients and 16 RT patients. Both types of antibodies tested were found more frequently in TC patients than in control persons or RT patients. GMTs for VCA and EA differed significantly /$p<0.01$/ in TC patients from those found in their controls. All SGLC patients possessed IgG antibodies against EA. No significant difference / $p > 0.05$/ between TC patients and the matched controls was found in the prevalence and titres of complement fixing antibodies against three other herpes viruses, namely herpes simplex virus, varicella-zoster virus and cytomegalovirus. In a further analysis, TC patients were grouped according to the histological type of their tumors, and serological findings were compared. Anaplastic carcinoma patients were found to possess significantly higher antibody levels against EBV EA and VCA /$p < 0.001$ and $p < 0.01$, respectively/ than the TC patients with other histological diagnoses.

TABLE 1

EBV IgG ANTIBODIES IN TC PATIENTS, THEIR MATCHED CONTROLS, SGLC PATIENTS AND RT PATIENTS[a]

Group	No. tested	VCA No./%/ positive	VCA GMT	EA No./%/ positive	EA GMT
TC patients	48	48 /100/	124.1	21 /44/	11.4
Matched controls	48	46 /96/	58.7	0 /0/	-
SGLC patients	6	6 /100/	89.1	6 /100/	48.4
RT patients	16	16 /100/	67.5	2 /12.5/	6.1

[a]Summary of different experiments

TABLE 2

RESULTS OF REASSOCIATION KINETICS TESTS ON TC AND SGLC BIOPSIES

Patient	Sex	Age /years/	Material code	Type of biopsy[a]	Histological diagnosis[b]	No. of EBV genomes per cell	Sensitivity of the test[c]
M.J.	F	64	TC 9	TC	E	3	2
H.B.	M	25	TC 11	TC	A	4 - 5	2
K.A.	F	57	TC 13	TC	E	15 - 18	2
P.V.[d]	F	53	TC 14	TC	E	neg.	2
			TC 15	TC	E	neg.	1.45
J.J.	M	81	TC 16	TC	E	neg.	1.45
M.F.[e]	M	77	TC 17p	TC	E	neg.	1.45
			TC 17m	LN	E	neg.	1.45
B.F.	M	68	TC 18	TC	E	neg.	0.1
K.J.	M	65	TC	TC	E	0.17	0.1
B.F.	F	78	TC 23	TC	A	3	0.1
S.J.	M	81	TC 30	TC	K	neg.	0.1
V.J.	M	69	T 2	SGLC	E	0.23	0.1

[a] TC = tonsillar carcinoma; LN = lymph node; SGLC = supraglottical laryngeal carcinoma

[b] E = epidermoid carcinoma without keratinization; K = epidermoid carcinoma with keratinization; A = anaplastic carcinoma

[c] EBV genome equivalents per cell genome

[d] In this patient, 2 biopsies /TC 14,15/ were available; the second was obtained one year after the first surgery

[e] In this patient, 2 biopsies were available; TC 17p originated from primary tumor, TC 17m from a metastase localized in regional lymph node

The results of reassociation kinetics experiments in twelve biopsy specimens originating from ten TC patients and in one specimen of SGLC origin are shown in Table 2. The sensitivity of the assay technique varied considerably in dependence on specific radioactivity of labeled probe DNA. EBV specific DNA sequences were detected in six specimens. According to the calculations made TC9 and TC23 contained approximately 3 EBV genome equivalents, TC14 4 to 5 genome equivalents. TC13 15 to 18 genome equivalents, TC21 0.17 genome equivalent, and SGLC-T2 0.23 genome equivalent per cell. EBV DNA was also detected in 4 out of the 16 tonsils obtained from RT patients. The EBV DNA content in these positive tissues was 1.46, 1.47, 1.97, and 2.17 EBV genome equivalents per one cellular genome.

Tables 3 and 4 summarize the results of in situ tests used for the detection of EBNA and EBV DNA in thin frozen sections from biopsy materials of 11 TC and 7 SGLC patients. Out of 7 TC tissues and 6 SGLC tissues tested for the presence of EBNA three were found positive in both cases. In preparations stained with haematoxylin-eosin cells in surface areas of the tumor foci were darker, i.e. more basophilic /Fig.1/, which reflects an increased metabolic activity. The highest EBNA positivity was usually detected in the same areas of the parallel neighbouring section /Fig.2/. None of the parallel sections was reactive with EBNA negative sera /Fig.3/. The nuclei in frozen sections of four tonsils obtained from RT patients were EBNA negative.

Positive in situ reaction /Fig.4/ with probe EBV DNA was found in 9 TC and 4 SGLC tissues. Two TC and 2 SGLC tissues did not react with EBV DNA. No hybridization with labeled HSV DNA probe was observed in any TC, SGLC, or RT tissue, the grains present were randomly distributed over the section area. EBV DNA probe hybridized neither with thin sections of tonsils of RT patients nor with biopsy materials free of tumor tissues.

TABLE 3

<u>IN SITU</u> DETECTION OF SPECIFIC MACROMOLECULES IN TC BIOPSIES

Patient	Age /years/	Sex	Material code	Histological diagnosis[a]	EBNA *in situ*	EBV NA[b] *in situ*	HSV NA[b] *in situ*
D.V.	68	M	TC 18	E	ND[c]	-	-
Č.T.	65	F	TC 19	E	ND	+	-
K.J.	65	M	TC 21	E	ND	+	-
V.J.	63	M	TC 25	E	±	+	-
S.J.	83	M	TC 30	K	ND	+	-
M.J.	60	M	TC 38	K	-	+	ND
A.R.	38	M	TC 44	K	-	+	-
Š.M.	64	M	TC 47	A	NS[d]	+	ND
V.J.	71	M	TC 56	E	+	-	-
Š.V.	38	M	TC 58	A	NS	+	-
U.M.	48	F	1143	K	+	+	-

[a]E = epidermoid carcinoma without keratinization; K = epidermoid carcinoma with keratinization;
A = anaplastic carcinoma
[b]NA = DNA and/or RNA
[c]ND = not done
[d]NS = non-specific

TABLE 4

IN SITU DETECTION OF EBV SPECIFIC MACROMOLECULES IN SGLC BIOPSIES

Patient	Age /years/	Sex	Material code	Histological diagnosis[a]	EBNA in situ	EBV NA[b] in situ	HSV NA[b] in situ
D.V.	68	M	T 1	K	$\pm$	+	ND[c]
V.J.	69	M	T 2	E	+	+	ND
B.M.	46	M	T 36	E	-	-	ND
M.M.	62	M	T 42	K	+	+	-
S.L.	57	M	T 52	K	-	+	-
Z.P.	45	M	T 55	K	-	-	-

[a]E = epidermoid carcinoma without keratinization; K = epidermoid carcinoma with keratinization
[b]NA = DNA and/or RNA
[c]ND = not done

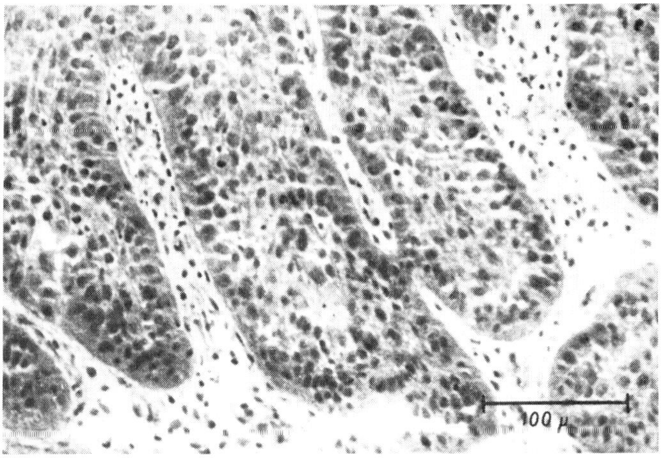

Fig. 1. Thin section of SGLC T 42 stained with haematoxylin - eosin.

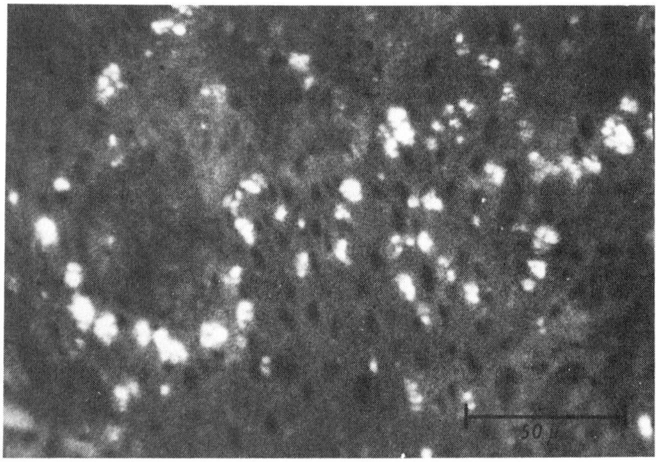

Fig. 2. Immunofluorescence micrograph of thin section of SGLC /T 42/ treated with EBNA antibody-positive human serum.

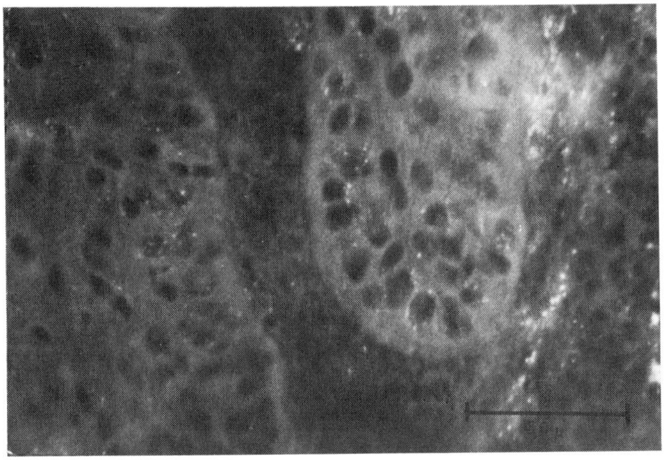

Fig. 3. Immunofluorescence micrograph of thin section of SGLC /T 42/ treated with EBNA antibody-negative human serum.

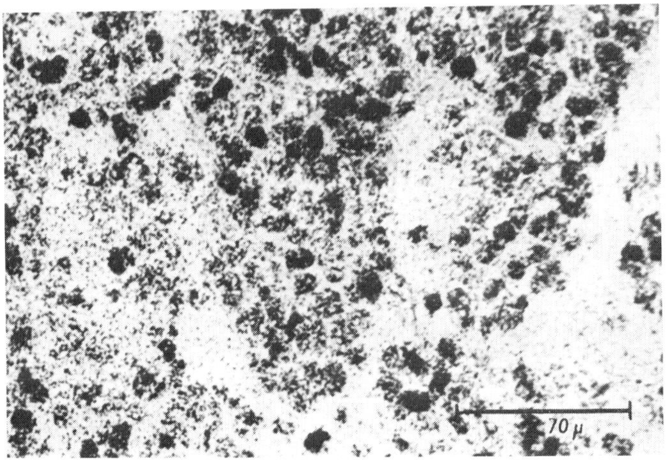

Fig. 4. In situ hybridization of ^{3}H-labelled EBV DNA with thin section of TC /CATO 30/.

DISCUSSION

Our results demonstrate a higher prevalence and higher titres of antibodies to the various EBV antigens in patients suffering from TC and SGLC, and thus extend our previous findings /17,20,32,34/. They also demonstrate the presence of EBV footprints directly in tumor tissues of some TC and SGLC patients. The frequency of biopsies positive for EBV DNA presence in *in situ* tests was higher than that found in reassociation kinetics experiments, though the reassociation kinetics was much more sensitive. This contradiction can be explained by inhomogeneity of biopsy material where a high proportion of non-tumor cells can be present. Indeed, only minor parts of preparation were usually found positive in *in situ* hybridization test and in a portion of surgically removed materials careful histological examination did not reveal any tumor tissue. EBV genome positive lymphocytes were most probably present in tonsils of non-tumor RT patients, which were reactive in the reassociation kinetics test. On the contrary, the strong *in situ* hybridization signal observed in positive TC or SGLC preparations indicated that high amounts of EBV DNA or RNA were present in carcinoma cells. Such intensity of the reaction, comparable with hybridization to TPA and n-butyrate-induced P3HR-1 cells, was never observed in non-tumor biopsies.

It is noteworthy that the highest VCA and EA antibody titres among the TC patients were seen in the anaplastic carcinoma patients /20/. This finding is reminiscent of similar results obtained in NPC patients /11,13/. However, an association of EBV DNA with TC and SGLC, in which foci of well differentiated tissues were found, was demonstrated by reassociation kinetics experiments and was further confirmed by findings of EBV DNA and EBNA by *in situ* methods. Thus, it seems that EB virus may be associated not only with anaplastic carcinoma but also with more differentiated carcinomas of Waldeyer ring provided that they originate outside the nasopharynx.

The observations reported here suggest that EBV is involved not only in NPC but also in some other carcinomas of Waldeyer ring. This cannot be considered as totally unexpected. The close contact between lymphocytes and epithelial cells within the Waldeyer ring could be essential for EBV penetration and subsequent transformation of epithelial cells /34,35,36/. This process could be influenced by some internal or environmental factors /36,37,38,39/. Whatever the mechanism of infection of epithelial cells is, it is apparently operative in any part of the Waldeyer ring.

These observations do not necessarily imply that EBV is etiologically involved in SGLC and TC. The same caution as in the case of NPC should be observed in interpreting the present findings.

REFERENCES

1. Evans,A.S., Niederman,J.C. and Mc Collum,R.W. /1968/ N.Eng. J.Med., 279, 1121.

2. Henle,G., Henle,W. and Diehl,V. /1968/ Proc.Natl.Acad. Sci. U.S., 59,94.

3. Henle,G. and Henle,W. /1970/ J.Infect.Dis., 121, 303.

4. Henle,G., Henle,W., Clifford, P., Diehl,V., Kafuko,G., Kirya,B., Klein,G., Morrow,R.H., Munube,G.M.R., Pike,M.G., Tuker,P.M. and Ziegler,J.L. /1969/ J.Natl.Cancer Inst.,43,1147.

5. Henle,G., Henle,W., Klein,G., Gunvén,P., Clifford,P., Morrow,R.H. and Ziegler,J.L. /1971/ J.Natl.Cancer Inst.,46,861.

6. zur Hausen,H., Schulte-Holthausen,H., Klein G., Henle, G., Henle,W., Clifford,P. and Santesson,L. /1970/ Nature /Lond./, 228, 1056.

7. Nonoyama,M., Huang,D.H., Pagano,J.S., Klein,G. and Singh,S. /1973/ Proc.Natl.Acad.Sci. U.S., 70, 3265.

8. Klein,G. /1975/ N.Engl. J.Med. 293, 1353.

9. De-Thé,G., Geser,A., Day,N.E., Tukei,P.M., Williams,E.H., Beri,D.P., Smith,P.G., Dean,A.G., Bornkamm,G.W., Feorino,P. and Henle,W. /1978/ Nature /Lond./, 274, 756.

10. De Schryver,A., Freiberg,S.J., Klein,G., Henle,W., Henle,G., de Thé,G., Clifford,P. and Ho,H.C. /1969/ Clin.Exp.Immunol., 5, 443.

11. Henle,W., Henle,G., Ho,H.C., Burtin,P., Cachin,Y., Clifford,P., de Schryver,A., de Thé,G., Diehl,V. and Klein,G. /1970/ J.Natl. Cancer Inst., 44, 225.

12. Wolf,H., zur Hausen,H. and Becker,V. / 1973/ Nature New. Biol., 244, 245.

13. Andersson-Anvret,M., Forsby,N., Klein,G. and Henle,W. /1977/ Int. J.Cancer, 20, 486.

14. Purtilo,D.T. /1981/ Cancer Genetics and Cytogenetics, 4,251.

15. Purtilo,D.T., Sakamoto,K., Saemundsen,A.K., Sullivan,J.L., Synnerholm,A.C., Anvret,M., Pritchard,J., Sloper,C., Sieff,C., Pincott,J., Pachman,L., Rich,K., Cruzi,F., Cornet,J.A., Collins,R., Barnes,N., Knight,J., Sanstedt,B. and Klein,G. /1981/ Cancer Res., 41,4226.

16. Hanto,D.W., Frizzera,G., Purtilo,D.T., Sakamoto,K., Sullivan,J.L., Saemundsen,A.K., Klein,G., Simmons,R.L. and Najarian,J.S. /1981/ Cancer Res., 41,4253.

17. Vonka,V., Šíbl,O., Suchánková,A., Simonová,I. and Závadová,H. /1977/ Int. J.Cancer, 19,456.

18. Wilmes,E., Wolf,H., Deinhardt,F. and Naumann,H.H. /1981/ in: Grundmann,E., Krueger,G.R.F. and Ablashi,D.V. /Eds./ Nasopharyngeal Carcinoma, Cancer Campaign Vol.5, Gustav Fischer Verlag, Stuttgart, P.145.

19. Oberender,H., Nowak,R., Donner,A., Kunkel,M., Gärtner,L. and Scholtz,H.J. /1983/ Acta virol., 27,277.

20. Břicháček,B., Suchánková,A., Hirsch,I., Šíbl,O., Řezáčová,D., Závadová,H. and Vonka,V. /1981/ Acta virol., 25,361.

21. Veltri,R.W., Mc Clung,J.E. and Sprinkle,P.M. /1976/ J.gen. Virol., 32,455.

22. Veltri,R.W., Heyl,L.W. and Sprinkle,P.M. /1977/ Proc.Soc.exp. Biol.Med., 156,282.

23. Simonová,I., Závadová,H. and Vonka,V. /1977/ Acta Virol., 21,184.

24. Řezáčová,D. and Barešová,Z. /1969/ Progr.immunobiol. Stand., 3,111.

25. Tanaka,A., Miyagi,M., Yajima,Y. and Nonoyama,M. /1976/ Virology, 74,81.

26. zur Hausen,H., O'Neill,F.J., Freese,U.K. and Hecker,E. /1978/ Nature /Lond./, 272,373.

27. Rigby,P., Dieckmann,M., Rhodes,C. and Berg,P. /1977/ J.molec. Biol., 113,237.

28. Pater,M.M., Hyman,R.W. and Rapp,F. /1976/ Virology, 75,481.

29. Frenkel,N., Locker,H., Cox,B., Roizman,B. and Rapp,F. /1976/ J.Virol., 18,885.

30. Schmitz, H. /1981/ J.Immunol.Meth., 42,337.

31. Reedman,B. and Klein,G. /1973/ Int.J.Cancer, 11,499.

32. Břicháček,B., Hirsch,I., Šíbl,O., Vilikusová,E. and Vonka,V. /1983/ Int. J.Cancer, 32,193.

33. Boguzsáková,L., Hirsch,I., Břicháček,B. and Vonka,V. /1983/ J.gen.Virol., 64,887.

34. Vonka,V., Suchánková,A., Břicháček,B., Hirsch,I. and Šíbl,O. /1979/ in: Bachmann,P.A. /Ed./, Mechanisms of Viral Pathogenesis and Virulence, Proceedings of the 4th Munich Symposium of Microbiology, WHO Collaborative Centre for Sellection and Evaluation of Data on Comparative Virology, Munich, pp.129-140.

35. Bayliss, G.J. and Wolf,H. /1980/ Nature /Wash./, 287,164.

36. Wolf,H., Bayliss,G.J. and Wilmes,E. /1981/ in: Grundmann et al. /Eds./, Cancer Campaign, Vol.5, Nasopharyngeal Carcinoma, Gustav Fischer Verlag, Stuttgart - New York, pp. 101-109.

37. Bayliss,G.J. and Wolf,H. /1981/ Proc.Natl.Acad.Sci. U.S., 78,7162.

38. Ito,Y., Kawanishi,M., Hirayama,I. and Takabayashi,S. /1981/ Cancer Lett., 12,175.

39. Zeng,Y. Zhong,J.M., Mo,Y.K. and Mia,X.C. /1983/ Intervirology, 19,201.

The Role of Viruses in Human Cancer. Volume II
G. Giraldo and E. Beth editors

EBV ASSOCIATED MEMBRANE ANTIGENS ON VIRIONS, PRODUCER CELLS AND TRANSFORMED LYMPHOCYTES

DAVID A. THORLEY-LAWSON
Department of Pathology and Medicine, Tufts University School of Medicine, 136 Harrison Avenue, Boston, MA 02111 U.S.A.

INTRODUCTION

Epstein-Barr virus, a ubiquitous member of the herpesvirus group, has been associated with several human diseases. Like other herpesviruses, infection during early childhood results in limited or no detectable clinical symptoms (1). The virus persists for life and reactivates frequently in normal individuals manifest by the maintenance of steady levels of both humoral and cellular immunity to the virus and the frequent appearance of infectious virus in the mouth (2). Thus, in the normal healthy adult seropositive individual an effective balance is struck, where the virus is able to replicate and be shed in the mouth and thereby, infect other individuals through mouth to mouth contact, however, the viremia is controlled by neutralizing antibody and the infected B lymphocytes are removed from the circulation by cell mediated immunity. The later form of immunity is particularly crucial as it is known that infected B lymphocytes are immortalized, being able to proliferate indefinately in culture (3,4) and to grow as lymphomas when placed back into the autologous host in animal models using new world primates, particularly cotton-top marmosets (5). Thus, the circulating infected B lymphocytes have the potential to develop into a life threatening lymphoproliferative disease.

This scenario can be viewed as the "normal" situation, with host and parasite virus well adapted to each other. However, as with the other herpesviruses, disturbance of this relationship can result in a wide spectrum of severity of disease which, in some instances, can be fatal. The most common situation, that may be reguarded as mildly aberrant, is those individuals, presumably as a consequence of Western hygiene, who do not become infected until adolescence or adulthood. It is generally accepted that EBV infection at this time can give rise to infectious mononucleosis (IM) (6). This disease results in the secretion of large amounts of virus in the mouth, in the infection of high numbers of B lymphocytes in the peripheral blood (as high as 5% of the B cells)(7,8) and seroconversion. The disease can be mild, lasting perhaps a few days or extremely debilitating for many months, however, it is rarely if ever fatal. Upon recovery from the disease state the individual behaves as a normal healthy seropositive as described above.

The rare cases of fatal infectious mononucleosis are generally thought to arise in individuals who are immunodeficient (9). The study of this syndrome within families has implicated an X-linked gene responsible for an immunodeficient response to EBV (10).

The other situation where EBV infection can result in life threatening lymphoproliferation is in individuals who are immunosuppressed in association with organ transplantation. It is now well documented that immunosuppressive drugs can result in reactivation of EBV as reflected in heightened antibody titers and virus shedding (11). In the case of cyclosporin A it has been shown that this reactivation can lead to lymphoma with the tumor cells consisting of EBV infected B lymphocytes (12,13). Although cyclosporin A is known to inhibit cell mediated immunity to EBV infections in vitro,(14,15) it is thought to do so through a nonspecific mechanism, i.e. the inhibition of the expansion of immune T cells, that affect the immune response generally. It is unclear at the present time why this particular drug should preferentially inhibit the immune response to EBV in vivo.

The cases described above represent clear cut examples of the role of EBV in specific human diseases. Less clear cut is the role of EBV in two other malignancies, Burkitt's lymphoma (BL) and nasopharyngeal carcinoma (NPC). BL is at high incidence only in a few locations of the world, notably tropical Africa. The high incidence of BL is associated with the presence of EBV in the tumor cells (16,17). However, unlike the examples described above, the BL lymphoma is monoclonal in origin (18). Furthermore, there are rare cases of EBV negative BL which occur throughout the world and this, has led to the speculation that EBV potentiates the tumor in high incidence areas in conjunction with other regional co-factors, both genetic and environmental in origin. Nasopharyngeal carcinoma of the undifferentiated form appears to be always associated with EBV in the tumor cells in both high and low incidence areas (19). This is the only tumor of non-lymphoid origin associated with EBV and may reflect a malignant derivative of the epithelial cells that constitute the usual site of replication. No epithelial tumor has ever been derived in animal models. The high incidence areas of NPC may also reflect the involvement of local genetic and environmental factors.

In summary, therefore, EBV is responsible for a common debilitating disease, IM, which in rare cases can result in fatal lymphoma. In addition it probably potentiates, if not plays, a major causative role in two human cancers, one of which, NPC, constitutes a major world health problem. For these reasons, a detailed study of the mechanisms that normally control EBV in vivo is called for, so that effective means may be developed to intervene against the virus.

It has been the interest of our laboratory to define, characterize and isolate the antigenic moieties on the surface of EB virions and infected B lymphocytes which are responsible for generating immunity with the long-term goal of developing means, such as a subunit vaccine, to intervene against the virus.

Membrane antigens (MA) on Virions and Virus Producer Cells - Targets of Neutralizing Antibodies

Our approcah to the isolation and identification of MA was to produce a high-titered rabbit antiserum against purified virions. This serum had the following properties (20).

1. It neutralized EBV infection *in vitro*.

2. It gave membrane immunofluorescence (MIF) with a small population of cells in EBV-producing cultures. Separation of the MIF-positive cells revealed that they were the only ones actively replicating the virus.

3. The membrane antigens on intact cells could be identified by means of a quantitative two-step binding radioimmunoassay using ^{125}I-labeled protein A as the second step. Inhibition of this assay could be used as a quantitative measure for the isolation of membrane antigens.

Using the protein A assay it was possible to demonstrate for the first time, the presence of EBV-associated membrane antigens on highly purified plasma membranes which could be effectively solubilized by a range of detergents and partially purified by ricin and lentil lectin affinity columns (21).

This material retained antigenic activity, as it was capable of generating high-titered neutralizing antibodies when injected *in vivo* (22). Analysis by immunoprecipitation of detergent-solubilized, [^{35}S] methionine-labeled, purified plasma membranes, revealed that the membrane antigen consisted of four polypeptides with molecular weights of 350,000, 220,000, 140,000, and 85,000 (Fig. 1) (21). These polypeptides were precipitated from all of 14 EBV-producing cell lines tested and from superinfected RAJI cells. Their production could be increased by treatment of the cells with TPA and n-butyrate (P3HR-1 only) and was decreased by treatment with PAA, with concomitant changes in the number of virus-producing cells. These polypeptides could not be precipitated from a wide range of EBV-positive nonproducer cell lines or EBV-negative cell lines. Thus, they are EBV-specific late polypeptides. Using three independent methods, it was possible to demonstrate that the 350,000, 220,000, and 85,000 MW polypeptides, but not the 140,000 MW polypeptide, were glycoproteins because (a) the 350,000, 220,000, and 85,000, but not the 140,000, MW proteins bound to ricin and lentil lectin affinity columns; (b) the 350,000, 220,000, and 85,000, but not the 140,000 MW proteins could be labeled metabolically with [^{3}H]glucosamine and [^{3}H] mannose; (c) the 350,000, 220,000, and 85,000, but not the 140,000, MW proteins were sensitive to tunicamycin, an inhibitor of N-asparagine-linked glycosylation (23). For these reasons we have adopted the terminology gp350, gp220, p140, and gp85 to describe the four major components of EBV-MA.

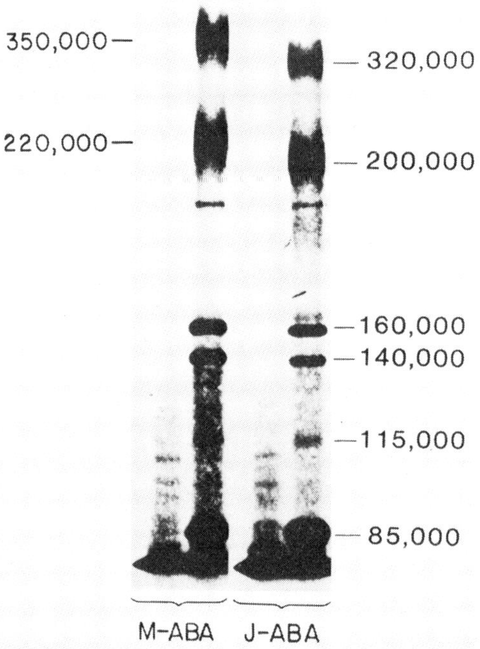

Fig. 1. Immunoprecipitation (4% SDS-PAGE) of Triton X-100-soluble Epstein-Barr (EBV) antigens from [^{35}S]methionine-labeled M-ABA (marmoset) and J-ABA (human) cells transformed with the same EBV isolate. The samples were run on the same gel and are shown side by side for clairty. These cells were induced by 12-0-tetradecanoyl-phorbol-13-acetate. M-ABA cells plus (A) preimmune or (B) immune serum; J-ABA cells plus (C) preimmune or (D) immune serum.

In subcellular fractionation studies it was possible to demonstrate that all the components of the membrane antigen are present on the membranes throughout the cell, with the possible exception of the gp85 polypeptide, which is poorly expressed in the nucleus. Last, we have been able to demonstrate that these four proteins are also the major constituents of [^{35}S] methionine-labeled virion envelopes (23).

Thus, in summary, we may conclude that gp350, gp220, p140, and p85 are truly membrane components in the plasma membrane of EBV-producing cells and the envelopes of virions because (a) they copurify with plasma membranes and viral envelopes; (b) they require detergents to be solubilized from both plasma membranes and virions; (c) at least three of the proteins are glycoproteins; and (d) at least three of the proteins (the glycoproteins) may be labeled in intact cells by lactoperoxidase and ^{125}I (D.A. Thorley-Lawson, unpublished observations).

In addition to the four major components, we have also detected several other components, including a glycoprotein of molecular weight 115,000 (gp115) in producer cells (Fig. 1) and a component of molecular weight 56,000 (p56), which is most readily detected in virions. Furthermore, we have detected minor glycoprotein components with molecular weights in the range of 120,000-190,000 that are precursors of the two large glycoproteins.

We have studied a large number (14) of independent EBV isolates and have found no consistent or significant variations in the patterns of the MA polypeptides, with two exceptions. First, gp350/300 and gp250/200 have slightly different molecular weights when synthesized in marmoset compared to human cells (23). This difference is best illustrated using the human J-ABA and marmoset M-ABA pair of cell lines (Fig. 1), which were derived at the same time with the same EBV isolate (24). The second case is the B95-8 cell line, which expresses very low levels of gp250/200 (23). B95-8 is, therefore, a variant in that it is the only virus strain that we have tested with a defective glycoprotein expression. Marmoset and human cell lines transformed in vitro with B95-8 virus are also atypical in their expression of gp250/200 indicating that the defect is in the viral DNA itself, not in the host cells. Interestingly, this defect must have been acquired by the virus during the establishment of the B95-8 cell line, as the parental isolate, the 833L cell line, is normal.

Because of the number of polypeptides in the MA, it would be difficult to isolate each component by standard biochemical techniques. We decided, therefore, to separate them by producing monoclonal antibodies against each component.

We have derived a series of monoclonal antibodies against EB virions. Of those that will immunoprecipitate polypeptides, most recognize the two large glycoproteins gp350/300 and gp250/200 (25) even in the presence of SDS, which separates the two polypeptides. This means that gp350/300 and gp250/200 share antigenic determinants and that they are structurally related. To study this relationship further, we have compared the proteolytic peptide map of the two glycoproteins. Such studies reveal that the glycoproteins share a large number of peptides in common (26). However, in pulse-chase experiments there was no suggestion that they derive from a common precursor. Thus they must derive either from a gene duplication or through differential RNA splicing. This point has been resolved by genetic mapping studies performed in collaboration with Drs. Mary Hummel and Elliott Kieff (27). Immunoprecipitation of the in vitro translation products of whole, EBV specific, RNA with the specific rabbit anti-gp350/220 serum, reveals two major polypeptides of 135 kd and 100 kd respectively, that map to the BamHI L fragment of the genome. The

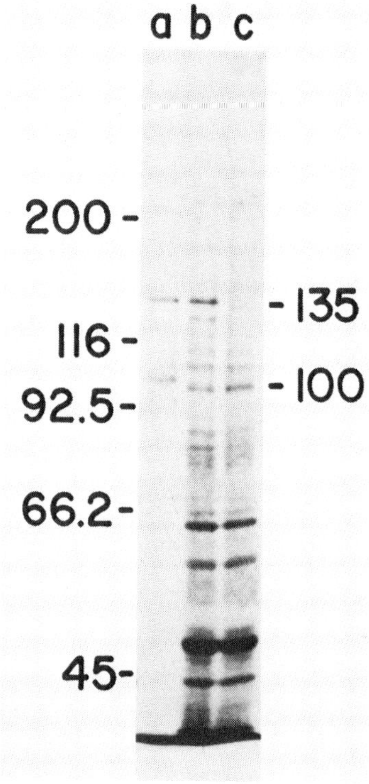

Fig. 2. Comparison of the BAMHI L 135 kd in vitro translation product with the precursor to gp350 made in vivo. A) BamHI L-specific RNA was selected by hybridization, translated in vitro and the polypeptides were immunoprecipitated with anti gp350/220. B and C) B95-8 cells were pulse-labeled with ^{35}S-methionine for 15' in the presence of tunicamycin, lysed and immunoprecipitated with (B) rabbit anti gp350/220 or C) normal rabbit serum. Numbers on the left indicate the sizes (in kd) of protein standards; numbers on the right indicate the sizes of polypeptides specifically immunoprecipitated by anti gp350/220.

135 kd polypeptide is identical in size to the unglycosylated precursor of gp350/300 synthesized in B95-8 cells after pulse labelling with [^{35}S]methionine in the presence of tunicamycin and identified with the rabbit anti-gp350/220 serum (Fig 2). Similarly, the 100kd polypeptide is of the expected size for the unglycosylated precursor of gp250/200. The 135 kd and 100 kd polypeptides are encoded by 3.6 and 2.8 kb RNAs, respectively, and these RNA's overlap considerably in their sequence suggesting that they are derived by differential splicing from the same gene. This observation provides an explanation for the similarity in peptide maps and the antigenic cross reactivity of the two glycoproteins.

Since, the polypeptide sizes for gp350/300 and gp250/200 are 135 kd and 100 kd, respectively about half of the apparent molecular weight of these glycoproteins is due to carbohydrate. More intriguing still, the molecular weights of these polypeptides in the presence of tunicamycin, which inhibits N-linked, but not O-linked, glycosylation are 300 kd and 180 kd, respectively. This means that there is a discrepancy of 165 kd and 80 kd, respectively, between the unglycosylated forms (135 kd and 100 kd MW) and the forms lacking N-linked sugar (300 kd and 180 kd), implying that a large amount of O-linked sugar is present (26). Additional evidence for the presence of O-linked sugar is:

a) The form synthesized in the presence of tunicamycin binds to ricin lectin implying that galactose residues are present as would be expected for O-linked glycans.

b) This form may be labelled with [^{3}H]-glucosamine and its molecular weight shifts after neuraminidase treatment,further indicating the presence of sugar moieties.

c) This form does not bind to lentil lectin although the mature form does, a finding consistent with the absence of N-linked sugar.

d) The sugar moieties are released by mild alkali treatment.

The presence of such large amounts of both O- and N-linked carbohydrate on a polypeptide may be unique in herpesviruses and rare among glycoproteins in general.

One crucial observation made with the monoclonal antibodies to the two large glycoproteins is that they will give both membrane immunofluorescence while on cells and a classical viral capsid antigen (VCA) like stain in the cytoplasm of acetone-fixed virus-producing cells (25). This means that classical VCA stain has no value as a test for serum antibodies against viral capsid components if those sera also contain anti-MA antibodies.

One of the monoclonal antibodies (C1) against the two large glycoproteins will neutralize the virus in vitro. We have confirmed the presence of these glycoproteins on the virion by performing immunoelectron microscopy on

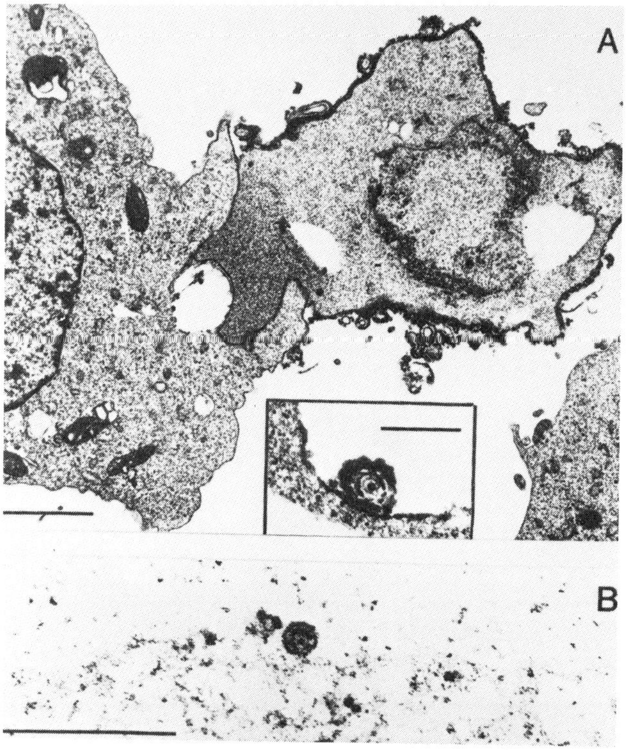

Fig. 3. (A) Electron micrographs of a producer cell in which the gp350/300 gp250/200 antigen is revealed using the monoclonal antibody C1 followed by a peroxidase-conjugated rabbit anti-mouse Ig. The cell in the upper right portion of the figure has virus particles in its nucleus and a dense peroxidase reaction product over its surface, whereas the nonproducer cells to the left and lower right have no indication of positive reaction. The magnification bar is 2 m. The inset shows a virion and a portion of plasma membrane from a different cell. Both the viral envelope and cell membrane are stained by the immunochemical reaction. The magnification bar in the inset is 0.5 m. (B) Electron micrograph of an EB virion on the surface of protein A Sepharose beads coated sequentially with rabbit anti-mouse Ig, C1 monoclonal antibody, and EB virions. The bead is represented by the darker region in the lower part of the electron micrograph. Virions were found sparsely but regularly around the perimeter of the C1 beads because the electron micrographs were of ultrathin sections. No virions were found on beads with control antiibodies or on C1 beads incubated with virions in the presence of detergent. The magnification bar is 0.5 m.

EBV-producing cells using the C1 antibody (Fig. 3) (28). In this experiment, C1 antibody stained both the plasma membrane of EBV-producing cells and the envelope of virions. The stain was specific because every cell observed with internal virions stained with the antibody and every cell without internal virions did not stain. No single exception was seen. These studies show in the most convincing way that gp350/300 and gp250/200 are EBV-specific structural components of the viral envelope, which are also expressed independently (i.e., not in virions) on the plasma membrane of EBV-producing cells.

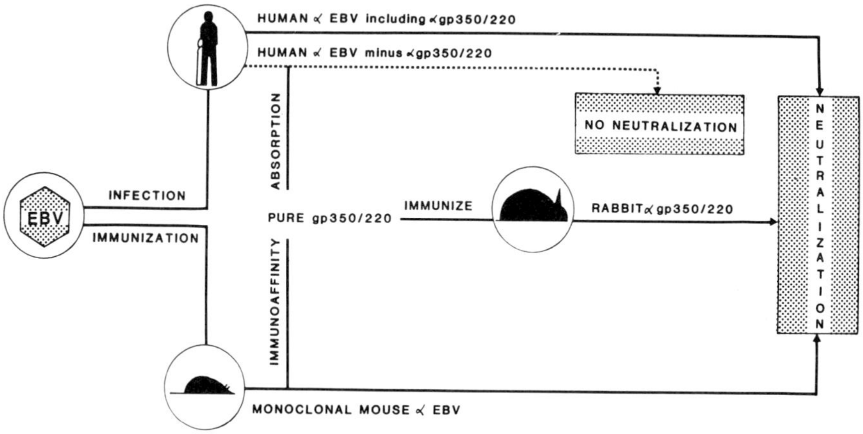

Fig. 4 Schematic representation summarizing our findings on gp350 and gp 220 and their role in stimulating neutralizing antibodies. Epstein-Barr virus (EBV) infection of humans results in the generation of virus-neutralizing antibodies. Production of monoclonal antibodies after immunizing mice with purified EBV allowed the isolation of monoclonal anti-EBV antibody that would neutralize the virus in vitro. This antibody could then be used to isolate the appropriate molecules (gp350/220) in a form that would both generate neutralizing antibodies in vivo and absorb the neutralizing antibodies from the sera of EBV-infected humans.

To purify the molecules, crude plasma membranes are extracted with detergent, and the solubilized glycoproteins are purified by ricin lectin affinity and C1 antibody-affinity chromatography. The purified material is antigenically active as it may be immunoprecipitated, and, most important, upon injection into rabbits, it will generate a high-titered virus-neutralizing antibody that specifically immunoprecipitates only gp350/300 and gp250/200 (28). Last, we have used the purified glycoproteins, either in solution or, more effectively, coupled to Sepharose 4B, to absorb, human sera. In this manner, it is possible to demonstrate that when the anti-gp350/300 and gp250/200 antibodies are absorbed from human sera most of the virus-neutralizing antibodies are also absorbed. This indicates that the two large glycoproteins can generate virus neutralizing antibodies in vivo, both in animals and humans, and is the main componenet responsible for generating virus-neutralizing antibodies during infection. A summary of these findings is presented schematically in Fig. 4.

EBV Associated Antigens on Infected B Lymphocytes

When EBV from throat washings or the supernatants of virus producing lymphoblastoid cell lines is added to peripheral lymphocytes, the virus will infect only the B lymphocytes (29,30,31,32). This will result in the subsequent outgrowth of continuously proliferating B lymphoblasts. Similarly, infected B cells from the peripheral blood of IM patients will spontaneously grow out. In this discussion, the term "transformation" will be used to define this process. EBV is unique, however, as it fixes the B cells at the lymphoblastoid stage and does not allow differentiation to a plasma cell. The B cell, now fixed at the proliferative stage, will grow indefinately and for this process the terms "immortalization" or "growth transformation" will be used. These B cells contain multiple copies of the viral genome, a small portion of which is transcribed into 3-5 RNA species (33). One of these RNA's encodes a nuclear protein designated EB nuclear antigen (EBNA) (34,35), which is present in all infected cells and is detected using sera from EBV seropositive individuals.

The first evidence for EBV associated antigens of the surface of transformed cells came from studies of cell mediated immunity. It was originally noticed that B lymphocytes would transform move effectively in vitro if the cultures were depleted of T lymphocytes and this phenomenon was termed "suppression" (36). It was later shown that the level of initial infection was much lower in the presence of T cells due to the production of interferon which inhibited transformation (37). Any cells that did transform would then proliferate for 7-10 days and subsequently cease. This phenomenon was termed "regression" (38). The "regression" event has subsequently been shown to be due to the induction of both MHC restricted (39) and unrestricted cytotoxic T cells (40). The MHC restricted T cells are of particular interest as they will also produce interferon in response to the autologous, EBV infected B cell, suggesting that they play many roles in controlling EBV infection (40). Not only do they directly destroy virus transformed cells, but they also produce interferon which can prevent the transformation of newly infected cells and is also known to be a potent activator of natural killer cell activity (41).

The observation that cell mediated immunity could be readily detected against EBV transformed lymphocytes suggested strongly that there were EBV associated antigenic changes on the surface of these cells. As a consequence, several laboratories attempted to identify these antigens by producing monoclonal antibodies with specificity for EBV transformed lymphocytes.

Detection of the BLAST-1 and BLAST-2 Antigens

The BLAST-1 and 2 antigens were detected by producing monoclonal antibodies with specificity for EBV transformed cell lines. One antibody has been produced against the BLAST-1 antigen and 7 against BLAST-2. The distribution of these antigens on a wide variety of cell types is very similar. They are found only on B cells transformed *in vitro* with EBV or poke-weed mitogen, *in vivo* with EBV or antigen, and on neoplastic B cells from chronic lymphocytic leukemia and poorly differentiated lymphoma. They are not found on cells of T, null, and myeloid lineage, whether obtained from peripheral blood, lymph nodes, neoplasms or lymphoblastoid cell lines. Thus, these antigens are B cell differentiation antigens, specific for the lymphoblastoid stage (8,12). The EBV association of these antigens comes from their level of expression. The antigens are expressed at a much higher level on EBV infected cells than on any other cell type studied and this is demonstrated in Fig. 5. The difficulty in detecting the level of expression of these antigen on normal B lymphoblasts has led to previous erroneous claims that the antigens are EBV specific (43,44,45). These antigens are of special interest, as they constitute the only EBV associated antigenic changes so far detected using monoclonal antibodies. Using these markers we may conclude that, although EBV infects small, resting B lymphocytes, the transformed cells express antigens characteristic of the lymphoblastoid stage. Thus, EBV infection, like mitogens and antigens, acts upon the small B cell and drives it to differentiate. EBV is unique, however, as it causes the B cells to become fixed at the stage where B cells normally proliferate--the lymphoblastoid stage, characterized by the expression of BLAST-1 and BLAST-2. It is likely that the virus takes control of the transformation process, including the expression of the BLAST antigens, causing them to be overproduced. Whether this overproduction is itself sufficient to immortalize a B lymphocyte or is part of a more complex process remains to be investigated.

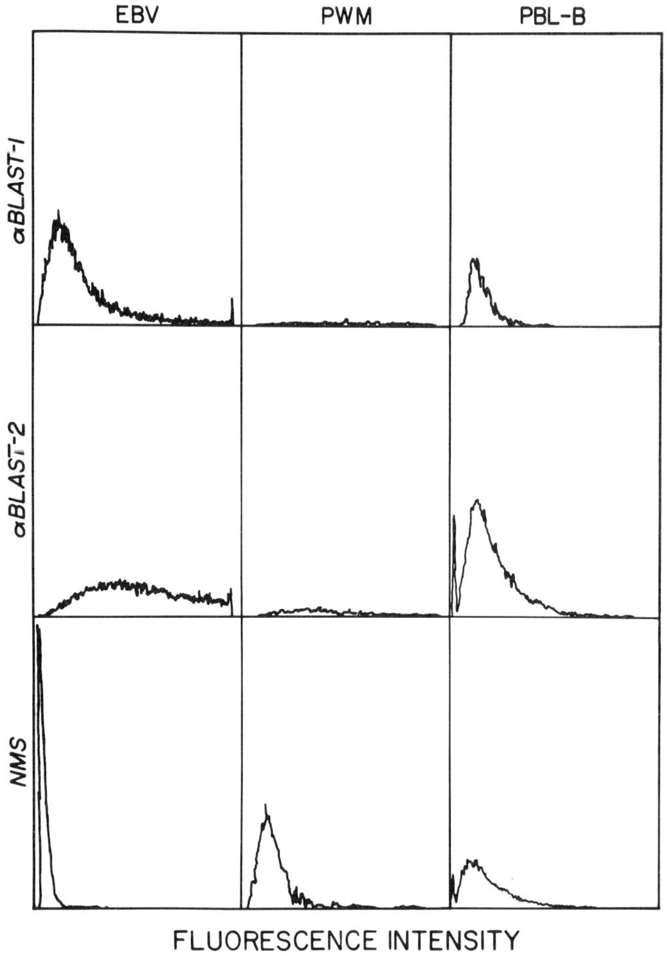

Fig. 5. Cytofluorgraphic analysis for expression of BLAST-1 and BLAST-2 on B lymphocytes transformed with either poke-weed mitogen (PWM) or Epstein-Barr virus (EBV) or untreated. N.B. The fluorescence on the PWM treated cells was much weaker than the EBV transformed cells. It was necessary to turn up the gain in order to see the fluorescence and this is reflected in the level of staining seen with the NMS control on both populations.

Biochemical Analysis of BLAST-1 and BLAST-2

The first evidence that BLAST-1 and BLAST-2 were distinct antigens comes from immunoprecipitation studies. BLAST-1 has a molecular weight of 44-45,000 and BLAST-2 of 47,000 (Fig. 6). Despite the similarity in their sizes the antigens were distinct, as no crossreactivity was seen. Thus, anti-BLAST-1

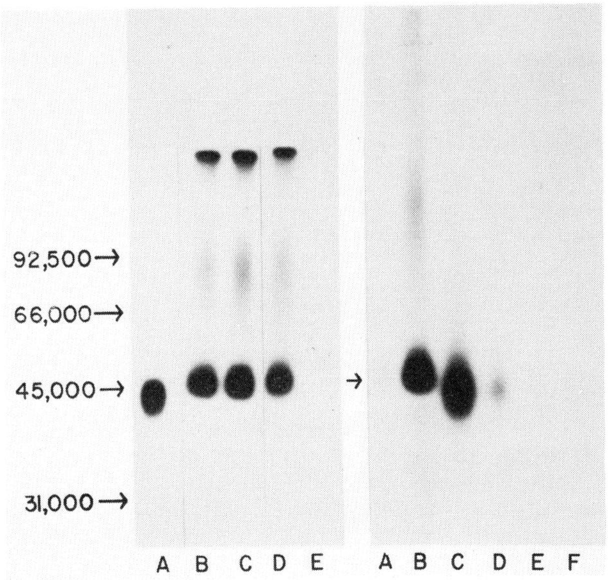

Fig. 6. (left panel) Immunoprecipitation of the antigens recognized by four independently derived monoclonal antibodies against polypeptides induced on B lymphocytes by EBV infection. Note: CS #1 is one of a series of five antibodies derived in the same laboratory and known to recognize the same molecules

A - anti-BLAST-1
B - CS #1
C - B532
D - MHMC
E - Negative ascites

Fig. 6 (right panel) Immunoprecipitation of BLAST-1 and BLAST-2 antigens after preclearing with either anti-BLAST-1 or anti-BLAST-2 (CS #1).

A - Precleared with anti-BLAST-2 followed by anti-BLAST-2.
B - As A) but precleared with anti-BLAST-1
C - Precleared with anti-BLAST-2 followed by anti-BLAST-2
D - As C) but precleared with anti-BLAST-1
E - As A) but precipitated with negative ascites
F - As B) but precipitated with negative ascites

could not preclear BLAST-2 and vice versa. Furthermore, the partial proteolythic maps of the antigens were different. Lastly, the biosynthetic processing of the antigens is distinct. In [^{35}S]methionine pulse-labelling experiments in the presence of tunicamycin both antigens appear to derive from 43,000 dalton precursors. The BLAST-1 antigen is associated with a second chain of 65,000 daltons [^{35}S-methionine, but not ^{125}I labelled], acquires carbohydrate, putatively of O- but not N-linked origin (lectin binding in the presence of tunicamycin) and is extensively sialated (neuraminidase sensitive). The BLAST-2 antigen, on the other hand, acquires N-linked sugar (tunicamycin sensitive), no detectable O-linked glycosylation and no sialic acid. The BLAST-1 antigen is stable at the cell surface whereas BLAST-2 is shed, as a stable 35,000 dalton polypeptide, into the culture supernatant.

Role of BLAST-1 and BLAST-2

At the present time, no functional role for these antigens is known. Clearly, however, it is reasonable to speculate that they could play a role both in the normal transformation process and in the aberrant process that leads to immortalization upon EBV infection.

The antigens could also play a role in stimulating cell mediated immunity either because of their inappropriate level of synthesis or because the inappropriate production leads to the expression of new "non-self" antigenic epitopes, as yet undefined, on the molecules.

CONCLUSIONS

Much progress has been made in the last few years in defining the immune response to EBV and the appropriate target antigens. We can now say with certainty that gp350/300 and 250/200 are the major targets for neutralizing antibodies and the prime targets for development of a subunit vaccine. Our recent experiments on the genetic mapping of these molecules should now pave the way for the development of genetically engineered, immunogenic peptides from the glycoproteins. This in turn should provide ample material for vaccination experiments. What is lacking, at present, is a readily accessible animal model system. The best one, at present, is with the cotton-top marmoset; however, the rarity of this animal will severely limit the number and type of experiments which can be performed.

A wealth of evidence now exists for antigenic change at the surface of B cells transformed and immortalized by EBV. These changes may be detected by both cellular immune responses and by mouse monoclonal antibodies. Whether the virus itself encodes a cell surface transformation antigen and what the nature of the target is for cellular responses remain two of the most important unanswered question in EBV research.

ACKNOWLEDGMENTS

I would like to thank the many collaborators without whom this work would not have been possible. This research was supported by grant numbers AI15310 and CA28737 from the National Institutes of Health.

REFERENCES

1. Henle, G., Henle, W.: J. Infect. Dis. 121:303-310, 1970.
2. Golden, H.D., Chang, R.S., Prescott, W., Simpson, E., Cooper, T.Y. J. Infect. Dis. 127:471-473, 1973.
3. Henle, W., Diehl, V., Kohn, G., zur Hausen, H., Henle, G. Science. 157:1064-1065, 1967.
4. Pope, J.H., Horne, M.K., Scott, W. Int. J. Cancer. 4:255-260, 1969.
5. Shope, T., Dechairo, D. and Miller, G. Proc. Natl. Acad. Sci. USA. 70:2487-2491, 1973.
6. Henle, G., Henle, W. and Diehl, V. Proc. Natl. Acad. Sci. USA. 59:94-101, 1968.
7. Robinson, J., Smith, D. and Niederman, J. Nature. 287:336-335, 1980.
8. Thorley-Lawson, D.A., Schooley, R.T., Bhan, A.K. and Nadler, L.M. Cell. 30:415-425, 1982.
9. Bar, R.S., DeLar, J., Chausen, K.P., Hurtubrise, P., Henle, W., and Hewetson, J.F. New Engl. J. of Med. 290:363-367, 1974.
10. Purtillo, D., Cassel, C.K., Yang, J.P.S., Harper, R., Stephenson, S.K., Landing, B.H., Vawter, G.F. Lancet. 935-941, 1975.
11. Strauch, B., Andrews, L-L., Siegel, N., Miller, G. Lancet. 1:234-237, 1974.
12. Crawford, D.H. et al. Lancet. 1:1355-1356, 1980.
13. Nagington, J. and Gray, J. Lancet, 1:536-537, 1980.
14. Bird, G.A., McLachlan, S.M. and Britton, S. Nature 289:300-301, 1981.
15. Palacios, R. 160:321-329, 1981.
16. Hausen, H., Schulte-Holthausen, H., Klein, G., Henle, W., Henle, G., Clifford, P., Santesson, L. Nature. 228:1056-1058, 1970.
17. Nonoyama, M., Pagano, J. Nature. 242:44-47, 1973.
18. Fialkow, P.J., Klein, G., Gartler, S.M., Clifford, P. Lancet 1:384-386, 1970.
19. Andersson-Anvret, M., Forsby, N., Klein, G., Henle, W. Int. J. Cancer. 20:486-494, 1977.
20. Thorley-Lawson, D.A. Cell 16:33-42, 1979.
21. Thorley-Lawson, D.A. and Edson, C.M. J. Virol. 32:458-467, 1979.
22. Thorley-Lawson, D.A. Nature. 281:486-488, 1979.
23. Edson, C.M. and Thorley-Lawson, D.A. J. Virol. 39, 1981.
24. Crawford, D.H., Epstein, M.A., Bornkamm, G.W., Achong, B.G., Finerty, S., and Thomps`on, J.L. Int. J. Cancer. 24:294-302, 1979.
25. Thorley-Lawson, D.A. and Geilinger, K. Proc. Natl. Acad. Sci. 77:5307-5311, 1980.
26. Edson, C.M. and Thorley-Lawson, D.A. J. Virol. 46:547-556, 1983.
27. Hummel, M., Thorley-Lawson, D.A. and Kieff, E. J. Virol. (submitted for publication, 1983.
28. Thorley-Lawson, D.A. and Poodry, C.A. In Virol. 43:730-736, 1982.
29. Schneider, U., zuur Hausen, H. Int. J. Cancer 15:59-66, 1975.
30. Yata, J., Desgranges, C., Nakagawa, T., Favre, M.C., de-The,, G. Int. J. Cancer. 15:377-384, 1975.
31. Menezes, J., Jondal, M., Leibold, W., Dorval, G. Infect. Immun.

13:303-310, 1976.
32. Thorley-Lawson, D.A. and Strominger, J.L. Virol. 86:423-431, 1978.
33. VanSenten, V, Cheung, A., and Kieff, E. Proc. Natl. Acad. Sci. USA. 78:1930-1934. 1981.
34. Reedman, B.M. and Klein, G. Int. J. Cancer. 11:499-520, 1973.
35. Summers, W.P. Grogan, E.A., Shedd, D., Robert, M., Liu, C.R. and Miller, G. Proc. Natl. Acad. Sci. 79:5688-5697, 1982.
36. Thorley-Lawson, D.A., Chess, L. and Strominger, J.L. J. Exp. Med. 146:495-508, 1977.
37. Thorley-Lawson, D.A. J. Immunol. 126:829-833, 1981.
38. Rickinson, A.B., Moss, D.J. and Pope, J.H. Int. J. Cancer. 23:610-617, 1979.
39. Misko, I.S., Moss, D.J. and Pope, J.H. Proc. Natl. Acad. Sci. USA. 77:4247-4250, 1980.
40. Slovin, S.F., Schooley, R.T. and Thorley-Lawson, D.A. J. Immunol. 130:2127-2132, 1983.
41. Trinchieri, G., and Santoli, D. J. Exp. Med. 147:1314-1333, 1977.
42. Thorley-Lawson, D.A. and Schooley, R.T. Proc. Natl. Acad. Sci. (submitted for publication), 1983.
43. Kintner, C. and Sugden, B. Nature. 294:458, 1981.
44. Slovin, S.F., Frisman, D.M., Tsoukas, C.D., Royston, I., Baird, S.M., Womsley, S.B., Larson, D.A. and Vaughan, J.H. Proc. Natl. Acad. Sci. 79:2649-2653, 1982.
45. Rowe, M. Hildreth, J.E.K., Rickinson, A.B. and Epstein, M.A. Int. J. Cancer, 29:373, 1982.

The Role of Viruses in Human Cancer. Volume II
G. Giraldo and E. Beth editors

THE MOLECULAR BIOLOGY OF HUMAN CYTOMEGALOVIRUS AND ITS RELATIONSHIP TO VARIOUS HUMAN CANCERS

ENG-SHANG HUANG, IVSTAN BOLDOGH, JOHN F. BASKAR, ENG-CHUN MAR
Cancer Research Center, Departments of Medicine, Microbiology and Immunology, University of North Carolina at Chapel Hill, North Carolina.

INTRODUCTION

Human Cytomegalovirus (CMV), a member of herpes group viruses, is ubiquitous and is an important human pathogen. This virus has been reported to be associated with a variety of clinical syndromes ranging from asymptomatic infections to cytomegalic inclusion disease, intrauterine fetal death, prematurity, congenitial maldevelopment, mental retardation, infectious mononucleosis and interstitial pneumonia in organ transplants or immunocompromised patients.[1,2] As with other herpes group viruses, human CMV is capable of establishing a long term latent infection following primary infection, and with subsequent recurrence under immuno-compromised situations. The virus can be reactivated despite the presence of a high level of humoral antibody titer in latently infected patients. Congenital infection can also occur in utero in spite of a substantial level of maternal immunity. The reactivation of a latent virus is often associated with preganancy, allograft transplantation and immunosuppresion due to malignant diseases or chemotherapy. Since human CMV stimulates host cell macromolecule synthesis and transforms hamster embryonic cells (4) and human fibroblasts (5,6,), it is speculated that this virus may have ocogenic potential. More recently, human CMV has been associated with Kaposis's sarcoma and pneumoncystic pneumonitis which are the two rare diseases occurring with increasing frequency among homosexual and heterosexual men (7-9).

Besides saliva and urine, human CMV is frequently isolated from human cervix, blood, milk and even from semen (10-12). We have been surprised by our recent observation that this virus could be isolated from semens of 11 out of 29 homosexual men in the Research Triangle Area, North Carolina (our unpublished data). In addition, CMV is also associated with a certain number of chronic cervicitis (13). With such a wide representation and involvement in genital infections, it is reasonable to assume that the occurrence of congenital embryonic abnormality and mental retardation could be due to the consequences of sexually transmitted CMV infections. The ubiquitious distribution of CMV infection and its asymptomatic interaction with its host frequently lead scientists to overlook CMV's oncogenic potential and the consequence that followed the reactivation of latent infection. In this section we would like to address specifically the molecular biology of CMV and its oncogenic potential including some possible evidences of CMV association with various human cancers.

MOLECULAR BIOLOGY OF HUMAN CMV

General properties. Human CMV is a double-stranded DNA virus with a genome of approximately 150 million (235 kilobase pairs)(14-15).Morphologically it is extremely hard to distinguish CMV from other herpes group viurses. Electron microscopic observation shows a naked virus consisting of a spherical core about 60nm in diameter, surrounded by a 90 to 110nm icosahedral capsid with 162 capsomeres(17).The naked virion particles are in turn surrounded by an envelope with a single or double membrane acquired from nuclear or cytoplasmic membrane of infected cell during maturation. In addition, some membrane-bound particles called "dense bodies" are often associated with the virus in the cytoplasmic inclusion and also in the extracellular fluid of infected cultures. The dense body has a size of 300nm and consists of a spherical particle surrounded by a membrane similar to that of the viral envelope(18-20). The dense body does not contain viral DNA, but shares most of the polypeptices contained in the purified virion.[18] SDS-polyacrylamide gel electrophoresis reveals that purified CMV virions contain at least 33 polypeptides with molecular weights ranging from 11,000 to 290,000 daltons (21-24).There are at least 8 glycopolypeptides repeatedly detected, and antisera to these glycopolypeptides contain antibodies that neutralize virus infectivity in vitro (19,22).

Molecular arrangement of CMV genome. Human CMV genome is one of the largest among DNA viruses. It is 50% larger than that of the herpes simplex virus (14-16). The density of CMV DNA is slightly higher than that of human host-cell DNA and is about 1.716gm/cm^3; which corresponds to G + C content of 56 to 58%. Melting curve analysis reveals a bisigmoidal curve having two distinguishable melting transition points at 86°C and 94°C; corresponding to G + C content of 43% and 79% respectively (25). By a partial denaturation analysis of CMV DNA, it has been found that viral DNA consists of 2 covalently linked L and S components with invertable orientation which gave four possible isomeric arrangements resembling that of herpes simplex virus (26,27) (see Fig. 1). Electron microscopic observation of self-annealed denatured whole length single stranded DNA or DNA fragments shows that CMV DNA (Towne strains) contains internal inverted repetitious sequences of both terminal-end sequences (TR_l and TR_s) (28,29). These internal inverted repetious sequences divide CMV DNA molecule into a large L and a smaller S segment as detected in partial denaturation map (25). The L segment contains

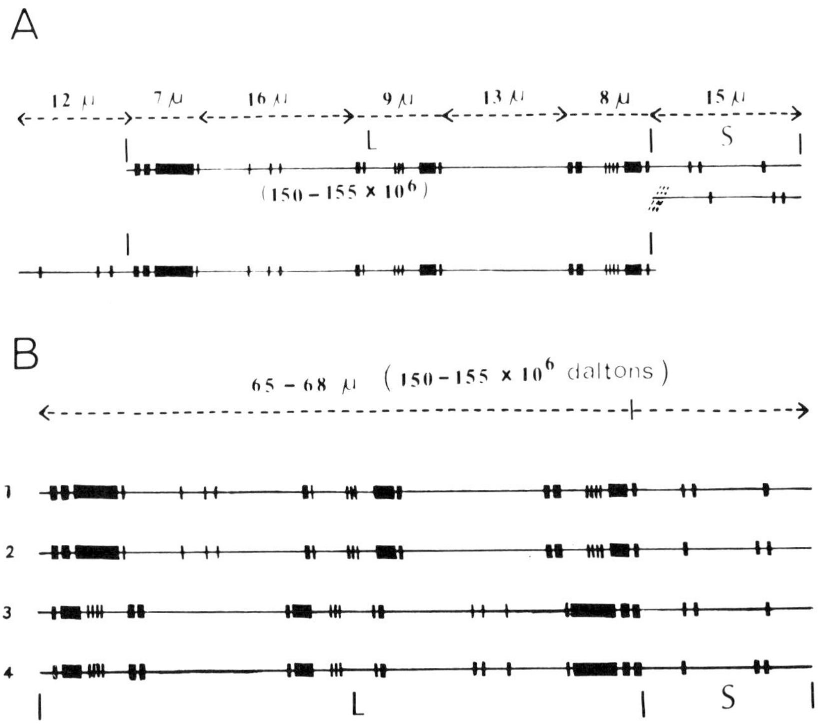

Fig. 1 Summary of observed features of partially denatured DNA molecules. (A) L and S subfragments of CMV DNA (B) Four possible isomeric arrangements. The heavy blocks indicate the denatured areas, at pH 11.36.

a unique region of 170kb boxed by inverted repeats of 11kb, and the S-segment with a unique region of 37kb surrounded by inverted repeats of 2-2.5kb (28,29) (see Fig. 2).

Due to the inverted rightward and leftward orientations of L and S components and the equal representation of 4 isomeric molecules, the restriction fragments are generated from terminal ends and L-S junction of DNA molecules (25,28). Restriction fragment patterns and physical maps of AD-169, (16,30), Towne (28), and Davis (31), strains of human CMV have been analyzed and mapped recently. Data from these analysis confirm the presence of four equimolar isomeric populations which differ only in the relative orientation of the L and S components of DNA molecules.

An intrastrained heterogeneity of viral DNA in the form of one or more additional steps of 750bp at the S end of DNA molecules (around 70%) and occasionally also of 300bp at the L end of DNA molecules (around 10%) have been found in

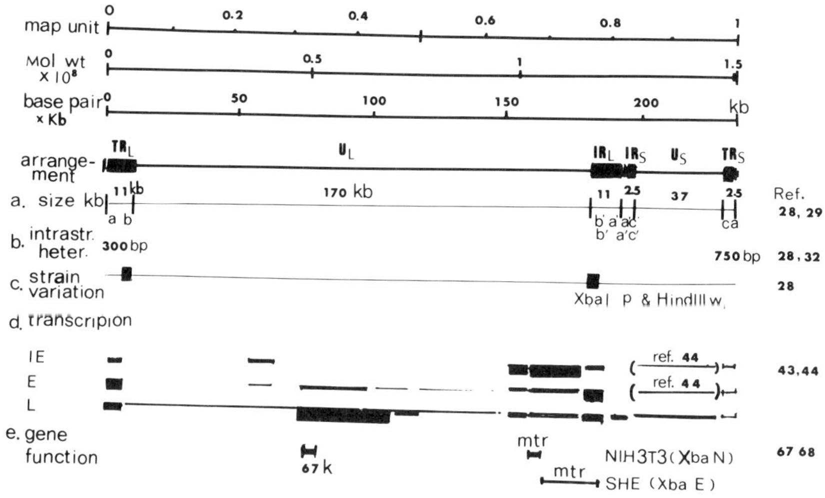

Fig. 2 Diagram of the molecular arrangement, gene transcription and gene functions of human cytomegalovirus.

the Towne strain CMV DNA (28,32). This kind of heterogeneity has also been found in herpes simplex virus in which one or more extra copies of a 250 bp direct terminal redundancy (the "a" sequence) occur in up to 40% of the L ends and joint fragments. This "a" sequence has been implicated to play an important role in the isomerization of viral genome(33,34). Human CMV DNA molecule is expected to have a similar "a" sequence arrangement, because Geelen and Westrate were able to find the unit length of the circular DNA molecule after a brief exonuclease III treatment followed by reannealing of full-length CMV DNA (35).

Genetic heterogeneity among different strains of human CMV. Great antigenic heterogeneity among human CMV strains have been demonstrated previously by various seroimmunology techniques (36-38). Similarly, the heterogeneity among CMV isolates also appears at the level of viral DNA sequences. Restriction enzyme digestion of DNA from various clinical isolates yields distinct DNA fragment patterns for each isolate. Although considerable matching of fragment

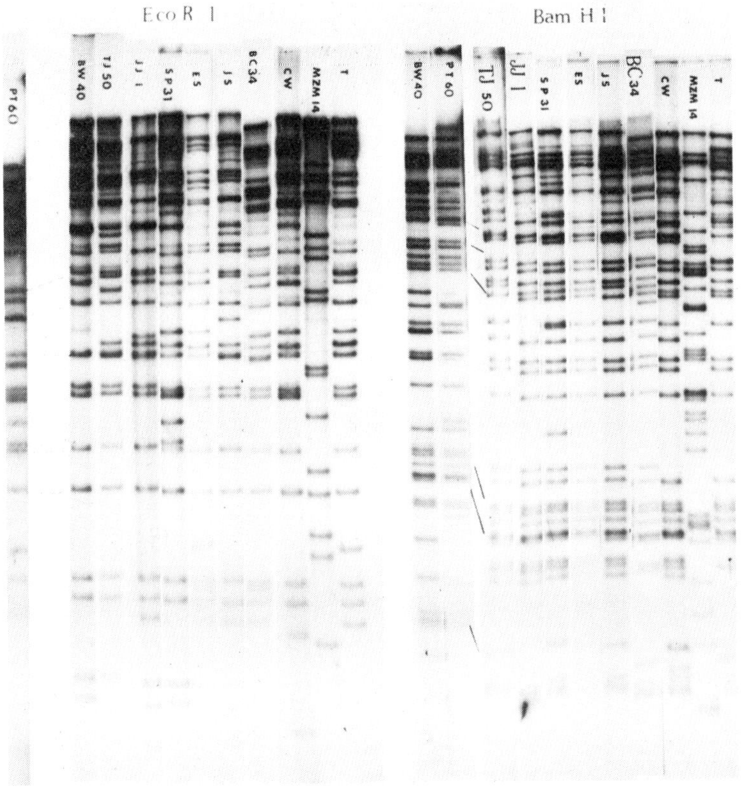

Fig. 3 DNA restriction finger prints of various CMV isolates. Enzymes used for analysis were EcoRI and Bam HI as indicated. MZM 14 is a chimpanzee CMV used for comparison.

patterns and comigrating fragments are found among them, no two epidemiologically unrelated isolates have identical fragment patterns (29) (Figure 3, gel MEM 14 is a Chimpanzee CMV). By DNA-DNA reassociation kinetics analysis of CMV inter-strain DNA sequence homology, we found that ten CMV strains shared at least 80% homology with the protoytype AD-169 CMV DNA (36). By reciprocal DNA hybridization, Pritchett also demonstrated that CMV strains Towne and AD169 share approximately 90% of DNA sequence homology (38). Though an extensive restriction fragmentation and nucleic acid hybridization observations, Lafemina and Hayward (38) concluded that the genetic variability and heterogeneity is associated primarily with the large repeat sequence, since both strains Davis and AD169 CMV genome are deleted by 2000-4000 bp at the extremely internal portion of both L repeats relative to Towne strain DNA (28,37). The deletions at these areas may, therefore, account for the genetic differences encountered among various CMV isolates (Fig. 2b). The size and the organization of the S inverted repeats in three CMV isolates examined by Lafemina and Hayward were generally very similar (28).

The degree of homology exists among various human CMV isolates is greater than that between HSV type I and type II, which is about 50% homology (39). Human CMV strains isolated from genital tract could not collectively be distinguished from strains isolated from other sites either in sequence homology or in restriction pattern as is the case with HSV I and II.

Transcription and translation of human CMV genome. Although human CMV is widespread, it is remarkably species-specific. The interaction between this virus and its infected cells is therefore different between permissive and non-permissive hosts. In permissive human fibroblasts, viral gene expression resembles herpes simplex virus and is cascadely regulated and sequentially expressed at both transcription and translation levels (40-43). In the absence of *de novo* protein synthesis (in the presence of cycloheximide), the viral immediate early transcription (RNA originated from a region of L unique (0.66-0.77 map units), with three size classes of RNA, 4.8-4.6, 2.2 and 1.9kb, were present on the polyribosomes in intermediate to high abundance (41-44). There were also viral RNA in low abundance originating from the long repeat sequence (0.01-0.35 and 0.795-0.825 map units) and a region in the long unique section (0.201-0.260 unit).

If the *de novo* protein synthesis allows to occur, a transition from highly restricted transcription to relatively extensive transcription occurred, and additional size classses of viral mRNA in high abundance are found associated with polyribosome (42-44). These abundance classes of viral mRNAs at intermediate abundance were mapped in two regions of long unique section (0.325-0.460 and 0.685-0.770 map units). Stable viral RNAs that were associated with polyribosomes in high abundance originated from the long repeat sequences, but not from the long unique section (43). At late phase of infection, and after virus DNA replication the abundant transcripts were from the regions of long unique sequences (0.325 to 0.460, and 0.660 to 0.685 map units) while the intermediate abundance RNA was from the long repeat sequences (43 , see Fig. 2-d). Together with the polypeptides synthesis and *in vitro* translation data, Wathaen and Stinski concluded that initiation of viral RNA synthesis is not dependent upon the synthesis of viral-specific protein, but the *de novo* viral protein synthesis and expression of immediate early viral genes is necessary for the transcription of the early viral genes in the long repeat and adjacent sequences. The expression and translation of viral genes in most of the long and short unique sequences appears to require viral DNA replication (42,43).

Besides the cascade type of regulation in gene transcription, the post-transcriptional control mechanism was also noticed (43,44). DeMarchi has found that during the immediate-early, early, and late phases of human CMV infection in permissive human fibroblasts, transcripts accumulated from approximately 20, 75, and 90% of the sequences of genome, respectively. But not all of the sequences which accumulated during the immediate-early and early phases were represented on polysomes (44). In study of four immediate-early gene transcripts, De Marchi has noticed that certain classes of transcripts strongly associated with polysomes immediately after synthesis, but some classes of RNA transcript retained in nuclei and was never transported to cytoplasm for further translation. Some species of RNA was spliced but some did not. She, therefore, postulated that the expression of human CMV genome transcripts are regulated at the level of transcript accumulation, transportation, stability of mRNA, splicing and affinity to polysome (44).

Wathen and Stinski (43) have also made similar observations that CMV RNA from the region of 0.685-0.770 map units (Xba E fragment) was transcribed preferentially and abundantly at immediate early and early times; however the association of RNA transcript from this region with polysome is higher at immeidate-early but low at early time after infection. Meanwhile, the transcript from the region 0.325-0.460 map units (Xba B) was abundantly transcribed at early times but little or not was associated with polyribosomes. In addition, the region of 0.075 to 0.325 map units of the extensive section of long unique region was transcribed at early times but little or no RNA was found to be associated with polysome. Therefore, post-transcription control apparently plays a very important role in CMV gene expression.

In abortively infected nonpermissive cells certain viral gene functions which normally are expressed in the early phase of productively infected human embryonic lung (HEL) fibroblasts are not expressed. In rabbit kidney (RK) DeMarchi found that the rate of accumulation of human CMV-specific RNA was 6-8 fold lower than in HEL during 24 hr post-infection. Most of the "early" transcripts which accumulated in infected permissive cells were also found in nonproductive RK cells, but some of the early transcripts (approximately 7% of genome) were significantly under represented in RK cells. Besides, the association of the early transcripts from one region of genome to polysome in cytoplasm of nonproductive infected RK cell was found in very low extent as compared to that in the permissive HEL hosts. Based on these observations, it is obvious that CMV gene transcription is regulated both at transcription and post-transcription levels.

In studying the immediate-early gene transcripts, Stinski's group has found the splicing phenomenon existing in one of the major immediate-early gene transcripts (1.95 kp). The essential elements including CAT sequence, TATA box, termination codon and polyadenylated site have been identified. Besides, four repeat sequences with potential tertiary structures as that of SV40 72 bp repeats were also demonstrated (Stinski, M.F., personal communication; presented at the Pathogenesis of Human Cytomegalovirus Conference, Philadelphia). DeMarchi also observed that one of the immediate-early transcripts found in cyclohexamide-treated cells might be spliced during processing (44).

In the early study of the late CMV gene transcripts, we have found that one of the most abundant transcripts with the size of 1.8 kb was located at the middle of the L uinique region (approximately 0.330 to 0.345 map units). The gene function of this most abundant late transcript was studied in COS cells, a SV40 transformed monkey kidney cell, by transfection of COS cells with a recombinant clone pHD 713SV2 derived from PSVOH (pBR322 carried SV40 origin of DNA replication and 72 bp repeats) and carried a 2.9 kb subfragment encoded for this abundant transcript. Western blotting analysis indicated that the gene of abundant transcript was for a 67 K major protien (M.G. Davis, et al, submitted for publication). DNA sequence analysis revealed that this abundance late gene coded for 67K polypeptide did not have obvious splicing. The first group of CAT sequence and TATA box were only 50 and 25 bp, respectively, upstream from the capping site, while the first initiation codon is among 90 bp downstream from capping region. The termination codon and polyadenylation site were approximately 1700 bp downstream of 5' RNA end (M.G. Davis, et al. submitted for publication).

Immediate early, early and late proteins. By cyclohexamide pre-treatment of host cells and ^{35}S methionine pulse labeling experiment, Stinski observed that at least three intracellular virus-specific proteins ICSPs, 75, 72 and 68 were synthesized at relatively high concentrations and several other ICSPs were present in relatively low concentratons at immediate-early phase of virus infection (40). At two hours post-infection, three additional ICSPs (ICSPs 59, 56 and 53), were detected at relatively high concentrations and four others at relatively low concentrations were also detected (ICSP 39, 27, 21, and 19). In addition, in the presence of proteolytic inhibitor, one protein (ICSP 135) which was previously not identified was also detected (19). The functions of each immediate early protein are still unclear. But it is believed these polypeptides (α protein) are needed for the synthesis and the expression and transcription.

By immunoprecipitation of immediate early antigen induced in CMV infected human lung fibroblast with anti-IEA positive human convalescent sera, Michelson, et al (45) were able to detect two polypeptides of 76K and 82K withing 90 min. after infection. These two polypeptides were located predominantly within the nucleus, although some of the 76K polypeptides were also found at perinuclear regions. Polypeptides of similar molecular weight were also found in CMV-infected nonpermissive rabbit skin fibroblasts (45). Lafemina and Hayward (32) also demonstrated that in CMV-infected Balb/c-3T3 cells a single major protein corresponding to the 68K immediate-early polypeptide could be detected within two hrs after cyclohexamide reversal. This 68K protein was overproduced in Balb/c-3T3 cells to such an extent as a major component of the nuclear fraction.

In vitro translation of polysome-associated virus-specific mRNA revealed approximately ten proteins (19). These virus specific-early (E) proteins are expressed in the cell for a prolonged period of time. They might play a very important role in regulation or participation in viral DNA synthesis. Virus specific late proteins requires the presence of viral DNA replication (23). The majority of the late proteins presumably are structural moieties of virion; these include capsid protiens, glycoprotiens, etc.

ONCOGENICITY OF HUMAN CMV

It is extremely difficult to present solid and direct evidence to conclude the oncogencity of human CMV in in human host, especially concerning the causal association between CMV and human cancers. Its widespread distribution and frequent asymptomatic interaction with humans make the interpretation of the significance, by the means of biostatistical and epidemiological analysis, more difficult. To facilitate the analysis of CMV oncogenicity, we have purposely categorized and interpreted the observations made by various investigators by the following format: (a) biochemical and biological interactions of CMV with virus-infected host, (b) oncogenic transformation of mammalian cells by virus or by virus DNA fragments in vitro, and (c) molecular epidemiological analysis of the association of CMV with various human malignancies. We believe that these analyses may not provide definitive conclusion for the causal association of CMV with cancer, but it at least will strengthen our knowledge toward understanding the role of CMV infection in malignant processes.

A. Biochemical interaction of CMV with infected-host

Stimulation of host cell nucleic acid synthesis. It is known that the ability of a virus to induce host cell macromolecule synthesis is correlated with its transformation and oncogenic potential. In human CMV system, St. Jeor, et al (46) first demonstrated that CMV was able to stimulate cellular DNA synthesis in both permissive HEL and in non-permissive monkey kidney (vero) cells. Their study indicated that stimulated cellular DNA was synthesized by semi-conservative mechanism and that viral coded function expressed after infection was essential for the stimulation of cellular synthesis. Boldogh, et al (47), also demonstrated that the ability of human CMV to stimulate cellular DNA synthesis was dependent on the expression of very early gene function and was relatively resistant to UV-irradiation. Besides the stimulation of cellular DNA synthesis, Tanaka, et al (48-50), also reported that cellualr mRNA, tRNA, ribosomal RNA and mitochondria DNA synthesis were greatly enhanced in CMV-infected permissive WI-38 and in non-permissive guinea pig embryonic fibroblasts. They also concluded that virus-specific protein synthesis was required for the stimulation of host cell macromolecules synthesis.

In non-permissive rabbit kidney (RK) cell, DeMarchi, et al (51,52), found that the induction of cellular DNA synthesis occurred only in those cells that synthesized viral-specific antigens. In contrast, in permissive cultures, synthesis of viral antigens and induction of cellular DNA synthesis are mutually exclusive; cells productively infected with CMV were not induced to synthesis cellular DNA, and the induction of cellular DNA synthesize occurred only in those cells in which viral structural antigens synthesis could not be detected. DeMarchi also found that defective virus particles are quite effective in stimulating cellular DNA synthesis (52).

Based on observations described above, it is obvious that human CMV has an important characteristic of stimulating cellular DNA synthesis both in permissive and non-permissive hosts. Similarly, host cell macromolecules synthesis which proceded the cellular DNA synthesis or which followed the cellular DNA synthesis should therefore coincidently be enhanced.

Stimulation of host cell enzymes synthesis. Kamata, et al (53) discovered that one of the chromatin-associated factors, which induced in human embryonic lung cells at very early stages of CMV infection, stimulated template activity of cell chromatin. This factor (factor 1) coincided with a major component of immediate-early antigen which was detectable within 1 hr after infection. As the consequence of derepression of host cell chromatin, numerous cellular enzyme activities have been found markedly enhanced in CMV-infected permissive and non-permissive cells. These include thymidine kinase (54,55), DNA polymerase (56,57), DNA dependent RNA polymerase (58), ornithine decarboxylase (59), plasminogen activator (60), exonuclease and topoisomerases (our unpublished observation).

In CMV-infected fibroblasts there was a rapid and high level of stimulation of cytosol and mitochondria thymidine kinase (TK), (55). The stimulation of these TK activities was found as early as 12 hrs post-infection and reached the plateau at 25 to 48 hrs. Characteristics of these enzymes with respect to electromobility, phosphodonor specificity, pH optima , salt inhibition and thermostability did not show significant distinction from that of normal cellular enzymes. There was also no detectable novel TK with an elcectromobility distinguishable from that of host enzymes in infected cells. These observations suggested that virus stimulated TK activities in CMV-infected cells were all of cellular origin. In contrast, a virus-specific TK with biochemical characteristics distinguishable from that of host enzymes was induced in HSV-1 infected cells. This HSV-1 specific TK can be separated from host cell enzymes by various affinity chromatographies, and is able to specifically and preferentially phosphorylate several purine and pyrimidine analogs, such as acyclovir, 5'-amino-2'5'-deoxy-5-iodouridine and 5'-allyl-deoxyuridine, etc., which in turn inhibit HSV DNA replication (61,62). In the case of human CMV systems, where no virus-specific TK has been detected, (55), the virus DNA replication is therefore relatively resistant to these pruine and pyrimidine analogs (our unpubished data).

Infection with WI-38 human fibroblasts with CMV also led to the stimulation of host cell α and β DNA polymerases syntheses. Besides, a novel virus-specific DNA polymerase was also induced (56,57). The characteristics of these virus-stimulated α and β form cellular DNA polymerases did not show significant distinction from that of mock-infected cells. In contrary, the virus-specific newly synthesized DNA polymerase exhibited different chromatographic behavior, salt sensitivity, template-specificity and PAA sensitivity sufficiently enough to distinguish it from that of cellular enzymes. The role of newly stimulated hosts cell α and β DNA polymerases is still unclear, but it is expected to be associated with the stimulation of host cell DNA synthesis. Whether these polymerases are also involved in the initiation or elongation viral DNA synthesis or act as core enzyme for the formation of viral-specific DNA polymerase (holoenzyme) has yet to be investigated.

DNA topoisoerases (TOPO) are a group of enzymes that change the helical turn numbers of supercoiled DNA molecules through the transient breakage and religation of the phosphodiester bond of DNA molecules (63). These enzymes are recently believed to be involved in site-specific as well as site-non-specific recombination, disentanglement of catenated DNA circles, facilitation of RNA transcription, and knotting and unknotting single-stranded DNA molecules (63.)

In the eukaryotic system, type 1 TOPO has been found to be associated with chromatin and is able to use both positive and negative supercoiled DNA as substrate without requiring ATP and Mg++ as co-factors' while eukaryotic type 2 enzyme resembles the prokaryotic type 2 TOPO and requires Mg++ and ATP as cofactors. In HEL cells the enzyme levels of both type 1 and type 2 topisomerases are greatly enhanced upon infection with CMV (our unpublished data). These enzymes are frequently co-purified with cellular α form and virus-induced DNA polymerases. The TOPO activity associated with virus-induced DNA polymerase did not require Mg++ and ATP as cofactor, and was resistant to phosphenoacetate (PAA) at a concentration of 50 μg/ml, while virus-specific DNA polymerase was sensitive to PAA at concentration of 10 μg/ml. The distinction in PAA sensitivity indicates that the active center for virus-induced DNA polymerase and TOPO activity should be different. Whether these two enzyme activities are derived from a single enzyme or are two co-purified enzymes should be further investigated.

Ornithine decarboxylase is an enzyme involved in the frist rate limiting reaction in the biosynthesis of protamine. The activity of this enzyme and the rate of enzyme synthesis are frequently linked to DNA synthesis. The increase of the enzyme activity was found to be associated with and to precede with DNA synthesis. The level of this enzyme in normal stationary phase cells usually is relatively low, but increases substantially upon the infection of cells with tumor viruses. In human CMV infected embryonic fibroblasts, a great degree of of stimulation of this enzyme activity was found 12 hrs post-infection (59). The degree of stimulation was dependent on the multiplicity of infection and was not inhibited or fed back by the addition of protamine, but was reversibly inhibited by PAA treatment (59). It was therefore concluded that CMV-induced stimulation of ornithine decarboxylase synthesis was independent of cellular DNA synthesis and might be related to virus-induced DNA polymerase or viral DNA synthesis.

Finally, infection of human hamster cells with UV-irradiated human CMV also led to the stimulation of plasminogen activator synthesis (60). The plasminogen activator is an enzyme which is able to convert plasminogen into plasmin, and the plasmin in turn is able to hydrolyze the fibrin. This enzyme appears to be closely related to malignant transformation. Transformation of cells by DNA and RNA tumor viruses frequently lead to the enhancement in the synthesis of plasminogen activator. The stimulation of plasminogen activator in human CMV-infected cells does not require virus DNA synthesis. It is therefore suggested that the stimulation of the plasminogen activator synthesis is an early gene function of human CMV (60).

B. Morphological transformation of mammalian cells in vitro by ctyomegalovirus

Albrecht and Rapp (4) first demonstrated the ability of UV-irradiated human CMV to transform hamster embryonic fibroblasts in vitro. Following the exposure of hamster embryonic cells to UV-irradiated human CMV, foci of transformed hamster cells were found. Of 16 foci selected, only one survived for three subsequent subcultures; it was called CX-90-3. In contrary, cells exposed to untreated CMV showed signs of degeneration and subsequent death. Two distinct cell types were observed in cultures derived from CX-90-3 after 8 passages of in vitro subcultivation. One (CX-90-3A) approached crisis at passage 10 and was terminated at passage 13. Another, CX-90-3B, showed loss of contact-inhibition and continued to grow well in vitro (4). Cloned CX-90-3b cells were found to be tumorigenic in golden Syrian hamsters. Tumors induced in hamsters were poorly differentiated malignant fibrosarcomas and could be continuously subcultured both in vitro and in vivo. Virus-specific antigens were demonstrated both in the cytoplasm and on the cell membrane of CX-90-3B cells using human covalescent sera. The status of the viral genome in CX-90-3B cells and in induced tumor cell lines is unclear. After a long period of subcultivation in vitro, the viral genome was undetectable in both CX-90-3B and its induced tumor cell line by DNA-DNA reassociation kinetics analysis with a sensitivity of 0.1 viral genome per cell (6).

Boldogh, et al (64), also demonstrated the transformation of hamster embryonic fibroblast cells using UV-irradiated human CMV. The morphologically transfromed cell lines and the cell lines derived from the induced tumors did bear CMV-specific cytoplasmic and membrane antigen but no infectious virus or virus-specific nuclear antigens.

Spontaneous release of human CMV from cell lines derived from human tissues has been reported on at least two occassions (65,66). A virus strain called Major (or Mj) was spontaneously released from a prostate cell line derived from a child. This prostate cell line grew in vitro to passage level well above the expected life span. Geder, et al (5), have speculated that cytomegalovirus gene function might play a certain role in permitting the life span of this prostate cell line extension beyond the normal level. They further studied the transforming ability of the Mj strain CMV by low multiplicity (0.001 PFU per cell) and persistent infections of human embryonic lung HEL cells with this virus. After a crisis period, foci of morphologically transformed cells appeared. Two transformed cell lines were then established. designated as CMV-Mj-HEL 1 and 2. Both cell lines were tumorigenic in athymic nude mice (5). CMV-specific membrane antigens could be detected in these transformed cells by immunofluorescent test with human covalescent serum or by cytoxicity tests using spleen cells from hamsters bearing isographs of CMV transformed cells. Perinuclear and paranuclear fluorescences were also observed in most of the transformed cells when anticomplement immunofluorescence technique were applied. Viral DNA had been found in transformed CMV-j-HEL cell at 0.3 genomes equivalent per cell at early passage (p48), but become undetectable after prolonged cultivation in vitro (our unpublished results, in collaboration with Dr. F. Rapp).

C. Morphological transformation of mammalian cells by viral DNA fragments

Morphological transfromation of mammalian cells by CMV DNA fragments has been achieved both in human fibroblasts (6), in NIH 3T3 cells (67), and in Syrian hamster embryo (SHE) cells, (68). The efficiency of transformation by DNA fragments is extremely low and is far less than what we expected. In human HEL fibroblasts we were able to obtain transformation foci when total Xba I or Hind III digested DNA fragments were used for transformation. Unfortunately, until now, we were not able to transform human HEL cells by a single cloned Xba DNA fragment; the study of double or combined fragments transformation is now underway.

To prevent the complication and to prevent the lytic infection, we initally used Xba digested Towne strain human CMV DNA fragments for our transformation study. The subconfluent (70-80%) HEL cells were transfected with Xba I CMV DNA fragments by calcium phosphate precipitation and DMSO shock as described previously (69). On the fith day post-transfection, the transfected cells were reseeded in low density in MEM with 3% fetal calf serum. Cultures were monitored for morphologically transformed loci for 6-7 weeks. Mechanically sonicated CMV DNA, calf thymus DNA or salmon sperm DNA, and DMSO treatment were included in our control experiment. A tumor promoter agent, 12-O tetradecanoyl-phorbol-13-acetate (TPA), was also used in our CMV transformation study to test whether CMV-initated transformation requires a co-carcinogen or a promoting factor to facilitate the transformation event (6). Some transfected cultures and various controls were then treated with TPA at 25ng/ml for 24 hours on 18 or 35 days after transfection.

In the absence of TPA treatment, the Xba I fragment-transfected culture yielded morphologically transformed foci at a frequency of approximately 1 per 10^6 cells when HEL cell was transfected with 2.5×10^{-6} μg per cell of Xba I DNA fragments. The TPA-treated culture yielded 5 to 6 transforming loci per 10^6 cells. No transforming foci have been found in cells transfected with sonicated CMV DNA either in the presence or in the absence of TPA. TPA and DMSO treatment alone did not induce any morphologically transformed foci in

HEL cultures (6).

CMV-specific antigen and viral specific mRNA could be detected in transformed foci by ACIF and by in situ DNA-RNA cytohybridization (Fig. 4, d-g). Three continuous transformed cell lines designated as BH19, BH21, and BH46 were established from cultures transfected with Xba I fragments and subsequently treated with TPA. These transformed cell lines were able to grow in soft agar and they were also able to grow to high densities with short doubling times in medium containing a low serum concentration (2% fetal calf serum). These transformed cell lines are morphologically epithelial, combined with short fibroblast, (see Fig. 4b) and are able to induce fibrosarcoma in ahtymic nude mice. The tumorigenicity of these transformed cells was increased significantly by subsequently passages in nude mice. Metastases consisting of poorly differentiated fibrosarcoma could be found in lung, liver and spleen of these animals. The cell lines derived from these tumors still bears some virus specific polypeptide that appeared in the original transformed cells. The karyotype of these tumor cells still retains characteristics similar to that of the original transformed cells.

The TPA treatment appears crucial for CMV DNA fragments to immortalize the morphologically transformed human cells. Without TPA treatment cells from the CMV transformed foci frequently cease multiplication after a couple of subcultivations. The mechanism of the establishment of CMV-induced transformation, therefore, appears similar to that of other carcinogen involved multiphase process (70); in which human CMV is capable of initiation of malignant transformation but the maintenance of this malignant state might require the presence of a co-carcinogen, such as a tumor-promoting agent.

In contrast to CMV DNA fragment transfomration of human fibroblast, we were able to establish CMV transformed cell lines by low multiplicity infection and extensive subcultivation of HEL cells with strains, BT1757 and Towne CMV without treatment. Cell lines derived from BT1757 (2E, 4D, 5B, GH and 7E), and from Towne (LH1, LH2, and LH5) share similar biological characteristics as those derived from DNA fragment transformed cells, and are all tumorigenic in athymic nude mice. This suggests that the continuous induction of cellular mitotic activity or cellular macromolecules synthesis by frequent subcultivation might function as tumor promoter in assisting CMV in the immortalization of HEL cells.

Nelson et al (67), have recently transfected NIH 3T3 cells with cloned AD169 strain CMV DNA fragments to identify the transforming region of human CMV. From the mapping and transfection experiments they have found that the transforming region was in the 2.9K subfragment of the Hind III E fragment with map units between 0.123 and 0.14 on the DNA molecule of the AD169 strain CMV.

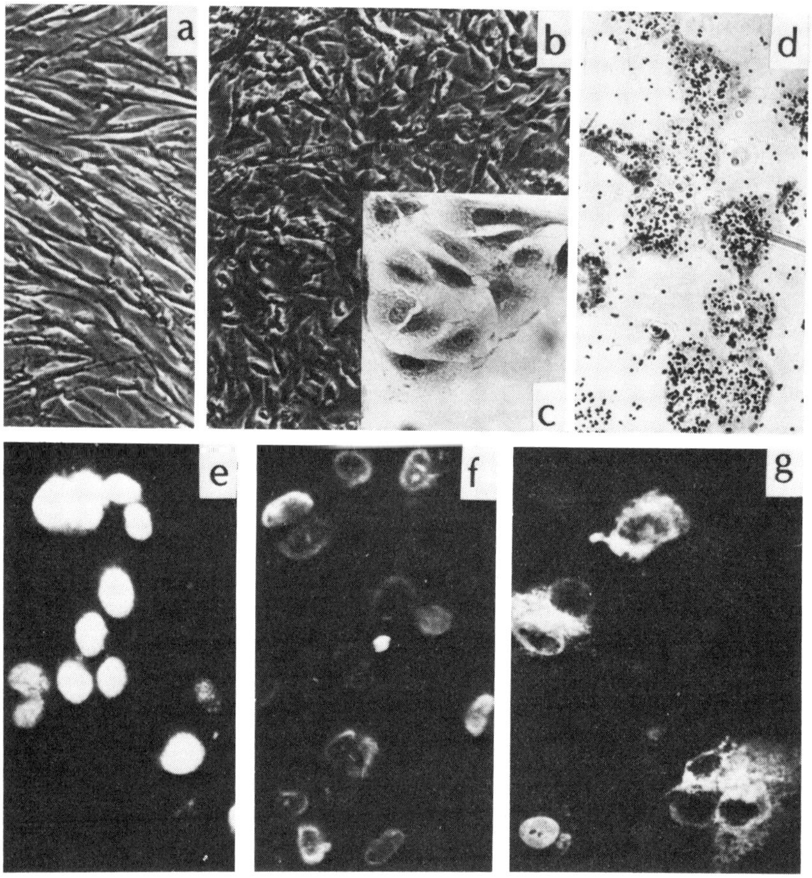

Fig. 4 Photographs of CMV DNA fragments (Xba I restricted transformed human embryonic lung (HEL) cells. (Data from ref. 6). (a) normal HEL cell; (b) morphologically transformed HEL cells derived from transformed foci; treated with TPA; (c) Haematoxylin-eosin stained transformed cells, X125; (d) viral specific RNA autoradiography. (e-g) CMV-specific antigens as detected by anti-complement immunofluorescence test.

From mapping and transfection experiments, they have found that the transforming region was in the 2.9 K subfragment of the Hind III E fragment with map units between 0.123 and 0.14 on the DNA molecule of the AD169 strain CMV. Transformed murine cells selected by 1.2% methylcellulose had high replicating efficiency and were tumorigenic in Balb/c nude mice. However, in their preliminary experiment, Nelson et al (67), were not able to detect viral DNA sequences homologous to the transforming Hind III E fragment with the sensitivity of 0.5 copy Hind II E fragment per cell. Treatment of the transfected cells with TPA was not necessary in this experiment. It was suggested that the promotion events fulfilled by TPA in HEL cells must reside endogenously in NIH 3T3 cells (67). This might also explain the high frequency of transformation of NIH 3T3 by numerous tumor viruses.

Through the study of a series of deletion mutant of pCM 4000 recombinants constructed by Exonuclease III and S1 nuclease sequential digestion of 2.9 kb Hind III-Xba I transforming fragment, Nelson, et al, were able to designate and to define the transforming region down to about the size of 172 bp; between 489 and 317 bases from Hind III site of PCM 4000. This implies that morphological transformation of NIH 3T3 cells can be achieved by small pieces of DNA without any obvious open reading frame (Nelson, Fleckenstein, John Galloway and McDougall, personal communication, Herpesvirus workshop, Oxford 1983, p.73). DNA sequence of this small piece of transforming sequence reveals a stem-like hair pin structure resembling that of bacteria insertion element within this short transforming DNA. A similar type of DNA sequence arrangement also exists in HSV I and II transforming DNA segments (D.A. Galloway and J.K. McDougall, personal communication).

In Syrian hamster embryo (SHE) cell systems, Clanton, et al (63), were also able to morphologically transform SHE cells with cloned Towne strain Xba I E fragment. This Towne Xba I E fragment was found homolgous to Bgl II transforming fragment N and C of herpes simplex type II, but lack of homology to the 2.9 kilobase subfragment of Hind III E is demonstrated by Nelson, et al (68). This Towne strain Xba I E fragment was also found able to transform NIH 3T3 cells. The resulting transformed cells were also oncogenic in athimic NIH nude mice (68).

D. Molecular epidemiological analysis of the association of cytomegalovirus with various human malignancies

As discussed above, the major setbacks in the epidemiological analysis of the association of CMV with human cancers is the ubiquitous distribution of CMV and the high prevalence of CMV antibodies in asymptomatic control population. It is almost impossible to come to the conclusion whether CMV has a causal association with a particular human cancer as analyzed by biostatistical approaches. Therefore the data presented here from our studies and that of others merely reflect the epidemiological association other than etiology or causal association. To date, human CMV has been implied to be associated with Kaposi's sarcoma (71-76), prostatic adenocarcinoma (77,78), adenocarcinoma of the colon (79-81), and cervical cancer (82).

a. Kaposi's Sarcoma(KS). Kaposi's sarcoma is a multifocal idiopathic pigmented hemorrhagic sarcoma with an extremely obscure nature (83,84). This tumor is characterized by the appearance of multicentric, violaceaus tumor masses over the skin and sometimes in internal viscera of adult males. Extensive involvement of the peripheral lymphnode with or without cutaneous lesion is often found in children (9). The detailed histophathology and clinical manifestation of KS were described elsewhere (83-86).

In general, the incidence of classical KS is extremely high in Negroes in some equatorial countries of Africa. It is one of the most common solid malignant tumors in Kenya and constitutes close to 2.5% of all malignant tumors (86). The prevalence rate is higher among Italian or Jewish descendants as compared to other races in Europe and America (83,84). In Africa KS attacks both younger and older populations with relatively aggressive syndromes, while in the United States, it appears to attack mainly older people with less aggressive clinical syndrome. Based on the epidemiological consideration, besides the classic endemic KS, there are two more groups of KS that have been recently discussed. One group arises as the result of aggressive immunosuppressive therapy in organ transplantations or as the result of disorders of the immune system with broader spectrum of age and sex ratio distribution. Another group of KS occurs predominantly in homosexual or bisexual men with no known predisposing of immunosuppressive disease or therapy. The KS of this group frequently accompanies with another rare disease, the pneumocystic pneumonitis, and is extremely fatal. Human CMV has been consistently found to be the most frequent infectious agent associated with these two rare diseases (7).

In searching for viral agents associated with Kaposi's sarcoma, Dr. Giraldo and his colleagus demonstrate the herpes type virus particles and subsequently isolated human CMV (strain K9V) from tissue culture cell line derived from classic type of African KS (71,72). Through an extensive seroepidemiological study, Giraldo and his colleagues have subsequently showed a significant association between CMV infection and KS in European and American KS patients (73,74). Statistically, no significant association between KS and other herpes viruses infections could be demonstrated. In collaboration with Dr. Giraldo, we examined classic KS tumor biopsies for CMV-specific nucleic acid sequences, macromolecules by nucleic acid hybridization and anticomplement immunofluorescence test respectively. In one set of the study, viral DNA and or viral DNA and or viral RNA could be detected in five out of ten tumor biopsies and viral specific antigens located in the nuclei could be demonstrated in 80% of the specimens (76). In contrast, no herpes simplex type II and

Epstein-Barr virus DNA sequence or virus specific macromolecules could be detected in these specimens. To date, we have examined 17 classic type KS specimens, among them 6 biopsied tumors carried CMV genomes at a level of 0.25 to 1 genome equivalent per cell. The existence of CMV DNA at 0.25 to 1 genome per cell in KS as described above does not represent the real value in a neoplastic cell. Since DNA was extracted from total biopsied specimens and a dilution factor due to the DNA from normal cell population should be considered in viral genome detection. In addition, the entire CMV DNA was used to prepare radioactive hybridization probe. It is therefore believable that CMV positive rate should go higher when a defined DNA fragment carried transforming gene is used for genome detection.

In the recent study of homosexual men with KS, Drew and his colleagues (7), found that CMV could be isolated from body secretion, semen or blood in 7 out of 9 patients. Viral cultures of KS biopsied specimens were negative, but CMV specific antigens and viral RNA could be demonstrated by immunofluorescence and by in situ nucleic acid hybridization tests in 6 out of 9, and 2 out of 3 tested, respectively. Normal tissue specimens from 3 KS patients were negative for CMV specific antigens (7).

In an independent observation Fenoglio and her colleagues also demonstrated the existence of CMV RNA in KS cells derived from a biopsy specimen of the skin lesions of a 35 year-old white Jewish homosexual man who had undergone surgery and chemotherapy for an embryonic carcinoma of the testis subsequently developed Kaposi's sarcoma.

b. Prostatic adenocarcinoma and benign hypertrophy of prostate. Adenocarcinoma and benign hypertrophy of prostate are the most common illnesses of the American male. The etiology of these diseases is still unkonwn. Some studies suggest that these diseases may be related to the endocrine imbalance; but up to date no conclusions have been made. Some preliminary observations imply that herpes simplex virus and human CMV might play some roles in the induction of prostatic adenocarcinoma (78). Lymphocytes from patients with adenocarcinoma of prostate were cytoxic to CMV-infected and CMV-transformed cells which bore CMV-specific membrane antigens (77). The peripheral lymphocytes from 84% of patients with prostatic carcinoma (77). These cellular immunological data strongly suggest a possible association of CMV with the adenocarcinoma of prostate, Furthermore, a long-term persistence of an oncogenic CMV strain (Major) in a cell line derived from a prostatic tissue (65), makes this speculation more attractive. To obtain more conclusive information, we have investigated the existence of viral DNA and viral specific macromolecules in normal, benign hypertrophy and adenocarcinoma of the prostate.

Surgically removed specimens from 13 normal, 9 benign hypertrophy of prostate (BHP), and 10 adenocarcinoma of prostate (ACP), were analyzed for the presence of CMV and herpes simplex virus type II DNA and viral specific RNA and antigens by DNA-DNA reassociation kinetics analysis, in situ nucleic acid hybridization and ACIF, respectivley (78). Experiemental results showed 3 out of 9 (33%), BHP and 4 out of 10 (40%), carcinoma of prostate carried CMV DNA and/or CMV-specific macromolecules while only 2 out of 13 (15.4%) in normal prostates. HSV-II specific products were also found in 2 out of 10 (20%), prostatic tumors and 1 our of 13 normal prostate (8%). It is worth noticing that 60% of the prostatic tumor showed the existence of herpes group viral macromolecules (CMV or HSV2). These data lead us to believe that both oncogenic CMV and HSV may play some crucial role in the production of prostatic carcinoma.

c. Cervical cancer. Considerable number of seroepidemiological and biological data suggest a strong causal association between HSV2 and cervical cancer (87,88). HSV-2 DNA, mRNA and protein markers have been found in human cervical cancer, in exfoliative tumor cells and in cells undergoing neoplastic change (89-91). On the other hand the seroepidemiological data from some geographic areas did not always support the close association HSV-2 with cervical cancer (82). Therefore, the multifactorial etiology of cervical cancer was suggested (13,82). Human CMV has been frequently found in cervixes and vaginal discharge, especially in Asia where the association of HSV2 and cervical cancer is not very obvious. Virus was isolated from cervical tumor biopsies (our unpublished data) and from cell cultures derived from tumor biopsies (82). In view of the oncogenic potential of CMV as demonstrated in vitro, we have performed a survey of CMV DNA in cervical cancers and in normal cervixes from various geographic regions by nucleic acid hybridization.

In collaboration with Dr. E. Russell Alexander, we examined eight cervical cancers and six normal cervix specimens from Taiwan. CMV DNA was found in 7 out of 8 (88%) cancer specimens at the level between 0.4 and 6.6 genome equivalent per cell while 3 out of 6 (50%) normal cervixes showed positive at 0.2 to 1 genome per cell. Infectious CMV was isolated from one tumor specimen which carried CMV DNA at 6.6 genome per cell level. No HSV-II DNA sequence could be detected in either cervical cancers or normal cervixes from Taiwan (6).

As for specimens from Africa (in collaboration with Dr. G de-The'), we found CMV DNA sequence in 9 out of 19 cervical cancers and 5 out of 10 normal cervixes at 0.1 to 1 and 0.5 to 1 gneome equivalent per cell, respectively. HSV-2 DNA was found in 1 out of 12 cervical cancers and in 1 out of 4 normal cervixes (6). In specimens from Finland and the United States, human CMV DNA was detected in 1 out of 11 and 2 out of 13 cervical cancers, respectively. No HSV II DNA was detected in in either normal or cervical sepcimens from Finland and the United States. The CMV-DNA positive rate is significantly lower in specimens from Finland and the United States in comparison with that from Taiwan and Africa. This might merely reflect socioeconomic status of each population. Nevertheless, the frequency of the positive CMV rate in cerivcal cancer was higher than that of HSV-2, and the rate of CMV positive specimens in cervical cancers from Taiwan was also higher than those from normal cervixes. However, the number of specimen studies was still limited. It is essential to examine more speicmens in order to come up with conclusive data. Again, with the ubiquitious nature of CMV it is extremely difficult to presume any causal association especially with cervical cancer, nervertheless, we cannot overlook the possibility. Based on the extremely high frequency of positive CMV DNA in both cervixes and cervical cancers, we can at least conclude that human cervix might be a site for latent DMV infection.

d. Carcinoma of the colon. Through extensive epidemiological analysis, it was suggested that environmental as well as genetic factors might play some important roles in the initiation of colon cancer (92,93). However, the precisive nature of factors or genes involved is still undetermined. By cytological observation and virus isolation, human CMV has been detected in the intestinal wall of patients with ulcerative colitis, a disease that has been suspected to be associated with colon cancer (94-96). By nucleic acid hybridization techniques, in collaboration with Dr. J.K. Roche of Duke University, we have shown the presence of viral DNA sequence in the diseased bowel of patients with ulcerative colitis, familial polyposis and carcinoma of the colon (79-81). In contrast, results were negative for CMV genome in the bowel of Crohn's disease patients (79, 80); patients with Crohn's disease have little or no increase in cancer risk.

Although CMV was detected in the majority of patients with colon cancers, we have found that CMV was also frequently existent in some histological normal and non-neoplastic diseased tissues of colon cancer patients (98). These observations make the interpretation of the association of CMV with colon cancer extremely difficult. It is more likely that the CMV detected probably was latent or harbored in the intestinal tissues of the majority of the patients. The fact that this virus is widely distributed does not also rule out the possible oncogenic role since close association of virus with host cells may give the virus an excellent opportunity to induce neoplastic transformation of its host during its long-term persistence.

CONCLUSION

The oncogenic potential of human CMV is strongly suggested by its ability to stimulate the synthesis of host cell macromolecules, and to transform human as well as other mammalian cell in vitro. Viral transformed cells were found tumorengenic in athymic nude mouse. Although CMV DNA, RNA and virus-specific antigens were found frequently existing in Kaposi's sarcoma, prostatic adenocarcinoma, cervical and colon cancers, its causal and etiological role with these human cancers is still unclear. The possiblities of the preferential replication of CMV in these neoplastic tissues and the reactivation of latent virus in patients with malignancy have to be considered. Due to the widespread CMV infection, it is also very difficult to study the causal association of CMV with human malignancies by sero-or molecular-epidemiological approaches. Nevertheless, the conncection between CMV and some human malignancies is impossible to dismiss. The important complication of CMV infection in cancer patients deserves great attention.

ACKNOWLEDGEMENTS

We thank Drs. E. Russell Alexander, G. de-The', J. Roche, G. Giraldo, and T.I. Malinin for their collaboration and for providing valuable tumor specimens, and Shu-Mei Houng, Barbara Leonard, and Sheila McNeill for technical assistance and manuscript preparation.

This project is supported by grants from NIC (CA21773) and NIAID (AI12717).

REFERENCES

1. Weller, T.H.: *N. ENGL. J. MED.* 285:203-214, 1971.
2. Rapp, F.: In: *VIRUS-HOST INTERACTION: VIRAL INVASION, PERSISTENCE AND DISEASE* Frankel-Conrat and Wagner, (Eds.), Plenum Publishing Corp., N.Y., 1980, Vol. 16: 193-232.
3. Stagno, S., Reynolds, D.W., Huang, E.S., Thomas, S.D., Smith, R.J., and Alford, C.A., Jr.: *N. ENGL. J. MED.* 296:1256-1258, 1977.
4. Albrecht, T., Rapp, F.:*VIROLOGY* 55:53-61, 1973.
5. Geder, L., Lausch, R., O'Neill, F., Rapp, F.,: *SCIENCE* 192:1134-1137, 1976.
6. Huang, E.S., Boldogh, I., Mar, E.C.: In: *VIRUSES ASSOCIATED WITH HUMAN CANCER*, Philips, L.A. (Editor), Marcel Dekker, N.Y., 1983, pp.161-193.
7. Drew, W.L., Conant, M.A., Miner, R.C., Huang, E.S., Zeigler, J.L., Groundwater, J.R., Gullett, J.H.: *LANCET* ii:;25-127, 1982.
8. Fenoglio, C.M. Oster, M.W., Gerfo, P.L., Reynolds, T., Edelson, R., Patterson, J.A.K.: *HUMAN PATHOLOGY* 15:995-959, 1982.

9. Haverkos, H.W. and Curran, J.W.: *CA-A J. FOR CLINICIANS* 32:330-339, 1982.

10. Lang, D.J., Kummer, J.F. and Hartley, D.P.: *N. ENGL. J. MED.* 291:121-123, 1974.

11. Lang, D.J. and Kummer, J.F.: *N. ENGL. J. MED.* 287:756-758, 1972.

12. Lang, D.J. and Kummer, J.F.: *J. INF. DIS.* 132:472-473, 1975.

13. Alexander, E.R.: *CANCER RES.* 33:1486-1496, 1973.

14. Kilpatrick, B.A. and Huang, E.S.: *J. VIROL.* 24:261-276, 1977.

15. DeMarchi, J.M., Blaneknship, M.L., Brown, G.S. and Kaplan, A.S.: *VIROLOGY* 89:643-646, 1978.

16. Geelen, J.L.M.C., Walig, G., Wertheim, P. and Van Der Noordae, J.: *J. VIROL.* 26:813-816, 1978.

17. Montplaisir, S., Belloncik, S., Leduck, W.P. Onji, P.A., Martinearu, B. and Krustak, E.: *J. INF. DIS.* 125:533-538, 1972.

18. Sarov, I. and Abady, I.: *VIROLOGY* 66:464-473, 1975.

19. Stinski, M.F.: In *"THE HERPESVIRUSES"* Vol. 2, Roizman, B. (ed.), Plenum Publishing Corp., N.Y., 1983, pp.67-113.

20. Huang, E.S., and Pagano, J.S.: In: *COMPARATIVE DIAGNOSIS OF VIRAL DISEASES*, Vol, I., Kurstak, E. and Kurstak, C., (eds.), Acadmeic Press, N.Y., 1977, pp.241-285.

21. Kim, K.S., Sapienza, V.J., Carp, R.I. and Moon, H.M.: *J. VIROL.* 20:604-611, 1970.

22. Stinski, M.F.: *J. VIROL.* 19:594-609, 1976.

23. Stinski, M.F.: *J. VIROL.* 23:751-767, 1977.

24. Gupta, P., St. Jeor, S. and Rapp, F.: *J. GEN. VIROL.* 34:447-454, 1977.

25. Kilpatirck, B.A. and Huang, E.S.: *J. VIROL.* 24:261-276, 1977.

26. Sheldrick, P. and Berthelot, N.: *COLD SPRING HARBOR SYMP. QUANT. BIOL.* 39:667-678, 1974.

27. Delius, H. and Clements, J.B.: *J. GEN. VIROL.* 33:125-133.

28. Lafemina, R.L. and Hayward, G.S. In: *"ANIMAL VIRUS GENETICS"* Fields, B.N. and Jaenish, R., (eds.), Academic Press, N.Y., 1980, pp.39-55.

29. Huang, E.S., Huong, S.M., Tegtmeir, G.E. and Alford, C.: *ANN. N.Y. ACAD. SCI.* 354:322-346, 1980.

30. Spector, D.H., Hock, L. and Tamashiro, J.C.: *J. VIROL.* 42:558-582, 1982.

31. DeMarchi, J.M.: *VIROLOGY* 114:23-38, 1981.

32. Lafemina, R.L. and Hayward, G.S.: *J. GEN. VIROL.* 64:373-389,1983.

33. Mocarski, S.E., Post, L.E. and Roizman, B.: *CELL,* 22:243-255, 1980.

34. Smiley, Y., Fong, B.S. and Leung, W.: *VIROLOGY,* 113:345-362, 1981.

35. Geelen, J.L.M.C. and Westrate, M.W. In: *HERPESVIRUS DNA,* Decker, Y., (ed.), Martinus Niykoff Medical Publishers, 1982.

36. Huang, E.S., Kilpatrick, B.A., Huang, Y.T. and Pagano, J.S.: *YALE J. BIOL. MED.,* 49:29-43, 1976.

37. Weller, T.H., Hanshaw, J.B. and Scott, D.E.: *VIROLOGY* 12:130-132, 1960.

38. Pritchett, R.F.: *J.VIROL.* 36:152-161, 1980.

39. Kieff, E.D., Hoyes, B., Bachenheimer, S.L. and Roizman, B.: *J. VIROL.* 9:738-745, 1972.

40. Stinkski, M.F.: *J. VIROL.* 26:686-701, 1978.

41. DeMarchi, J.M., Schmidet, C.A., and Kaplan, A.S.,: *J. VIROL.* 35:277-297, 1980.

42. Wathen, M.W., Thompson, D.R. and Stinski, M.F.: *J. VIROL.* 38:446-459, 1981.

43. Wathem, M.W. and Stinski, M.F.: *J. VIROL.* 41:462-477, 1982.

44. DeMarchi, J.M.: *VIROLOGY* 124:in press, 1983.

45. Michelson, S., Horodniceau, F., Kress, M. and Trady-Panit, M: *J. VIROL.* 32:259-267, 1979.

46. St. Jeor, S., Albrech5, T.B., Funk, F.D. and Rapp, F.: *J. VIROL.* 13:353-362, 1974.

47. Boldogh, I., Gonczol, E., Gartner, L. and Vaczi, L.: *ARCH. VIROL.* 58:289-299, 1978.

48. Tankaka, S., Furukawa, T., Plotkin, S.A.: *J. VIROL.* 15:297-304, 1975.

49. Furukawa, T., Tanaka, S. and Plotkin, S.A.: *PROC. SOC. EXP. BIOL. MED.* 148: 211-214, 1975.

50. Furukawa T.,. Sakuma, S., Plotkin, S.A.: *NATURE* (London) 262:414-416, 1976.

51. DeMarchi, J.M., Ben-Porat, T. and Kaplans, A.S.: *VIROLOGY* 97:457-463, 1979.

52. DeMarchi, J.M. and Kaplan, A.S.: *VIROLOGY* 82:93-99, 1977.

53. Kamata, T., Tanaka, S., Watanabe, Y.: *VIROLOGY* 90:197-208, 1978.

54. Zavada, V., Erban, V., Rezacova, D. and Vonka, V.: *ARCH VIROL.* 52:333-339, 1976.

55. Estes, J.E. and Huang, E.S.: *J. VIROL.* 24:13-21, 1977.

56. Huang, E.S.: *J. VIROL.* 16;298-310, 1975.

57. Hirai, K., Furukawa, T. and Plotkin, S.A.: *VIROLOGY* 70:251-255, 1976.

58. Tanaka, S., Ihara, S. and Watanabe, Y.: *VIROLOGY* 89:179-185, 1978.

59. Isom, H.J.: *J. GEN. VIROL.* 42:265-278, 1979.

60. Yamanishi, K. and Rapp, F.: *J. VIROL.* 31:415-419, 1979.

61. Elion, G.B., Furman, P., Fyfe, J.A., deMiranda, P., Beauchamp. L., Schaeffer, J.J.: *PROC. NATL. ACAD. SCI. USA* 74:5716-5720, 1977.

62. Cheng, Y.C., Domin, B.A., Sharma, R.A. and Bobek, M.: *ANTIMICROB. AGENTS CHEMOTHER.* 10:119-122, 1976.

63. Pulleyblank, D.E. and Ellison, M.J.: *BIOCHEMISTRY* 21:1155-1161, 1982.

64. Boldogh, I., Gonczol, E. and Vaczi, L.: *ACTA MICROBIOL. ACAD. SCI. HUNG.* 25:269-275, 1978.

65. Rapp, F., Geder, L., Murasko, D., Lausch, R., Ladda, R., Huang, E.S., and Webber, M.M.: *J. VIROL.* 16:982-990-, 1975.

66. Williams, L.L., Blakeslee, J.R. and Huang, E.S.: *J. GEN. VIROL.* 47:519-523, 1980.

67. Nelson, J.S., Fleckenstein, B., Galloway, D.A. and McDougall, J.K: *J. VIROL.* 43:83-91, 1982.

68. Clanton, D.J., Jariwalla, R.J., Kress, C. and Rosenthal, L.J.: *PROC. NATL. ACAD. SCI. USA* 80:3826-2830, 1983.

69. Stow, N.D. and Wilkie, N.M.: *J. GEN. VIROL.* 33:447-458, 1976.

70. Bereblum, I.: *J. NATL. CANCER INST.* 60:723-726, 1978.

71. Giraldo, G., Beth, E., Coeur, P., Vogel, C.L. and Dhru, D.S.: *J. NATL. CANCER INST.* 49:1495-1507, 1972.

72. Giraldo, G., Beth, E., and Haguenau, F.: *J. NATL. CANCER INST.* 49: 1509-1513, 1972.

73. Giraldo, G., Beth, E., Kourilsky, .FM., Henle, W., Henle, G., Mike, V., et al.: *INT. J. CANCER* 15:839-848, 1975.

74. Giraldo, G., Beth, E., Henle, W. Henle, G., Mike, V., Safai, B., Huraux, J.M., McHardy, J., De-The', G.: *INT. J. CANCER* 22:126-131, 1978.

75. Girlado, G., Beth, E. and Huang, E.S.: *INT. J. CANCER* 26:23-29, 1980.

76. Boldogh, I., Beth, E., Huang, E.S., Kyalwazi, K.S. and Giraldo, G.: *INT. J. CANCER* 28:469-474, 1981.

77. Sanford, E.J., Dagen, J.E., Geder, L., Rohner, J.T. and Rapp, E.: *J. UROL.* 118:809-810, 1977.

78. Boldogh, I., Baskar, J.F., Mar, E.C. and Huang, E.S.: *J. NATL. CANCER INST.* 70:819-825, 1983.

79. Roche, J.K and Huang, E.S.: *GASTROENTEROLOGY* 72:228-233, 1977.

80. Huang, E.S. and Roche, J.K: *LANCET* i:957-960, 1978.

81. Roche, J.K, Cheng, K.SL, Huang, E.S. and Lang, D.L.: *INT. J. CANCER* 27:659-667, 1981.

82. Melnick, J.L., Lewis, R., Wimberly, I., Kaufman, R.H. and Adams, E.: *INT. VIROLOGY* 10:115-119, 1978.

83. Safai, B. and Good, R.: *CA-A CANCER J. FOR CLINICIANS* 31:2-12, 1981.

84. Kaposi, M.: *ARCH. FUR. DERMAT. SYPHILIUS* 4:265-273, 1972.

85. Taylor, J.F., Templeton, A.C., Vogel, C.L., et al.: *INT. J. CANCER*8:122-135, 1971.

86. Kungu, A. and Gates, D.G.:*ANTIBIOTICS CHEMOTHER*29:38-55, 1981.

87. Rawls, W.S., Adam, E. and Melnick, J.L.: *CANCER RES.*33:1477-1482, 1973.

88. Nahmias, .AJ., Naib, Z.M. and Josey, W.E.:*CANCER RES.* 34:1111-1117, 1974.

89. Frenkel, N., Roizman, B., Cassai, E., and Nahmias, A.,: *PROC. NATL. ACAD. SCI. USA* 69:3784-3789, 1972.

90. McDougall, J.K., Galloway, D.A., and Fenoglio, C.M.: *INT. J. CANCER* 25:1-8, 1980.

91. Dressman, G.R., Burek, J., Adam, E., Kaufman, R.H., Melnick, J.L., Powell, K.L. Purfoy, D.J.M.: *NATURE* 283:591-593, 1980.

92. Hanszel, W. and Correa, P.: *CANCER* 28:14-24, 1971.

93. Burdette, W.J.: *CANCER* 28:51-59, 1971.

94. Powell, R.D., Warner, N.E., LEvine, R.S. and Kirsner, J.B.: *AM. J. MED.* 30:334-340, 1961.

95. Levine, R.S., Wanner, N.F. and Johnson, C.F.: *ANN. SURG.* 159:35-48, 1964.

96. Farmer, G.W., Vincet, M.M. and Fuccillo, D.A.: *GASTROENTEROLOGY* 65:8-18, 1973.

© 1984 Elsevier Science Publishers B.V.
The Role of Viruses in Human Cancer. Volume II
G. Giraldo and E. Beth editors

THE SIGNIFICANCE OF INTERFERON IN SERUM OF PATIENTS WITH ACQUIRED IMMUNE DEFICIENCY SYNDROME (AIDS) AND OF PERSONS AT RISK

ELENA BUIMOVICI-KLEIN[1], MICHAEL LANGE[1], RICHARD J. KLEIN[2], MICHAEL H. GRIECO[1], AND LOUIS Z. COOPER[1]
[1]St. Luke's-Roosevelt Hospital Center, Departments of Pediatrics and Infectious Diseases, New York, NY 10019 (U.S.A.), and [2]New York University Medical Center, Department of Microbiology, New York, NY 10016 (U.S.A.)

INTRODUCTION

Since the first cases of AIDS were reported in 1981 (1,2), their number has increased steadily (Fig. 1). As of September 2, 1983 a number of 2,259 cases of AIDS have been reported in the USA. Of these, 917 (41%) are known to have died (3). For the limited purpose of epidemiologic surveillance, the Centers for Disease Control define a case of AIDS as a reliably diagnosed disease that is at least moderately indicative of an underlying cellular immunodeficiency in a person who has had no known underlying cause of cellular immunodeficiency and no other cause of reduced resistance reported to be associated with that disease.

The etiology of the syndrome is not yet known, but the groups at risk are homosexual or bisexual men (71%) and men or women who abuse intravenous drugs (17%). AIDS was also found among Haitians, hemophiliacs, heterosexual partners of persons in an AIDS risk category and persons without identified risk factors who had received blood or blood products within five years before their illness (3).

A laboratory test for premonitory signs in groups at risk of AIDS would be highly desirable. DeStefano et al. (4) have found that 63% of homosexual men with Kaposi's sarcoma (KS) have an acid-labile interferon-alpha (IFN) in their serum; this

FIGURE 1. Cases of acquired immunodeficiency syndrome (AIDS), by quarter of report — United States, second quarter 1981 — second quarter 1983

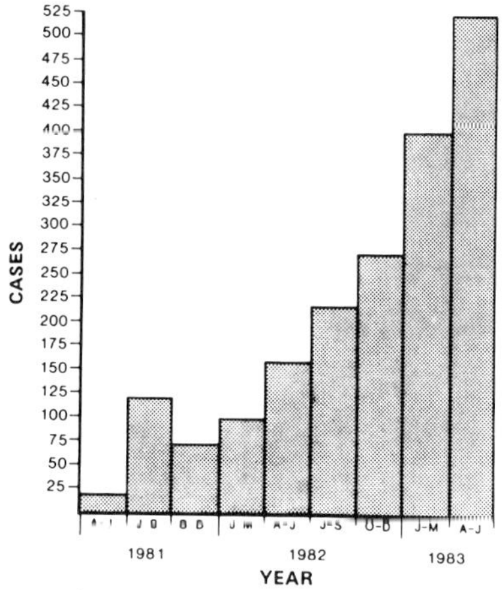

*Includes backlog of cases identified at beginning of CDC surveillance

Morbidity and Mortality weekly Report, v.32, p. 389. 1983

IFN was present in 29% of homosexuals with lymphadenopathy, in 8% of asymptomatic homosexuals and in none of the normal control subjects. The same type of IFN was also detected in the serum of hemophiliacs with AIDS (5). Since these data suggest that IFN-alpha in serum may have some diagnostic value, we have initiated a study involving patients with AIDS and persons at risk from the disease.

STUDY POPULATION AND LABORATORY METHODS

We have examined sera from 24 healthy heterosexual males, 76 homosexuals who were asymptomatic or presenting minor ailments not directly related to AIDS (recurrent herpes, slight weight loss, fever of unknown origin, transient diarrhea), 17 patients with AIDS and opportunistic infections and 8 patients with AIDS and KS only.

IFN determinations were done according to methods already described (4). In short, a micromethod in which human fibroblasts trisomic for chromosome 21 (strain GM-2504) are treated with serial serum dilutions and challenged with encephalomyocarditis virus, was used for IFN titrations. Positive identification of IFN activity was done by neutralization with anti-IFN-alpha or anti-IFN-gamma serum, and titers were compared with standard IFN preparations. In this study one IU was equivalent to 12.5 test units of IFN.

RESULTS

The determinations showed that 23 out of 24 normal control subjects had no IFN activity in their serum. IFN was detected in all 17 patients with AIDS (mean titer 1:123) and all 8 patients with AIDS and KS (mean titer 1:262). IFN was also detected in 52 out of 76 sera from homosexual controls (mean titer 1:31). All sera (titer 64 or greater) contained IFN-alpha, as determined by neutralization with anti-IFN serum, except for two sera in which IFN-gamma was detected (Table 1).

Table 1
IFN TITERS IN SERUM OF PATIENTS WITH AIDS AND HEALTHY HOMOSEXUAL AND HETEROSEXUAL MEN

Study group	Number of subjects with indicated serum IFN titers				
	4	8-32	32-128	128	Mean titer
Healthy heterosexuals	24	0	0	0	4
Healthy homosexuals	28	28	15	5[a]	31[b]
AIDS with OI	0	4	4[a]	9	123
AIDS with KS	0	2	0	6	262

[a]One subject with IFN-gamma in serum
[b]Subjects with titer ≥ 128 not included in mean

Interestingly, the 5 homosexuals without AIDS, but with the highest IFN titers (mean 436), developed AIDS within 2 to 17 months after examination of their sera:

(1) The man with a IFN-gamma titer of 192 had a titer of 256 one year later while still healthy, but AIDS developed with *Pneumocystis carinii* after another 5 months. At that time his IFN titer had dropped to 20.

(2) The patient with an IFN titer of 256 was admitted to hospital 6 months later with persistent cytomegalovirus infection from which he died after another 4 months. At 1 and 8 weeks after hospital admission his IFN titers were 256 and 512, respectively.

(3) One patient with an IFN titer of 512 acquired AIDS and *P. carinii* pneumonia 2 months after testing and died 2 months later.

(4) The other patient with an IFN titer of 512 was diagnosed as Kaposi's sarcoma within 5 months and high IFN titers have persisted.

(5) The last patient (IFN titer 1446) was diagnosed as Kaposi's sarcoma 8 months after testing and IFN titers were still high 12 and 17 months later.

COMMENTS

What is, therefore, the significance of IFN in patients with AIDS or in those at risk of developing the disease: Is IFN an end-product of the disease, a defense reaction against infectious agents, or does IFN play a role in the pathogenetic mechanism of AIDS?

IFN in serum might be induced by the elusive etiologic agent of AIDS or by multiple non-specific viral, bacterial and parasitic infections to which persons at risk are continuously exposed. The presence of IFN might also be the consequence of autoimmune components such as autoantibodies, autoantigens, or immune complexes. It is indeed known that patients with autoimmune disorders frequently have IFN in their serum (6). The presence of IFN in serum could then induce or further enhance the various immune abnormalities observed in patients with AIDS (inversions of T cell subset ratios, supression of cellular immune reactions, polyclonal hyperglobulinemia and formation of immune complexes). These immune disfunctions could eventually facilitate the establishment of opportunistic infections or the development of KS (Fig. 2). Of course this mechanism is only hypothetical and many alternatives can be suggested.

ORIGIN AND ROLE OF INTERFERON IN AIDS PATIENTS

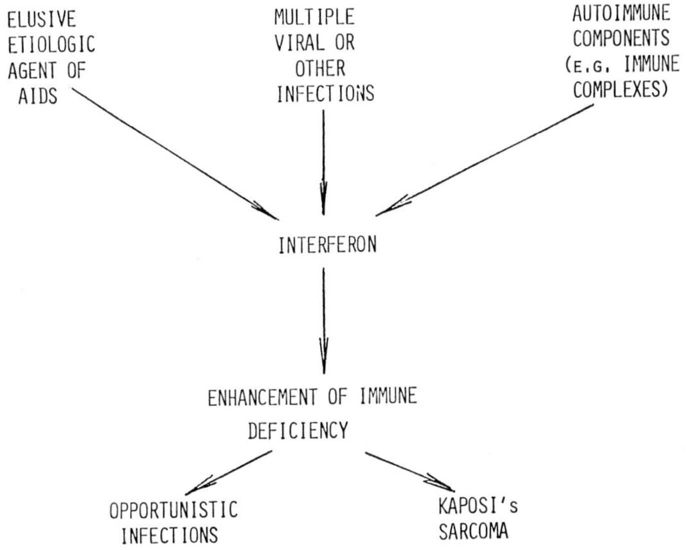

The data suggest that high IFN titers in sera of homosexuals at risk may be predictive of the opportunistic infections or Kaposi's sarcoma associated with AIDS. Such early detection might facilitate changes in life style among promiscuous homosexuals or allow for other more effective measures of prevention or therapy. IFN has been used as a therapy in Kaposi's sarcoma patients (7); however, the presence of serum IFN may influence the outcome of trials of IFN in this disease as it seems questionable whether the treatment will be effective in patients who have high IFN activity in their serum.

A preliminary note regarding our observations has been published recently (8).

ACKNOWLEDGMENTS

We thank Dr. J. Vilcek for critical evaluation of the data and donating anti-IFN serum, Mrs. D.H. DeStefano for providing us with cultures of GM-2504 cells, and Mrs. M.C. Maniscalco for excellent secretarial assistance.

REFERENCES

1. CDC. Pneumocystis pneumonia - Los Angeles. (1981) Morbid. Mortal. Weekly Rep. 30: 250.

2. CDC. Kaposi's sarcoma and Pneumocystis pneumonia among homosexual men - New York City and California. (1981) Morbid. Mortal. Weekly Rep. 30: 305-308.

3. CDC. Update: Acquired immunodeficiency syndrome (AIDS) -- United States. (1983). Morbid. Mortal. Weekly Rep. 32: 465-467.

4. DeStefano, E., Friedman, R.J., Friedman-Kien, A.E., Goedert, J.J., Henriksen, D., Preble, O.T., Sonnabend, J.A., and Vilcek, J. (1982) Acid-labile human leukocyte interferon in homosexual men with Kaposi's sarcoma and lymphadenopathy. J. Infect. Dis. 146: 451-455.

5. Eyster, M.E., Goedert, J.J., Poon, M.C., and Preble, O.T. (1983) Acid-labile alpha interferon: a possible marker for the acquired immunodeficiency syndrome in hemophilia. New Engl. J. Med. 309: 583-586.

6. Vilcek, J. (1982) The importance of having gamma. In: I. Gresser (Ed). Interferon 4/1982, Academic Press, pp. 129-154.

7. Krown, S.E., Real, F.X., Cunningham-Rundles, S., Myskowski, P.L., Koziner, B., Fein, S., Mittelman, A., Oettgen, H.F., and Safai, B. (1983) Preliminary observations on the effect of recombinant leukocyte A interferon in homosexual men with Kaposi's sarcoma. New Engl. J. Med. 308: 1071-1076,.

8. Buimovici-Klein, E., Lange, M., Klein, R.J., Cooper, L.Z., and Grieco, M.H. (1983) Is presence of interferon predictive for AIDS? Lancet ii: 344.

The Role of Viruses in Human Cancer. Volume II
G. Giraldo and E. Beth editors

DIAGNOSIS AND PREVENTION OF HUMAN CYTOMEGALOVIRUS INFECTIONS

MAX A. CHERNESKY
McMaster University Regional Virology Laboratory, St. Joseph's Hospital, Hamilton, Ontario, Canada.

INTRODUCTION

Both primary and reactivated cytomegalovirus infections are very common and subject a great degree of morbidity and mortality on human populations. Table 1 lists various reasons for a concerned effort towards diagnosis and prevention of human cytomegalovirus (HCMV) infections. HCMV is the most common cause of viral congenital infection affecting 0.2 - 2.5% of live births (1,2). Although only approximately 10% of congenitally infected infants have classical symptoms of cytomegalic inclusion disease about 16% of those "silently" infected may develop hearing loss (3,4,5). Spontaneous heterophile negative mononucleosis has been reported and transfusion of blood from sero-positive donors into seronegative recipients runs a risk of transfusion-associated mononucleosis. Transfusion of CMV infection is of particular consequence in neonates (6). Modern medicine with its approach to allograft transplantation; especially in kidney, marrow and cardiac donation; with immunosuppressive therapy, creates a problem of both primary and reactivated infections (7). In the health care field, the anxiety created during care of HCMV excreters by pregnant nurses could be alleviated by an appropriate vaccine. Recent information is accumulating which implicates HCMV as a sexually transmitted disease (8,9,10,11) and there is gathering evidence of a causal relationship in Kaposis sarcoma (12,13,14,15). HCMV may also play a role in acquired immunodeficiency syndrome. The ability of HCMV to transform cells (16,17) and reports linked to human oncogenesis (18,19,20) suggest an etiological role. Thus all of these factors provide a rationale for efforts in diagnosis and prevention.

TABLE 1
RATIONALE FOR DIAGNOSIS AND PREVENTION OF CYTOMEGALOVIRUS INFECTIONS

1.	Congenital
2.	Transfusion in non-immune patients
3.	Transplant patients
4.	Pregnant health care workers
5.	Sexually transmitted disease
6.	Oncogenic capacity

DIAGNOSIS

The laboratory diagnosis of HCMV infections may be made from a variety of specimens including throat washings, urine, blood, semen, cervical swabs, breast milk or tissues from lung, kidney or liver and a myriad of approaches (Table 2).

TABLE 2

SPECIMENS AND APPROACHES TO THE LABORATORY DIAGNOSIS OF HUMAN CYTOMEGALOVIRUS INFECTIONS

Specimens	Approaches
urine	Cell cultures
throat washings	Light microscopy
blood	Electron microscopy
semen	Antigen detection
cervical swabs	Viral DNA measurement
breast milk	IgM measurement
cerebrospinal fluid	Seroconversion
lung tissues	
kidney tissues	
liver tissues	

Any of these specimens may yield virus (and its characteristic cytopathology) in human fibroblast cell cultures (Figure 1).

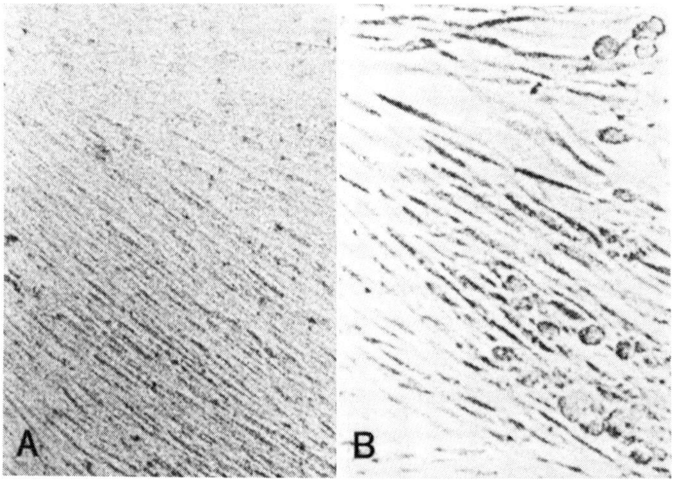

Fig. 1. Human foreskin fibroblast cell cultures. a. Control uninoculated, b. inoculated with urine from a congenitally infected infant.

Histopathology techniques are useful in providing early diagnosis of HCMV infection (Figure 2) by the observation of typical intranuclear "owl eye" inclusions.

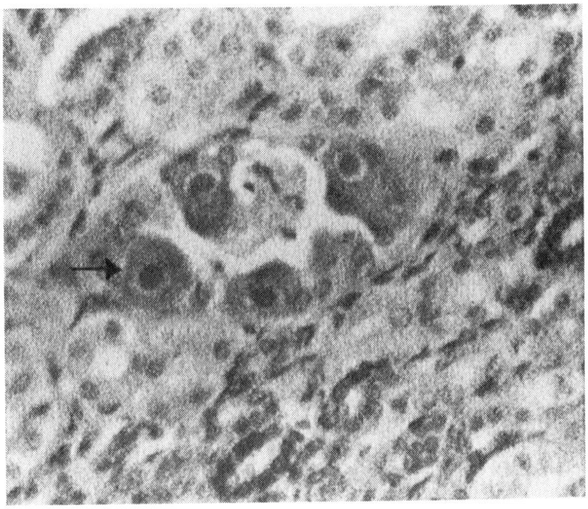

Fig. 2. Section of kidney from a patient with glomerulonephritis. Arrow points to typical "owl eye" intranuclear inclusion within epithelial cells.

Electron microscopy (EM) techniques have been used for detecting virus particles in clinical material such as urine (Figure 3) in infections where large quantities of virus are present (21,22).

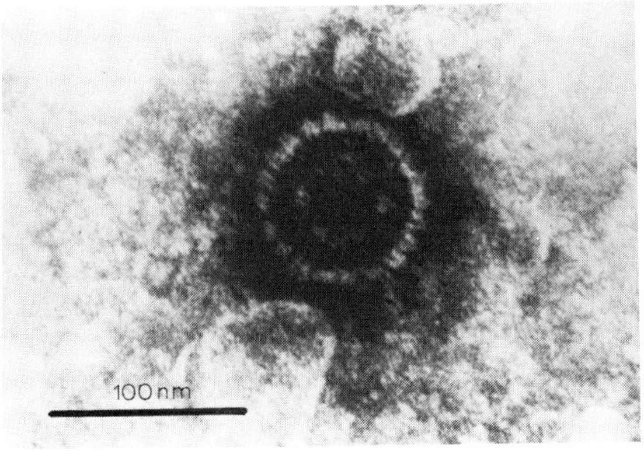

Fig. 3. Cytomegalovirus particle in urine of a congenitally infected infant. Stained with phosphotungstic acid.

In some cases enhancement procedures such as concentration or the use of aggregating antibody (Figure 4) may be required (23).

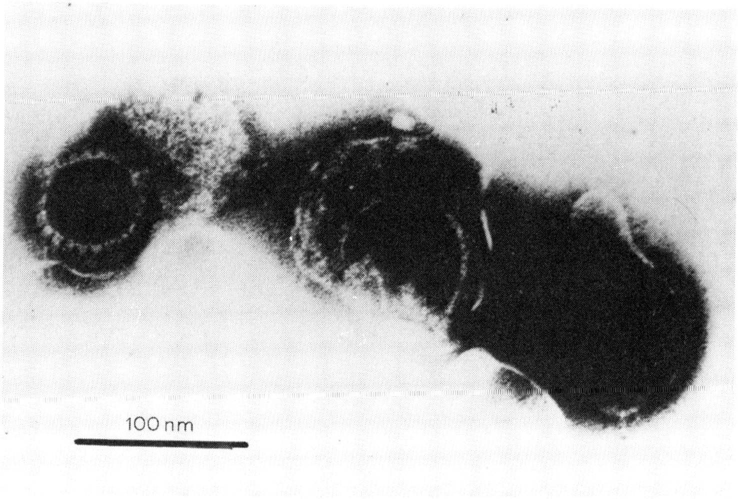

Fig. 4. Solid phase immune electron microscopy preparation of urine from a patient following kidney transplantation. Grid was coated with staphylococcal protein A and rabbit capture antibody. Stained with uranyl acetate.

More recently molecular tools have been applied to diagnosis. Restriction endonuclease (RE) digestion of viral DNA has allowed typing of HCMV isolates (24,25,26) and contributed to our knowledge of infection pathogenesis and natural history. Very sensitive and specific radioactive and enzyme labelled probes have been developed and used for the detection of CMV DNA both in situ and using dot hybridization on nitrocellulose filters (27-31). In a recent study by Chou and Merigan (27) analysing urine from patients with various forms of CMV infection these authors found a correlation between urine infectivity and intensity of DNA hybridization (Table 3). Positive hybridization occurred in specimens containing $10^{2.5}$ $TCID_{50}$/0.2 ml virus or higher.

TABLE 3

CYTOMEGALOVIRUS DNA HYBRIDIZATION AND TISSUE CULTURE INFECTIVITY IN URINE

Disease	Counts/Background	$Log_{10}TCID_{50}$/0.2 ml
Cytomegalic inclusion disease	5710/349	4.5
Retinitis	1727/257	2.9
Retinitis (inactive)	866/257	3.5
Asymptomatic primary infection	877/257	4.5
Retinitis	530/257	2.5
Retinitis	4365/275	4.2

Adapted from reference 27.

Sensitive serological approaches to diagnosis involve the detection of virus specific antigens in clinical specimens or CMV-specific IgM responses in serum (Table 4).

TABLE 4

SEROLOGICAL METHODS FOR THE DETECTION OF CYTOMEGALOVIRUS ANTIGENS AND MEASUREMENT OF IgM ANTIBODY RESPONSES

Antigen Detection
Indirect hemagglutination inhibition
Immunofluorescence
Immunoperoxidase
Solid phase enzyme immunoassays
IgM Measurement
Immunofluorescence
Solid phase enzyme or radio-immunoassays
- capture antigen
- capture anti IgM, (labelled antigen)

Although the antigen detection techniques of indirect hemagglutination inhibition (32) is simple to perform it tends to lack sensitivity required for performance on clinical specimens. The more sensitive microscopical techniques such as fluorescence and peroxidase have been described and used both on tissue culture and clinical specimens (33-35). Limited publications (36-39) have appeared using solid phase enzyme immunoassay (EIA) for HCMV antigen detection and this tends to be due to a lack of specificity on clinical specimens. For example, our experience with an indirect EIA employing polyclonal guinea pig and rabbit antisera was that sensitivity and specificity were high for the detection of tissue culture virus but were much less on urine specimens (Table 5).

TABLE 5

DETECTION OF HCMV ANTIGENS IN CELL CULTURE FLUIDS AND URINE BY ENZYME LINKED IMMUNOSORBENT ASSAY (ELISA)

Specimen	ELISA*		Total
	Positive	Negative	
Positive cell culture fluid	11	0	11
Positive urine**	12	8	20
Negative urine***	6	23	29
TOTALS	29	31	60

* indirect ELISA employing polyclonal g. pig and rabbit antiserum
** urine positive in cell culture
*** urine negative in cell culture

It is likely that newer approaches using monoclonal capture and/or detector antibodies (40) and biotinylated probes should enable improvements to these assays. IgM assays employing CMV antigens (41,42) or anti μ antibodies (43) on the solid phase have been described and each has inherent technical advantages as well as pitfalls. As more information is accumulated it appears that this approach to diagnosis may be viable in congenital (44-47) and transplant-associated infections (48-50). The anti μ technique of Schmitz et al. (51) employing enzyme-labelled CMV antigens to detect virus-specific IgM (Figure 5) appears to have the advantages of simplicity coupled with sensitivity and specificity. A large number of techniques are available for the measurement of other CMV-specific immunoglobulin classes (52-68). These methods are largely retrospective diagnostic tools but have a role to play in immunity screening which facilitates infection prevention especially in transfusion and transplantation situations.

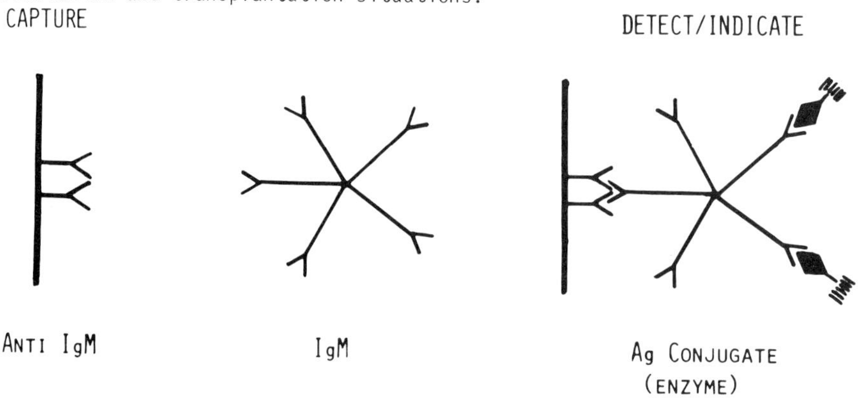

Fig. 5. Antibody capture IgM assay employing enzyme-labelled antigen.

PREVENTION

As well as those therapeutic and preventative approaches eluded to above; chemotherapy (69), interferon therapy (70), passive immunization with CMV immunoglobulins (71) and vaccination (72-75) have all been used on patients. Of these, it appears that passive or active immunization against HCMV is viable for the prevention of morbidity and mortality associated with CMV disease, and not necessarily for the prevention or termination of infection. There have been 2 approaches to the use of live attenuated HCMV vaccines. Elek and Stern (73) have used the AD169 strain passaged a minimum of 56 times in human embryo and foreskin fibroblasts, then administered to medical student volunteers, orally, intradermally or subcutaneously. Neff et al. (72) have performed clinical and laboratory studies with the AD169 vaccine in 43 adult male priests and seminarians. Good complement fixing (CF), neutralizing (N) and immune adherence (IH) antibody and cell mediated immunity (CMI) responses were induced. In the Stern series CF and N responses gradually declined to minimal levels during an 8 year followup, whereas good CMI responses persisted except in individuals during high dose immunosuppressive therapy. In this small series none of the volunteers have reactivated vaccine virus but some have become infected with wild strains. Three of the women have given birth and there was no evidence of CMV reactivation during pregnancy. None of the babies were infected with CMV and all infants were seronegative between 6 months and 1 year of age (Stern - personal communication).

The Towne strain vaccine of Plotkin and coworkers has been more extensively studied (76). This strain which was passaged 125 times in WI38 cells including 3 clonings was tested in cell cultures and animals as measurements of attenuation (Table 6) before clinical trials.

TABLE 6

PARAMETERS OF ATTENUATION UNDERTAKEN FOR THE TOWNE 125 STRAIN OF HCMV

1. Administered to mice treated with antilymphocytic serum.
2. Inoculated into neonatal hamsters.
3. Inoculation of ultraviolet irradiated virus into hamster embryo cells.
4. Maintenance of inoculated cultures at elevated temperatures.

The Towne vaccine DNA is different from other strains of HCMV by RE and possesses a characteristic resistance to trypsin. Towne vaccine induces N antibodies to itself and to AD169, and there is 90% DNA homology between the 2 strains, with the 10% difference found within Xba 1 restriction fragments (77). Clinical trials using subcutaneous administration of the Towne vaccine are underway in transplant recipients (78, Balfour - personal communication) and normal volunteers (74,75,79). In all of these studies seroconversion rates in seronegative vaccine-recipients varied between 78 and 100% (Balfour - personal communication) and some local reactions to vaccine was experienced which appeared to be dose related (Plotkin - personal communication). Antibody responses persisted at high levels for the first year and CMI responses developed within 3 weeks of immunization in normal volunteers (Figure 6) (75). In a study of 6 vaccine recipients over an 8 week period there were no significant changes in percentages or absolute numbers of lymphocyte subsets or con A responses following vaccination (80). The Towne strain has not been isolated from vaccine recipients following immunization.

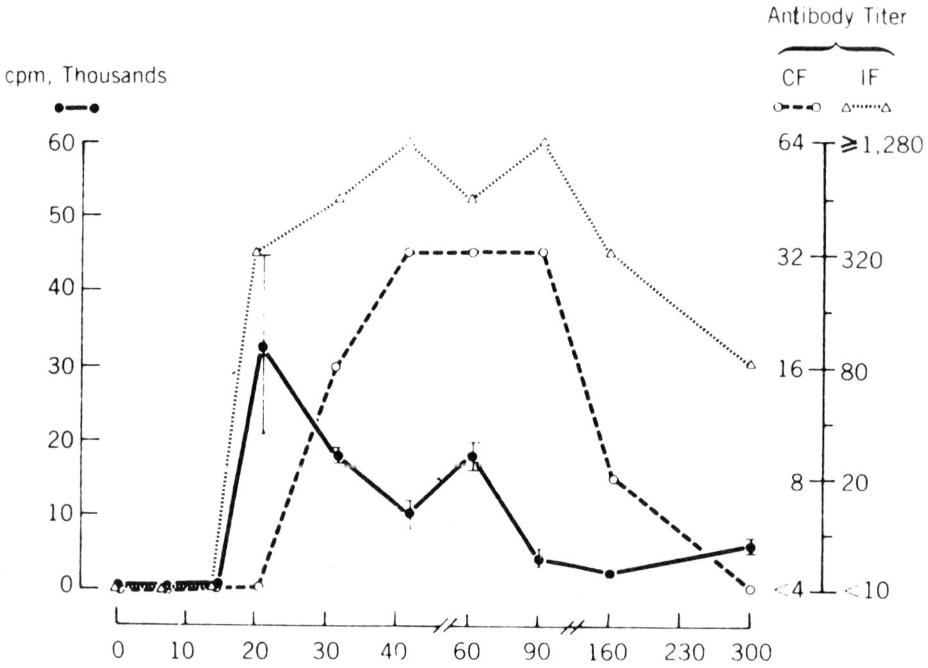

Fig. 6. Cell mediated (lymphocyte proliferation to optimal concentration of virus) and humoral (complement fixation - CF, and indirect immunofluorescence - IF) immune responses in a volunteer following administration of cytomegalovirus Towne vaccine. (Reproduced with permission from R.C. Gehrz et al. ref. 75)

Although vaccinated transplant patients have become infected with CMV, RE analysis of isolates indicated that they were different from the vaccine strain. Illnesses in vaccine recipients have been mild without mortality whereas disease in placebo-recipients has tended to be more severe and lethal (Balfour - personal communication). Thus it is becoming apparent that current immunization with the live attenuated Towne strain of HCMV may not protect against infection but appears to have some impact on morbidity and mortality in this particular population of patients. Although inactivated animal CMV vaccines have been produced and tried in animal models (81-83) their degrees of efficacy have not been encouraging. Biophysical methods and serological monoclonal antibody techniques (84-87) for isolating and purifying non-infectious HCMV antigens should allow the development of subunit vaccines for trial.

Passive immunization using high-titered HCMV immune globulin (IG) or plasma (IP) has been reported in marrow or kidney recipients (69,86). Results in the marrow patients indicated that CMV infection could be prevented by immunoprophylaxis in seronegative patients who were not given granulocyte transfusions. If infection takes place in immunized patients the incidence of interstitial pneumonia in this group is lower than in controls. In a small series of kidney allograft patients with varying degrees of clinical severity post transplant, 5 of 7 patients administered IG or IP responded within 24 hours (89). Thus the therapeutic and prophylactic roles of passive immunization need to be more accurately determined with larger trials.

In summary, justification for diagnosis and prevention of HCMV infection is apparent. Critics of immunization attempts with live-attenuated vaccine voice arguments of a lack of time-proven markers of attenuation and potential for oncogenicity (90,91); whereas the other potentially beneficial side of this sword of immunoprophylaxis is not only that of prevention of acute CMV disease but also of oncogenic outcome.

REFERENCES

1. Stagno, S., Reynolds, D.W., Tsiantos, A., Fuccillo, D.A., Long, W. and Alford, C.A. (1975) Journal of Infectious Diseases, 132, 568.
2. Larke, R.P.B., Wheatley, E., Saigal, S. and Chernesky, M. (1980) Journal of Infectious Diseases 142, 647.
3. Saigal, S, Lunyk, O., Larke, R.P.B. and Chernesky, M.A. (1982) American Journal of Diseases of Children 136, 896.
4. Melish, M.E. and Hanshaw, J.B. (1973) American Journal of Diseases of Children 126, 190.
5. Pass, R.F., Stagno, S., Myers, G.J. and Alford, C.A. (1980) Pediatrics 66, 758.
6. Yeager, A.S., Grumet, F.C., Hafleigh, E.B., Arvin, A.M., Bradley, J.S. and Prober, C.G. (1981) Journal of Pediatrics 98, 281.
7. Simmons, R.L., Matas, A.J., Rattazzi, L.C., Balfour, H.H., Howard, R.J. and Najarian, J.S. (1977) Surgery 82, 537.
8. Jordan, M.C., Wyatt, E.R., Noble, G.R., Stewart, J.A. and Chin, T.D.Y. (1973) New England Journal of Medicine 288, 932.
9. Lang, D.J. and Kummer, J.F. (1975) Journal of Infectious Diseases 132, 472.
10. Wenckebach, G.F.C. and Curry, B. (1976) Archives of Pathology and Laboratory Medicine 100, 609.
11. Drew, W.L., Mintz, L, Miner, R.C., Sands, M. and Ketterer, B.(1981) Journal of Infectious Diseases 143, 188.
12. Hymes, K.B., Greene, J.B., Marcus, A. et al. (1981) Lancet ii, 598.

13. Giraldo, G., Beth, E. and Haguenau, F. (1972) Journal of the National Cancer Institute 49, 1495.

14. Giraldo, G., Beth, E. and Haguenau, F. (1972) Journal of the National Cander Institute 49, 1509.

15. Drew, W.L. (1982) Lancet ii, 125.

16. Albrecht, T. and Rapp, F. (1973) Virology 55, 53.

17. Geder, L., Lausch, R., O'Neill, F.J. and Rapp, F. (1976) Oncogenic transformation of human embryo lung cells by human cytomegalovirus. Science 192, 1134.

18. Hashiro, G.M., Horikami, S. and Loh, P.C. (1979) Intervirology 12, 84.

19. Melnick, J.L. (1978) Intervirology 10, 115.

20. St. Jeor, S. and Weisser, A. (1977) Infection and Immunity 15, 402.

21. Montplaisir, S., Belloncik, S., Leduc, N.P., Onji, P.A., Martineau, B. and Kurstak, E. (1972) Journal of Infectious Diseases 125, 533.

22. Lee, F.K., Nahmias, A.J. and Stagno, S. (1978) New England Journal of Medicine 299, 1266.

23. Chernesky, M.A. (1979) in: Lennette, D., Specter, S. and Thompson, K. (Eds.), Diagnosis of Viral Infections: The Role of the Clinical Laboratory, University Park Press, Baltimore, pp. 125-142.

24. Spector, S. (1983) Lancet i, 378.

25. Wilfert, C.M., Huang, E.S. and Stagno, S. (1982) Pediatrics 70, 717.

26. Yow, M.D., Lakeman, A.D., Stagno, S., Reynolds, R.B. and Plavidal, F.J. (1982) Pediatrics 70, 713.

27. Chou, S. and Merigan, T.C. (1983) New England Journal of Medicine 308, 921.

28. Kafetos, F.C., Jones, C.W. and Efstratiadis, A. (1979) Nucleic Acids Research 7, 1541.

29. Brandsma, J. and Miller, G. (1980) Proceedings of the National Academy of Science U.S.A. 77, 6851.

30. Williams, L. (1980) Journal of General Virology 51, 435.

31. Huang, E.S. and Pagano, J.S. (1977) in: Kurstak, K. and Kurstak, C. (Eds.), Comparative Diagnosis of Viral Diseases I, Academic Press, New York, pp. 241-285.

32. Black, S.B., Raas, M., Mintz, L., Shinefield, H.R. and Drew, W.L. (1983) Journal of Laboratory and Clinical Medicine 101, 450.

33. Gerna, G, Vasquez, A., McCloud, C.J. and Chambers, R.W. (1975) Archives of Virology 50, 311.

34. Schmidt, N.J. and Gallo, D. (1980) Journal of Clinical Microbiology 11, 186.

35. Volpi, A., Whitley, R.J., Ceballos, R., Stagno, S. and Pereira, L. (1983) Journal of Infectious Diseases 147, 1119.

36. Yolken, R.H. and Stopa, P.J. (1980) Journal of Clinical Microbiology 11, 546.

37. Sundqvist, V.A. and Wahren, B. (1981) Journal of Virological Methods 2, 301.

38. McIntosh, K., Wilfert, C., Chernesky, M., Plotkin, S. and Mattheis, M. (1980) NIAID News, pp. 793-802.

39. Pronovost, A.D., Baumgarten, A. and Andiman, W.A. (1982) Journal of Clinical Microbiology 16, 345.

40. Goldstein, L.C., McDougall, J., Hackman, R., Meyers, J.D., Thomas, E.D. and Nowinski, R.C. (1982) Infection and Immunity 38, 273.

41. Schmitz, H., Doerr, H.W., Kampa, D. and Vogt, A. (1977) Journal of Clinical Microbiology 5, 629.

42. Kangro, H.O. (1980) British Journal of Experimental Pathology 61, 512.

43. Yolken, R.H. and Leister, F.J. (1981) Journal of Clinical Microbiology 14, 427.

44. Kangro, H.O., Griffith, P.D., Huber, T.J. and Heath, R.B. (1982) Journal of Medical Virology 10, 203.

45. Griffiths, P.D., Stagno, S., Pass, R.F., Smith, R.J. and Alford, Jr., C.A. (1982) Journal of Infectious Diseases 145, 647.

46. Griffiths, P.D. (1981) British Journal of Obstetrics and Gynaecology 88, 582.

47. Griffiths, P.D., Stagno, S., Pass, R.F., Smith, R.J. and Alford, Jr., C.A. (1982) Pediatrics 69, 544.

48. Pass, R.F., Griffiths, P.D. and August, A.M. (1983) Journal of Infectious Diseases 147, 40.

49. Rasmussen, L., Kelsall, D., Nelson, R. et al. (1982) Journal of Infectious Diseases 145, 191.

50. Sutherland, S. and Briggs, J.D. (1983) Journal of Medical Virology 11, 147.

51. Schmitz, H., von Deimling, U. and Flehmig, B. (1980) Journal of General Virology 50, 59.

52. Horodniceanu, F. and Michelson, S. (1980) Archives of Virology 64, 287.

53. Tegtmeier, G.E., Sweet, G.H. and Bayer, W.L. (1982) Journal of Medical Virology 10, 17.

54. Griffiths, P.D., Buie, K.J. and Heath, R.B. (1980) Archives of Virology 64, 303.

55. Yeager, A.S. (1979) Journal of Clinical Microbiology 10, 64.

56. Al-Nakib, W. (1980) Journal of Medical Virology 5, 287.

57. Cabau, N., Crainic, R., Duros, C. et al. (1981) Journal of Clinical Microbiology 13, 1026.

58. Coupland, B. (1982) Medical Laboratory Sciences 39, 91.

59. Czegledy, J., Gergely, L., Vaczi, L. and Valyi-Nagy, T. (1982) Acta Virologica 26, 73.

60. Chia, W.K. and Spence, L. (1979) Canadian Journal of Microbiology 25, 1082.

61. Booth, J.C., Hannington, G., Aziz, T.A.G. and Stern, H. (1979) Journal of Clinical Pathology 32, 122.

62. Pinku, A., Haikin, H., Friedman, M. and Sarov, I. (1982) Journal of Medical Virology 9, 111.

63. Cheung, K.S., Roche, J.K., Capel, W.D. and Lang, D.J. (1981) Journal of Clinical Laboratory Immunology 6, 269.

64. Levy, E. and Sarov, I. (1980) Journal of Medical Virology 6, 249.

65. Pozzetto, B. and Gaudin, O.G. (1981) Biomedicine 35, 187.

66. Torfason, E.G., Kallander, C. and Halonen, P. (1981) Journal of Medical Virology 7, 85.

67. Sarov, I., Levy, E., Aymard, M. et al. (1982) Clinical and Experimental Immunology 48, 321.

68. Tamura, T., Chiba, S., Chiba, Y. and Nakao, T. (1980) Infection and Immunity 29, 842.

69. Baublis, J.V., Whitley, R.J., Chien, L.T. et al. (1978) in: Pavan-Langston, D., Buchanan, R.A. and Alford, C.A. (Eds.), Adenine Arabinoside: An Antiviral Agent, Raven Press, New York.

70. Hirsch, M.S., Schooley, R.T., Cosimi, A.B. et al. (1983) New England Journal of Medicine 308, 1489.

71. Meyers, J.D., Leszczynski, J., Zaia, J.A. et al. (1983) Annals of Internal Medicine 98, 442.

72. Neff, B.J., Weibel, R.E., Buynak, E.B., McLean, A.A. and Hilleman, M.R. (1979) Proceedings of the Society for Experimental Biology and Medicine 160, 32.

73. Elek, S.D. and Stern, H. (1974) Lancet i, 1.

74. Plotkin, S.A., Farquhar, J. and Hornberger, E. (1976) Journal of Infectious Diseases 134, 470.

75. Gehrz, R.C., Christianson, W.R., Linner, K.M., Groth, K.E. and Balfour, H. (1980) Archives of Internal Medicine 140, 936.

76. Plotkin, S.A., Furukawa, T., Zygraich, N. and Huygelen, C. (1975) Infection and Immunity 12, 521.

77. Pritchett, R.F. (1980) Journal of Virology 36, 152.

78. Glazer, J.P., Friedman, H.M., Grossman, P.A. et al. (1978) Lancet i, 90.

79. Fleisher et al. (1982) American Journal of Diseases of Children 136, 291.

80. Carney, W.P., Hirsch, M.S., Iacoviello, V.R., Starr, S.E., Fleisher, G. and Plotkin, S.A. (1983) Journal of Infectious Diseases 147, 958.

81. Bia, F.J., Griffith, B.P., Tarsio, M. and Hsiung, G.D. (1980) Journal of Infectious Diseases 142, 732.

82. Medearis, Jr., D.N. and Prokay, S.L. (1978) Proceedings of the Society for Experimental Biology and Medicine 157, 523.

83. Jordan, M.C. (1980) Journal of Clinical Investigation 65, 798.

84. Pereira, L., Hoffman, M. and Cremer, N. (1982) Infection and Immunity 36, 933.

85. Pereira, L., Hoffman, M., Gallo, D. and Cremer, N. (1982) Infection and Immunity 36, 924.

86. Schmitz, H., Muller-Lantzsch, N. and Peteler, G. (1980) Intervirology 13, 154.

87. Tegtmeier, G.E. (1976) Yale Journal of Biology and Medicine 49, 69.

88. Winston, D.J., Pollard, R.B., Ho, W.G. et al. (1982) Annals of Internal Medicine 97, 11.

89. Nicholls, A.J., Brown, C.B., Edward, N., Cuthbertson, B., Yap, P.L. and McClelland, D.B.L. (1983) Lancet i, 532.

90. Lang, D.J. (1980) Reviews of Infectious Diseases 2, 449.

91. Osborn, J.E. (1981) Journal of Infectious Diseases 143, 618.

The Role of Viruses in Human Cancer. Volume II
G. Giraldo and E. Beth editors

WOODCHUCK HEPATITIS VIRUS-INDUCED HEPATOMAS CONTAIN INTEGRATED AND CLOSED CIRCULAR VIRAL DNAs

JESSE SUMMERS[1], C. WALTER OGSTON[1], GERALD J. JONAK[1], SUSAN M. ASTRIN[1], GAIL V. TYLER[2] AND ROBERT L. SNYDER[2]
[1]Institute for Cancer Research, Fox Chase Cancer Center, Philadelphia, PA 19111 (U.S.A.) and [2]Penrose Research Laboratory, Philadelphia Zoological Gardens, Philadelphia, PA 19104 (U.S.A.)

ABSTRACT

Thirty-four woodchucks naturally infected with the woodchuck hepatitis virus (WHV) and at least twenty uninfected woodchucks were maintained in captivity during a period of several years. Hepatocellular carcinomas were observed to occur exclusively in the WHV-infected group (16/34).

Hepatocellular carcinomas from nine of these infected woodchucks were analysed for the presence of WHV-specific DNA sequences by restriction enzyme analysis and gel electrophoresis and hybridization. All but one of the examined tumors contained viral sequences integrated at one or more specific sites in the cellular DNA. Separate nodules in the same tumorous liver had different restriction enzyme patterns of integrated viral DNA, suggesting different clonal derivations of the tumors. A species of viral DNA found in both tumor tissues and in non tumorous infected liver was identified as covalently closed circular DNA with a length equal to the partially single stranded virion DNA.

MATERIALS AND METHODS

Animals. Woodchucks were trapped from study areas in southeast Pennsylvania, Delaware, Maryland and New Jersey. Some animals were purchased from the Cocalico Woodchuck Farms (Denver, PA). Woodchucks were housed at the Penrose Research Laboratory, Philadelphia Zoological Gardens, until they developed hepatomas.

Tests for viral antigens. Serum samples from the animals under study were tested for antibody to the woodchuck hepatitis surface antigen (WH_sAg) by passive haemagglutination, or for circulating WH_sAg by haemagglutination inhibition (20). Circulating virus was detected by an assay for the virion-associated DNA polymerase activity, using the endogenous DNA as the template (17).

Pathology. Samples of tumor and normal tissues were taken at necropsy and either frozen at -70°C, or fixed in formalin for later histologic examinations.

DNA extraction. DNA was extracted from tissues in two ways. In earlier experiments, the 0.1 to 0.5 g tissue was homogenized in a Dounce homogenizer in 5 ml 10 mM Tris-HCl, pH 7.4, 10 mM EDTA, (TE) at 0°C, and immediately transferred to a tube containing 5 ml TE with 0.3 M NaCl + 1 mg/ml Pronase (Calbiochem, La Jolla, CA). The lysate was incubated 30 min at 37°C, and extracted with phenol:chloroform (1:1) until no precipitate was visible at the interphase. The aqueous layer was removed and the nucleic acids precipitated with 2 volumes of absolute ethanol. The precipitate was washed twice with ethanol, dried, and redissolved in 5 ml TE. Pancreatic RNAse was added (20 µg/ml final concentration) and the DNA incubated 30 min at 37°C. The solution was adjusted to 0.15 M NaCl, 0.1% SDS, and 500 µg/ml Pronase, incubated at 37°C for 30 min, extracted once with an equal volume of phenol:chloroform (1:1) and the DNA precipitated with two volumes of ethanol. DNA was stored in distilled water at 4°C. In later experiments (see text) a preliminary purification of nuclei was performed in order to minimize contamination of the cellular DNA with vegetative forms of viral DNA. The tissue was homogenized in a Dounce homogenizer with a loose-fitting pestle, in 10 ml TE + 0.15 M NaCl (TES). The large tissue fragments were allowed to settle and the supernatant fluid removed. Nuclei were pelleted by centrifugation at 1000 g for 5 min, and the nuclear pellet was washed twice with TES and once with 10 ml TE containing 0.15 ml NaCl + 0.1% Triton X-100. The washed nuclei were resuspended in 5 ml TES, digested with Pronase and the DNA was isolated as described above.

Restriction enzymes and digestions. Restriction endonucleases were obtained from New England Biolabs (Beverly, MA). Digestions of 3 to 5 µg DNA were performed in 0.02 ml of the appropriate buffer, and loaded directly on the agarose gels. Digestions of 25 µg DNA were performed in 0.1 ml of the appropriate buffer, stopped by the addition of 20 mM EDTA and 0.1% SDS, and the digested DNAs precipitated with 2 volumes of ethanol. The precipitated DNAs were dried and dissolved directly in electrode buffer for loading on the gels.

Gel electrophoresis. DNAs (3 to 5 µg) were loaded on horizontal 0.8% agarose gels (14 x 15 x 0.2 cm). Electrophoresis was carried out at 15 mA for approximately 10 hr. Larger amounts of DNA (25 µg) were separated by electrophoresis through vertical 1% agarose gels (10 x 12 x 0.9 cm) 30 mA for 16 hr.

Blotting and hybridization to WHV DNA. The DNA in vertical agarose gels was transferred to nitrocellulose filters as previously described (4). The horizontal gels were transferred to nitrocellulose by a modification (1) of the method of Southern (15). To facilitate transfer of DNA from the gels, and to make closed circular DNAs detectable, a mild depurination was carried out prior to alkaline treatment of the gel. Gels were equilibrated for 1 hr at room temperature in 250 ml 0.05 M acetic acid, adjusted to pH 4.2 with sodium hydroxide. The sodium acetate solution was discarded, a fresh solution added, and the gel incubated at 50°C for one hour. The DNA in the gel was then denatured and hydrolyzed at the apurinic sites by soaking in 0.2 N NaOH for at least 30 minutes. This treatment introduced one strand scission for every 2000 nucleotides in calibration experiments with closed circular DNA (data not presented).

A ^{32}P-probe specific for WHV DNA was prepared by radiolabeling *in vitro* a recombinant DNA plasmid pBH20 (7) containing the entire WHV genome (4). A pBH20-WHV DNA (200 ng), nicked by a depurination procedure similar to the one described above, was denatured, and incubated in a 0.02 ml reaction containing 0.1 M Tris HCl, pH 7.5, 1.75 mM $MgCl_2$, 100 μM each dATP, dGTP and dCTP, 200 p moles α^{32}P-TTP (approx. 400 Ci/mmole, Amersham, Arlington Heights, IL), 0.5 μg oligonucleotide primers prepared from calf thymus DNA (16), and 5 units *E. coli* DNA polymerase I (Boeringer-Mannheim, Indianapolis, IN). After 10-20 minutes (50-80% incorporation), the reaction was terminated by the addition of 0.1 ml TE with 0.15 M NaCl, 0.2% SDS, and 100 μg wheat embryo ribosomal RNA. The nucleic acids were precipitated with two volumes of ethanol, collected by centrifugation, and stored frozen in a small volume of distilled water.

The nitrocellulose filters were hybridized to 50 ng/ml ^{32}P-WHV DNA probe overnight, and washed as previously described (4). Autoradiograms were exposed at -70°C to Kodak XR-5 x-ray film with a DuPont Quanta II-F intensifying screen.

RESULTS

The hepatocellular carcinomas examined in this study were obtained from a colony of captive woodchucks housed at the Penrose Research Laboratory. One group of 34 animals were already infected with WHV at the time of their capture, and have remained infected during their captivity. Uninfected woodchucks either have antibody to the viral surface antigens (WHsAg) indicating a probable acute infection with WHV at sometime in the past, or have no evidence of exposure to WHV. As reported previously, woodchucks persistently infected with WHV develop hepatocellular carcinomas with a high frequency (11,13,14,17). To date, 16 of the 34 infected woodchucks have developed hepatocellular carcinomas. No hepatocellular carcinomas have been found in the uninfected control groups. A detailed report of the pathology studies on these groups of woodchucks will be published elsewhere (Tyler et al.).

Tumors from nine WHV-infected woodchucks were analyzed for the presence of WHV DNA. Table 1 summarizes the virological history and histological findings on these animals at necropsy. All nine had been persistently infected with WHV

TABLE 1
WOODCHUCKS WITH HEPATOMAS THAT WERE ANALYZED

Animal Number	Presence of WHV at death	Duration of WHV infection (months)	Degree of chronic active hepatitis[a]	Number of primary tumors
CW300	+	>24	4.5	2
CW307	+	>12	4.5	>8
CW309	+	>14	2.5	4
CW320	+	>21	3.0	2
CW325	-	>10	2.0	>2
CW326	+	>30	4.0	?
HW86	+	>17	2.0	>3
HW98	+	>10	4.0	1
HW99	+	>12	3.5	>4

[a]The degree of CAH was estimated as described in the text and in the legend to Figure 1.

and eight had WHV infections at the time they were sacrificed. The livers of these animals on histological examination showed a form of hepatitis similar in many respects to chronic active hepatitis (CAH) in man (10). CAH is characterized by hepatocellular necrosis, bile duct proliferation, and an inflammatory exudate which extends into the parenchyma of the lobules. Grades of CAH were assigned on the basis of the amount of necrosis, bile duct proliferation, and inflammatory exudate and fibrosis (scarring) observed in liver sections. A grade of +1 indicates little or no necrosis in the parenchyma and segmental necrosis of hepatocytes lining the portal triads, i.e., the limiting plate hepatocytes. Inflammatory cells occupy the portal triads and a few inflammatory cells are found a small distance beyond the limiting plates, usually in areas where limiting plate hepatocytes were dead or degenerating. An example of a +1 CAH is shown in Fig. 1A. A grade of +5 was assigned to livers in which a significant portion of the hepatocytes have been replaced by fibrous connective tissue. In these livers, fibrosis along with proliferated bile ducts, numerous inflammatory cells, and degenerating hepatocytes extends from one portal tract to the next or to a central vein, thus collapsing the architecture of the liver. An example of +4.5 CAH is shown in Fig. 1B.

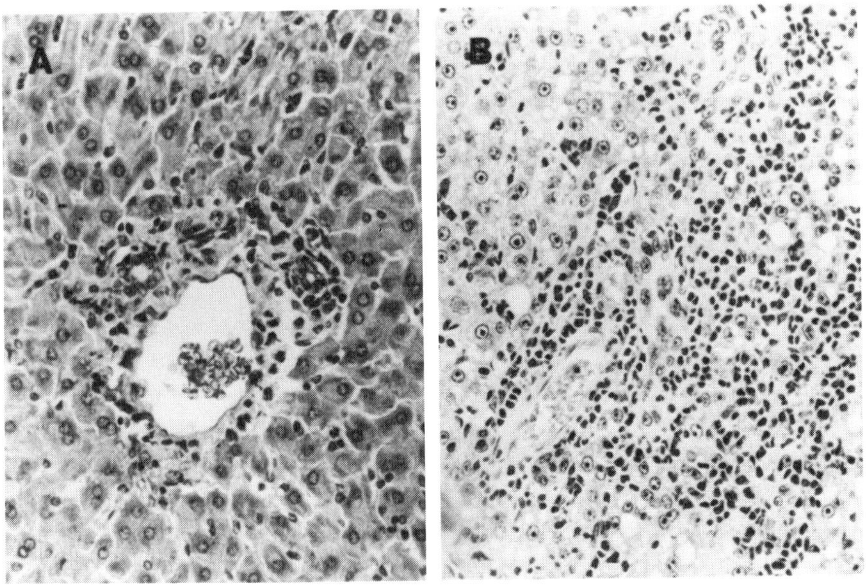

Fig. 1. Chronic active hepatitis of varying severities in WHV-infected woodchucks. (A) CAH designated +1 showing inflammation limited to the portal area. (B) CAH designated +4.5 showing inflammatory cells and necrosis extending past the limiting plate and involving a major portion of the parenchyma.

Hepatocellular carcinomas in the animals studied were generally found as multiple nodules ranging in size from less than 0.5 cm in diameter to 5.0 cm in diameter, and occurring in more than one lobe. A thin fibrous capsule surrounded each tumor, thereby demarcating the tumor from surrounding liver tissue. No evidence of extrahepatic metastases or invasion of hepatic veins by tumor cells was observed in any animal.

We analyzed tumors from these nine animals for the presence of WHV-specific DNAs as an initial step in investigating the role of the virus in the initiation of these hepatocellular carcinomas. In initial experiments, total DNAs extracted from six tumors were digested with endo R Bam HI or endo R Sac I, separated on agarose gels, transferred to nitrocellulose filters, and analyzed by hybridization. Cloned WHV DNA, excised from the λgt WHV recombinant (4) with Eco RI, was included as a marker and hybridization standard equivalent to one WHV genome per cell (0.5 pg WHV DNA per μg cellular DNA). Most tumor DNAs contained low molecular weight vegetative forms of viral DNA, migrating with and faster than the linear viral DNA standard. In all but one tumor, however, higher molecular weight fragments containing viral sequences were seen in one

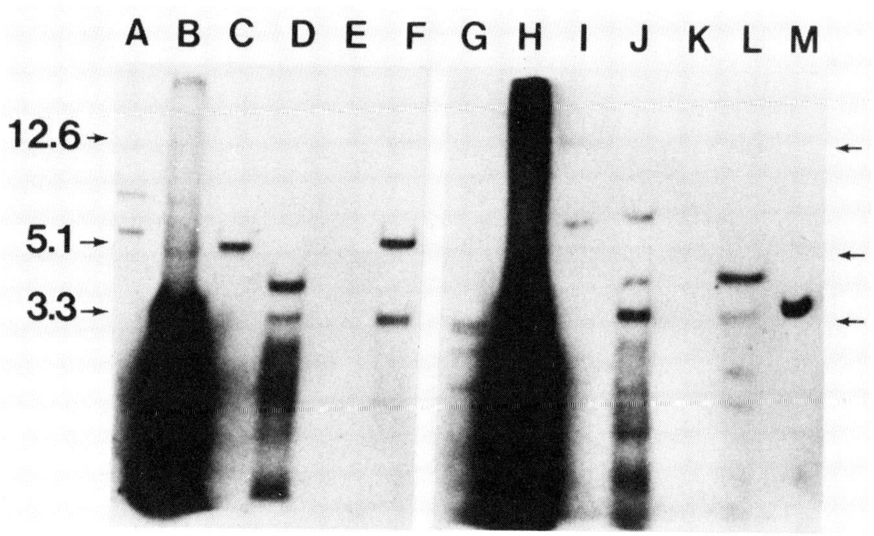

Fig. 2. WHV DNA sequences in hepatoma DNAs. DNA (25 µg) from each tumor was cleaved with either Sac I (A-F) or Bam HI (G-L) and separated by electrophoresis through a 1% vertical agarose gel. The DNA was denatured, transferred to a nitrocellulose filter and hybridized to a ^{32}P probe of cloned WHV DNA as described in MATERIALS AND METHODS. A hybridization standard (1 genome per cell) of cloned WHV DNA (Eco RI cleaved) was also run (well M). The positions of molecular length marker DNAs (expressed in kbp) are shown at the margins. The DNA samples were extracted from hepatomas from animals CW300 (A,G); CW307 (B,H); CW309 (C,I); CW320 (D,J); CW325 (E,K); and CW326 (F,L).

or both restriction digests. The hybridization of these fragments was generally in the range of that seen with the one genome standard, although often much less. Hybridization to the faster migrating species was much greater than that to the standard, indicating that they exist as many copies per cell. Since these low molecular weight DNAs containing viral sequences exist in even greater amounts in infected non-tumor tissue (see Fig. 3), we presume them to be intermediates involved in the vegetative replication of the virus. DNA from the tumor of CW325 did not hybridize to amounts of probe detectable in the autoradiographic exposure shown in Fig. 2. Longer exposure of the autoradiogram (5 times that shown in Fig. 2) revealed a 5.5 kbp Sac I fragment containing viral information (not shown). No Bam HI fragment containing viral sequences was seen in the longer exposure. The DNA from this tumor was slightly degraded, and this degradation may account for our inability to detect the small amount of viral sequences in this tumor DNA using other

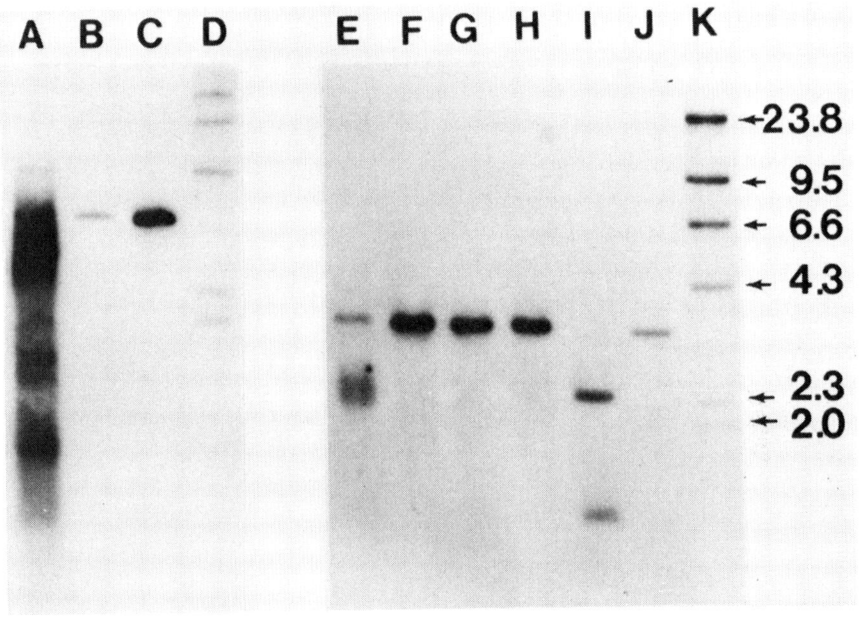

Fig. 3. WHV DNA in infected woodchuck liver. Total DNA (A) and enriched nuclear DNA (E-I) were analyzed by electrophoresis through a 1.5% (A-D) or a 0.8% horizontal agarose gel, transferred to nitrocellulose filters, and hybridized with a ^{32}P-WHV probe as described. (A) 0.2 μg total liver DNA; (B) 5 pg WHV DNA; (C) 50 pg WHV DNA; (D) ^{32}P-λ Hind III markers; (E-I) 1.0 μg enriched nuclear DNA; (E) undigested; (F) digested with Eco RI; (G) digested with Bam HI; (H) digested with Sac I; (I) digested with Eco RI, Sac I; (J) WHV DNA, 5 pg; (K) ^{32}P-λ Hind III fragments.

restriction endonucleases. It is clear, however, that very little viral information is present in the DNA of this tumor.

Unintegrated forms of viral DNA in our preparations were often of sufficient quantity to interfere significantly with detection of integrated WHV DNA. In order to alleviate this problem we carried out a preliminary isolation of nuclei and washing procedure on the additional tumors analyzed. This procedure was tested on infected non-tumor liver in order to identify which forms of free viral DNA remained after this treatment. As shown in Fig. 3, total DNA extracted from infected liver is very heterogeneous in size, consisting of multiple DNA species, whose structures have not been determined. Comparison of densitometer tracings of the tracks containing total liver DNA with the hybridization standards indicates that the infected cells contained approximately 2000 genome equivalents per cell. DNA extracted from an enriched nuclear preparation contains much less viral sequences (Fig. 3, lane E). The predominant species is slightly heterogeneous in mobility, migrating on the gel at an average position corresponding to a linear molecule of 2.5 kbp, and is not readily detected if the depurination step is omitted before transfer of the DNA to the nitrocellulose filter. Cleavage of this DNA with Bam HI, Eco RI, or Sac I all generate a single discrete viral-specific DNA which comigrates with the cloned viral DNA marker. The rapidly migrating species in undigested DNA is, we conclude, a covalently-closed circular DNA that contains a single site for Bam HI, Eco RI and Sac I. This interpretation is confirmed by a double digestion with Eco RI and Sac I, which generates two fragments of 1000 and 2300 kbp, which add up to the length of the intact genome. The amount of closed circular DNA present in this DNA sample from the enriched nuclear preparation was determined by densitometry to be equivalent to approximately 50 molecules per cell. A minor species of viral DNA in the undigested sample was seen migrating at the position of 3.4 kbp. This 3.4 kbp band is probably relaxed circular DNA, which is only slightly separated from linear DNA of the same length on this 0.8% gel. In the digested samples no viral-specific restriction fragments larger than the linear viral marker DNA could be detected on exposures long enough to detect 0.1 genome equivalents per cell.

We employed the preceding nuclear fractionation step prior to DNA isolation in examining the tumor DNAs from three additional woodchucks. The results of these analyses are shown in Figs. 4 and 5. In Fig. 4, DNA from three separate nodules from the liver of one woodchuck were analyzed before and after restriction of the DNA with Kpn I, Xho I or Bam HI. Closed circular DNA is seen migrating at the 2.5 kbp position in all three tumors, in an amount corresponding to about 10 genomes per cell compared with the hybridization standards. Kpn I and Xho I do not cleave this viral DNA as judged by lack of conversion to the linear form (although some endonucleolytic "nicking" by the Kpn I enzyme preparation is seen, since the circular form migrating at the 3.4 kbp position appears in these digestions). In two of the tumors (1 and 3), high molecular weight fragments appear after cleavage with Kpn I and Xho I. Since neither of these enzymes cleaves within the viral sequences (they do not cleave the closed circular DNA), one may presume that the number of high molecular weight bands are equal to the number of common sites within the tumor at which viral information is integrated. Thus one fragment greater than 3.3 kbp is generated by

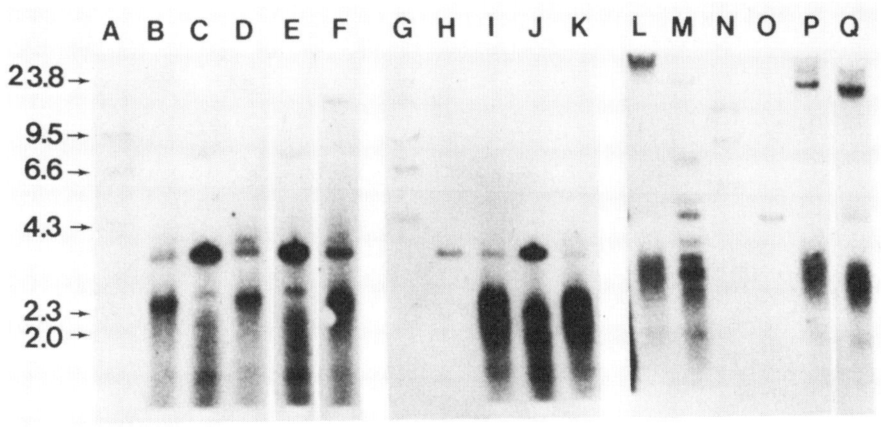

Fig. 4. WHV DNA in three hepatomas from an infected woodchuck. Enriched nuclear DNAs were isolated from three large tumor nodules in the liver of an infected woodchuck, HW86. DNAs, either undigested or digested with Bam HI, Kpn I or Xho I were separated by electrophoresis through 0.8% agarose gels, transferred to nitrocellulose filters, and hybridized with a cloned ^{32}P-WHV DNA probe. ^{32}P-λ Hind III fragments were run as molecular length markers (A,G,N). DNA (3 μg) from tumor 1 (lanes B-F) was run undigested (B), or after digestion with Bam HI (C), Xho I (D), Bam HI, Kpn I (E), or Kpn I (F). DNA (3 μg) from tumor 2 (lanes I-K) was run undigested (I), or digested with Bam HI (I), or Kpn I (J). DNA (3 μg) from tumor 3 (lanes L,M,P,Q) were run undigested (L), or digested with Bam HI (M), Xho I (P), or Kpn I (Q). Hybridization standards of one genome per cell were run in wells H and O.

cleavage with both Kpn I and Xho I in tumor number 1, indicating a single common site within the tumor at which viral DNA is integrated. Tumor number 3 would have viral sequences integrated at two common sites. No common sites for integrated viral DNA could be seen in tumor number 2.

Additional analyses of tumors from two other wooodchucks are seen in Fig. 5. Four tumors from a single animal are shown before and after digestion with Kpn I and Bgl I, as well as a single tumor from a different animal. Three of the four tumors from woodchuck HW99 contain sites at which viral information is integrated, as judged by two enzyme digestions. No common sites were detected in tumor number 1 using Kpn I or Bgl I, but digestion of the DNA with Eco RI, or Sac I reveals single faint bands of 7.0 and 5.5 kbp, respectively (not

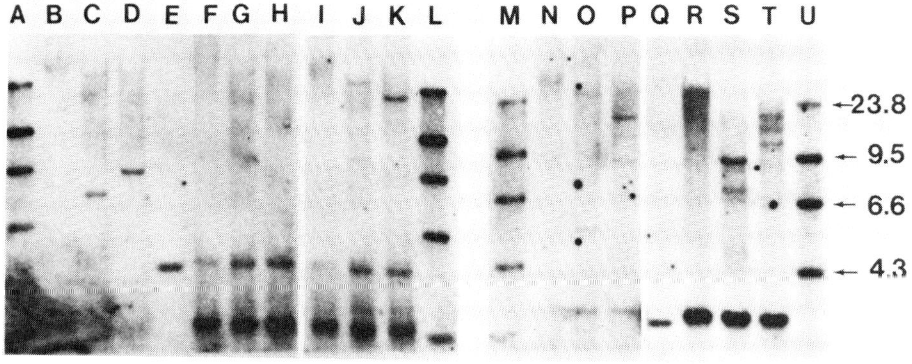

Fig. 5. WHV DNAs in hepatomas from two infected woodchucks. Enriched nuclear DNAs were isolated from four tumor nodules in the liver of one woodchuck (HW99) and from the single nodule in the liver of a second woodchuck (HW98). The DNAs, either undigested or digested with Bgl I or Kpn I, were separated by electrophoresis through 0.8% agarose gels, transferred to nitrocellulose filters, and hybridized with a cloned ^{32}P-WHV DNA probe. ^{32}P-λ Hind III fragments were run as molecular length markers (A,L,M,U). DNA (3 μg) from each tumor was run either undigested, or digested with Bgl I or Kpn I and are displayed in the figure in that order. HW99 tumor 0 (B-D); HW99 tumor 1 (F-H); HW99 tumor 2 (I-K); HW99 tumor 3 (N-P); HW98 tumor (R-T). A hybridization standard of one genome per cell was run in wells E and Q.

shown). The DNA from the hepatoma of HW98 appeared to have viral information integrated at three or four common sites.

DISCUSSION

Hepatocellular carcinomas have arisen in 16 of 34 WHV-infected woodchucks under study, and in none of 20 control uninfected animals. Persistent WHV infection is very probably a factor in the etiology of these hepatic tumors. However, since the two groups of animals under study were not randomly selected and experimentally infected, but had acquired their infections in the wild, it cannot be ruled out that unknown factors are responsible for both persistent WHV infections and hepatocellular carcinoma.

While WHV DNA was not found to be integrated at a common site or sites in hepatocytes of one chronically infected liver, most liver tumors examined contained viral DNAs integrated at one or a few such sites. Integrated viral DNA was not detected in one of the tumors from an animal that had multiple tumors. However, since in some cases (CW325, HW99 tumors), integrated viral sequences were marginally detectable, we are reluctant to conclude that this tumor contained no common site of integrated viral DNA. The shortest viral sequence detectable in our hybridizations would correspond to roughly 0.05 to 0.1 genome equivalents per cell, or 150-300 nucleotide pairs. Weak hybridization, as was observed in some tumors, could indicate either that such a short viral-specific sequence was present in every cell of the tumor, or that longer viral sequences were present in a subpopulation of tumor cells. The presence of viral DNA integrated into a subpopulation of cells of a single tumor nodule could arise by an integration event occurring in a single cell of the tumor after tumor growth had already begun. In this case, the integration event would not have occurred in the initial transformed cell but sometime after transformation. Alternatively, a tumor initiated by a cell containing integrated viral DNA may be able to recruit and sustain the growth within the tumor of otherwise normal hepatocytes. Consequently, the tumor would be made up of cells containing integrated viral DNA and recruited cells with no integrated viral DNA. We are unable to distinguish among these alternatives at the present time.

We have assayed the DNAs of separate tumor nodules occurring within the liver of the same animal. In the two animals examined in this way restriction enzyme digestion patterns of the integrated viral DNAs were different for the DNAs of the different nodules. We have considered two possible interpretations of this finding. Either these nodules are metastases that are clonally derived from subpopulations of cells of an undetected primary tumor which contained different integration sites, or they are independently occurring primary tumors. We favor the latter interpretation since histologically, these separate nodules were judged to be primary tumors on the basis of their clear physical demarcation from the surrounding normal liver, and by their non-invasive appearance. In addition, no candidate primary tumor containing multiple integration sites related to those found in adjacent tumors was detected.

The role of viral DNAs in tumor formation or cell transformation has been intensively studied in a number of experimental tumor virus systems. In general, tumor viruses have been found to cause malignant transformation either by introducing a transforming gene into the infected cell, or by causing the infected cell to express at high levels a particular cellular transforming gene that is otherwise repressed (6,9). Both of these mechanisms, where demonstrated, require the persistence of viral DNA within the transformed cells. We have observed two forms in which WHV DNAs exist in WHV-induced hepatomas.

Unintegrated forms of viral DNA, presumably involved in replication, as well as covalently-closed circular DNAs were found commonly in the hepatomas examined and in all WHV-infected non-tumor tissues (data not presented). Whether any of these forms of DNA exist within the tumor cells themselves, or whether they are found only in non-tumor cells trapped within the matrix of the growing tumor cannot be resolved by these experiments. In addition, WHV DNA sequences integrated into high molecular weight DNA was found in all but one hepatoma. The presence of these integrated viral sequences is consistent with, but certainly does not prove, a mechanism of direct transformation by WHV followed by clonal growth of the transformed cell into a tumor nodule. The mechanism by which malignant transformation of hepatocytes might be mediated by viral DNA is a major object of this study. It is doubtful that WHV carries and introduces into the cell a transforming gene for hepatocytes since infected cells cannot be distinguished from uninfected ones on the basis of their morphology. Moreover, WHV infection *per se* does not seem to result in the death of the infected cell. Infected cells are thus able to interact with the infecting virus over a long period of time. The long duration of this interaction in the chronically infected cell may allow time for an infrequent transforming event to occur that may involve the viral DNA.

Integration of WHV DNA into the DNA of chronically infected cells is not an infrequent event, however. Although such integrated WHV DNA was not detected by the methods employed in this study, in subsequent experiments, WHV DNA integrated in non-tumor infected cells has been detected by cloning and amplification of the integrated genomes in bacteriophage lambda (Rogler *et al.*, in preparation). Examination of these clones indicates that viral DNA is found at a level of about one genome per cell and integrated at many different sites in chronically infected liver, and is, therefore, not detected by hybridization to Southern blots. These results emphasize that the presence of integrated viral DNA is not unique to WHV-induced hepatomas, and that integration of WHV DNA is not sufficient for transformation. A detailed examination, currently underway, of the structure and sites at which WHV is integrated in hepatomas and in non-tumor tissue may reveal whether this viral DNA plays any role in malignant transformation of hepatocytes.

ACKNOWLEDGMENTS

This work was supported by Public Health Service grants AI-16260, AI-15166, RR-05539 and CA-06927 from the National Institutes of Health and by an appropriation from the Commonwealth of Pennsylvania.

REFERENCES

1. Astrin, S. M. (1978) Proc. Natl. Acad. Sci. USA, 75, 5941.
2. Brechot, C., Pourcel, C., Louise, A., Rain, B. and Tiollais, P. (1980) Nature, 286, 533.
3. Chakraborty, P. R., Ruiz-Opazo, N., Shouval, P. and Shafritz, D. (1980) Nature, 286, 531.
4. Cummings, I., Browne, J., Salser, W., Tyler, G., Snyder, R., Smolec, J. M. and Summers, J. (1980) Proc. Natl. Acad. Sci. USA, 77, 1842.
5. Edman, J. C., Gray, P., Valenzuela, P., Rall, L. B. and Rutter, W. J. (1980) Nature, 286, 535-538.
6. Hayward, W. S., Neel, B. G. and Astrin, S. M. (1981) Nature, 290, 475.
7. Itakura, K., Hirose, T., Crea, R., Riggs, A., Heyneker, H. L., Boliver, F. and Boyer, H. (1977) Science, 198, 1056.
8. Marion, P. L., Salazar, F. H., Alexander, J. J. and Robinson, W. S. (1980) J. Virol., 33, 795.
9. Neel, B. G., Hayward, W. S., Robinson, H. L., Fang, J. and Astrin, S. M. (1981) Cell, 23, 323.
10. Popper, H. (1975) Amer. J. Path., 81, 609.
11. Popper, H., Shih, J. W.-K., Gerin, J. L., Wang, D. C., Hoyer, B. H., London, W. T., Sly, D. L. and Purcell, R. H. (1981) Hepatology (in press).
12. Shafritz, D. A. and Kew, M. (1981) Hepatology, 1 (in press).
13. Snyder, R. L. (1978) Longevity and disease patterns in captive and wild woodchucks, in: 1977-78 Regional Workshop Proceedings of American Association of Zoological Parks and Aquariums, Wheeling, WV.
14. Snyder, R. L. and Summers, J. (1980) Woodchuck hepatitis virus and hepatocellular carcinoma, Cold Spring Harbor Laboratory, Cold Spring Harbor, NY, pp. 447-457.
15. Southern, E. M. (1975) J. Mol. Biol., 98, 503.
16. Summers, J., O'Connell, A. and Millman, I. (1975) Proc. Natl. Acad. Sci. USA, 72, 4597.
17. Summers, J., Smolec, J. M. and Snyder, R. (1978) Proc. Natl. Acad. Sci. USA, 75, 4533.
18. Szmuness, W. (1978) Prog. Med. Virol., 24, 40.
19. Twist, E. M., Clark, H. F., Aden, D. P., Knowles, B. B. and Plotkin, S. A. (1981) J. Virol., 37, 239.
20. Werner, B. G., Smolec, J. M., Snyder, R. and Summers, J. (1979) J. Virol., 32, 314.

The Role of Viruses in Human Cancer. Volume II
G. Giraldo and E. Beth editors

HEPATITIS B VIRUS INFECTION AND HEPATOCELLULAR CARCINOMA - PERSPECTIVES FOR PREVENTION

ALAIN GOUDEAU, BERNARD YVONNET, FRANCIS BARIN, FRANCOIS DENIS, PIERRE COURSAGET, JEAN-PAUL CHIRON, IBRAHIMA DIOP MAR.
Institut de Virologie and Viral Oncology Study Group
Facultés de Médecine et de Pharmacie, 37032 TOURS Cedex, France,
Facultés de Médecine et de Pharmacie, DAKAR, Sénégal.

RELATION BETWEEN HEPATITIS B AND HEPATOCELLULAR CARCINOMA

In veterinary medicine there are numerous cancers for which a viral etiology has been clearly assessed. On the contrary, in human pathology the involvement of viruses in the genesis of cancers is far less documented. The association of Hepatitis B virus (HBV) infection with hepatocellular carcinoma (HCC) has been gaining more attention in the recent years. A growing body of epidemiological, anatomoclinical and virological evidences point to a filiation from chronic type B hepatitis to postnecrotic cirrhosis and HCC. Arguements in favour of this etiological implications can be summarized as follow :

- There is a striking correspondance between areas where HCC is a prominent cancer (first of cancers in male in West Africa, third in Asia) and area where HBV is endemic (fig. 1 A and 1 B).
- Retrospective case-control studies and prospective cohort studies have shown that HBV infection precedes and usually accompanies the development of HCC (1, 2) and that chronic HBs antigenemia represent a very high-risk factor in the outcome of PHC : the relative risk of carriers to controls exceeding 200 in some studies (3).
- HCC usually arises in a liver exhibiting a cirrhosis of the macronodular post-necrotic type (80 % of the cases) and/or lesions of chronic hepatitis (2, 4).
- HBV specific antigens and integrated viral DNA are present in HCC tissues (5, 6).
- HCC cell lines have been isolated. Some lines exhibit integrated sequences of HBV DNA and synthetize viral antigen (7).

In the genesis of HCC, it is likely that HBV is not the unique causative factor. Genetic status, hormonal balance and immunological response seem to play their individual role : there is for instance an increased incidence of liver cancers in males as compared to females with a sex ratio

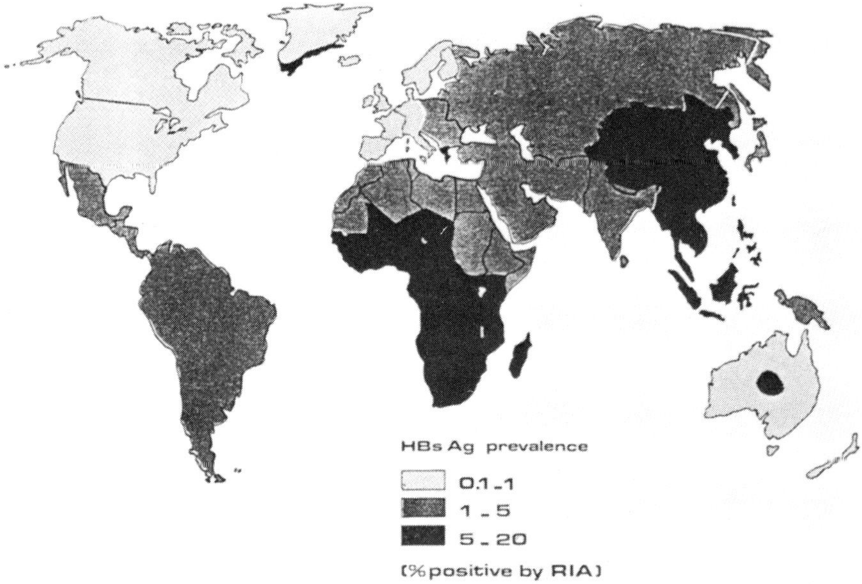
HBs Ag prevalence
0.1 - 1
1 - 5
5 - 20
(% positive by RIA)

Fig. 1A

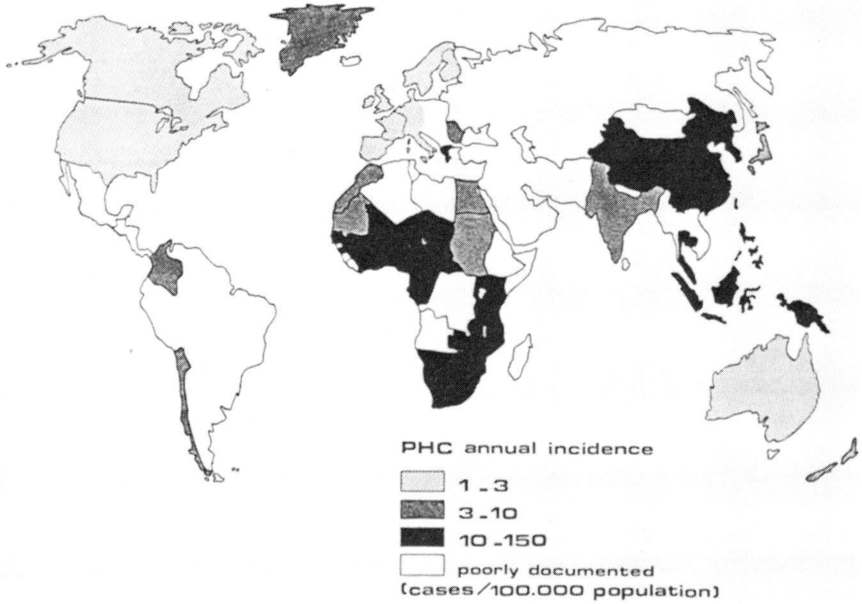
PHC annual incidence
1 - 3
3 - 10
10 - 150
poorly documented
(cases/100.000 population)

Fig. 1B

of 5-10 to 1 in African patients, although the prevalence of HBsAg carriers is identical in both sexes.

In the chain of cellular events that lead to HCC, HBV could act as a promoting factor of the chronic inflammatory process leading to progressive liver necrosis. Anarchic growth inside the regenerating nodules (doubling time of the hepatocytes being reduced from 300-400 days down to 10-20 days) will then increase the risk of oncogenic transformation. Additional environmental factors such as mycotoxins present in the food may account for HBsAg negative patients with HCC and for varying incidences of HCC in populations with identical HBsAg prevalence (8).

Further evidence of the exact virus/cell mechanisms involved in the genesis of HCC may arise from the study of other hepadnaviruses close to HBV that infect various vertebrates (woodchucks, ground-squirrels, ducks). These viruses can persist in their respective host and it has been shown that the woodchuck hepatitis virus was associated with chronic liver diseases and liver cancers (9).

Several controlled trials have established that hepatitis B vaccines offer substantial protection against chronic HBV infection (10, 11, 12). It has thus been hypothetized that mass immunization against HBV infections may prove usefull to prevent post-necrotic cirrhosis and HCC (13). If this hypothesis shall be shown correct then this will be the first conclusive evidence of a viral etiology for a human cancer (a Pastorian proof).

The French Hepatitis B vaccine (HB vaccine) was initially developped for the prevention of professionnal infections occuring in high-risk hospital settings such as hemodialysis units or laboratories. Clinical use of the HB vaccine began in the fall of 1975. Eight years later more than a million doses have been administered to hospital staff and patients in over thirty countries. Controlled trials carried out in France having established that the HB vaccine was safe, immunogenic and protective (12, 14), we proposed in 1978 to immunize children of endemic areas to prevent early infections.

PREVENTION OF HBV INFECTIONS IN CHILDREN FROM ENDEMIC COUNTRIES

Base-line studies

Epidemiology studies were carried out in Senegal, an endemic country of West Africa, to assess the transmission pathways of HBV infections (15). They showed that more than 90 % of the adult population is or has been infected by HBV and that one of six individuals is a chronic HBsAg carrier. Most of primary HBV infections occurs before teen-age (Fig. 2). A first peak is observed within the breast-feeding period resulting in a 17 % carrier rate in children at the age of two. A second peak is observed at 6-7 years of age when children enter primary school. At the age of 13, children evidence the same rate of HBV infection (90 %) and HBsAg chronic carriage (13 %) than adults of the same area. Although HBsAg and/or HBeAg can be occasionnally found in the blood of neonates from carrier mothers, lack of anti-HBc IgM argues against transplacental transmission of HBV infection (16). The vast majority of early HBV infections seem related to pernatal contaminations and/or close contacts between mother and child during the breast-feeding period ; children developping their HBV infections (usually asymptomatic) a few weeks after birth. These data were strong arguements to undertake a programme of children immunization to prevent perinatally transmitted Hepatitis B. The programme[1] which began in 1978 involved several feasibility, potency and efficacy field trials. These pilot studies were conceived to set the basis for future mass immunization campaigns.

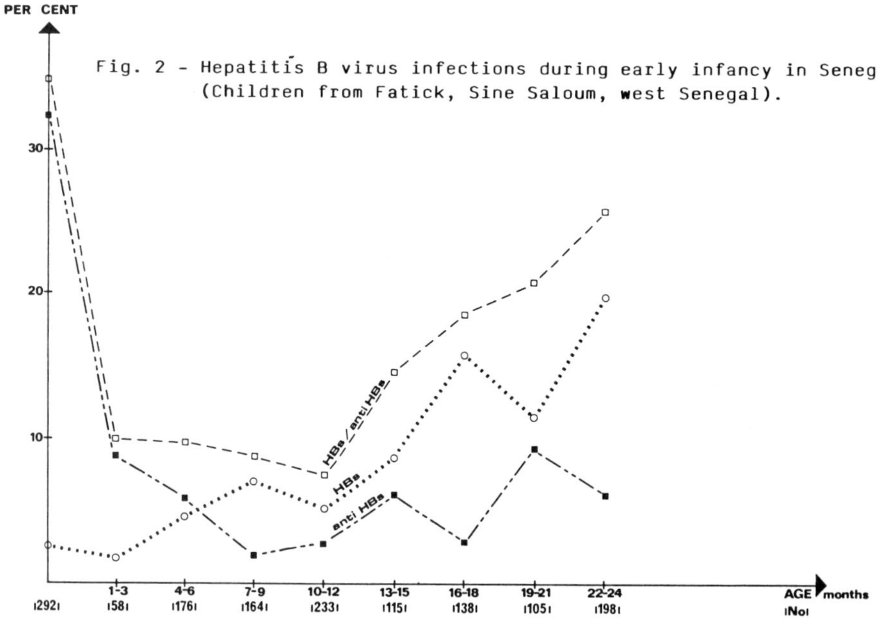

Fig. 2 - Hepatitis B virus infections during early infancy in Senegal (Children from Fatick, Sine Saloum, west Senegal).

1- The Franco-Senegalese programme "Prévention Hépatite-Hépatome" is supported by grant n° 28/77/423 project 223/DH/77 from the Ministère de la Coopération (France) and the Secrétariat d'Etat à la Recherche Scientifique et Technique (Sénégal).

Study populations and protocol

The first study was designed to investigate the logistics of immunization campaigns lead by small medical teams visiting villages on a regular basis (17). A target population of children aged 3-24 months was choosen because before two years of age, children are usually breast-fed and the mothers very cooperative to bring their child to the consultations.

A second study focused on the immunization of new-born infants in maternity units (18).

- Children age 3-24 months : a feasibility and efficacy controlled trial

Detailed base-line studies and design of this controlled trial have been described previously (15, 17). Villages where hepatitis B vaccine was to be used and the control villages were selected at random within the Niakhar district in West Senegal. Children of the HB vaccine group (N° 465) received 3 injections of vaccine (5 µg/dose) at 1 month intervals and a booster injection 1 year after. Children of the control group received a Diphteria-Tetanus-Polio triple vaccine (DTP) according to the same schedule (N° 219).

Medical surveillance of children was ensured by local paramedics and by a franco-senegalese team of residents living permanently in the centre of the vaccine trial area.

Blood samples were collected on the day of the first injections, the day of the third injection then two months later and at the time of the 12th month booster. Further samplings were collected two months and one year after the booster.

- New born infants : a potency study

Infection rate is very high within the first months of life in children of endemic areas and some of them may become infected prior to being immunized grieving the results of fastidious and costly programmes. Vaccination of neonates was then undertaken in maternity units of Dakar to assess that virtually every neonatal hepatitis B could be prevented by early prophylaxis.Newborn infants received 3 doses of 5 µg HBsAg vaccine at 1 month intervals ; the first injection being given within the first two days of life. Children were enrolled in the study whatever the HBV status of their mother. A blood sample was drawn at birth then at four and twelve months.

Hepatitis B vaccine and laboratory methods

Hepatitis B vaccine (HEVAC B°) was provided by Institut Pasteur Production, Paris. The vaccine is a preparation of highly-purified HBsAg of both subtypes **ad** and **ay** adsorbed on alum. One dose contains 5 µg of HBsAg. Laboratory tests were performed on location in the Dakar Faculty of Medicine and Pharmacy. HBsAg, anti-HBs and anti-HBc were tested by radioimmunoassays or enzymeimmunoassays from Abbott (AUSRIA or AUZYME, AUSAB-RIA or EIA, CORAB or CORZYME, HBe-RIA or EIA). Occasionnal doubtfull results were controlled by additive testing. Titration of anti-HBs positive sera was carried out according to the manufacturer recommandations and expressed in milli-International Units per milliliter (mIU/ml).

Results

Safety

During the regular consultations safety records of vaccinees were collected. No long-term local reaction nor systematic symptoms of intolerance were observed or reported by the mothers to the medical teams. Results showed that the HB vaccine was perfectly safe for new-born infants and older children, irrespective of their HBV status prior to immunization.

Potency - Children 3-24 months

Anti-HBs responses to the vaccine was studied according to HBV status before immunization (Table I). The trial involved 465 children aged 3 to 24 months of whom : 309 were seronegative prior to immunization, 44 were anti-HBs positive, 71 were anti-HBc positive alone and 41 were HBsAg positive :

Table I - Anti-HBs seroconversion after HB vaccine in seronegative and HBV positive children (12 months follow-up)

HBV status when immunized		Anti-HBs seroconversion N°	(%)
HBsAg	N° 41	4	(9.8 %)
anti-HBc alone	N° 71	64	(90.1 %)
anti-HBs	N° 44	43	(97.7 %)
Seronegative	N° 309	285	(92.2 %)
Total	N° 465	396	(85.2 %)

Seroconversion rate to anti-HBs was excellent in seronegative children : 92.2 % ; kinetics of anti-HBs response and titres compared with the one observed in healthy adults.

Isolated anti-HBc found in 71 children was in most of the case acquired from mother with high-titer antibody. Such passive anti-HBc can last for over a year in children of HBsAg or anti-HBs positive mothers. It did not interfer with anti-HBs response to the vaccine.

Anti-HBs found in 44 children at time of the first injection of HB vaccine could correspond either to passively-acquired anti-HBs or be a marker of passed infection. Anti-HBs response of such children should therefore be considered with caution.

Persistance of HBsAg was correlated to the activity of the infection (presence of HBeAg) and the child age.

Potency - New born infants

Results of the initial trial had proven that HB vaccine was very efficient when given to seronegative children but became useless for children already infected (19). Immunization of neonates aimed to prevent such early infection.

Anti-HBs response of neonates was satisfactory irrespective of their mother HBV status (Table II).

Table II - Anti-HBs response of new-born infants to the HB vaccine according to their mothers HBV status

Mother HBV status at time of delivery	N°	Children HBV status after 12 months follow-up				
		HBsAg	anti-HBs anti-HBc	anti-HBs alone	anti-HBc alone	Sero-negative
HBsAg	11	1	5	5	0	0
anti-HBc alone	2	0	0	2	0	0
anti-HBs	39	0	0	34	0	5
Seronegative	8	0	0	8	0	0
Total	60	1(1.7%)	5(8.3%)	49(81.7%)	0	5(8.3%)
			54(90.0%)			

At the age of 3 months, 58 of 60 infants (96.7 %) had anti-HBs antibody and 54 (90.0%) at one year. Passively-acquired anti-HBs did not seem to interfer with the response to the HB vaccine : 35 newborns were anti-HBs positive, their response to the HB vaccine was similar to the one of the 25 anti-HBs negative neonates in terms of seroconversion rate and antibody titers.

All infants born to mothers who were seronegative, anti-HBs alone or anti-HBc alone seroconverted to anti-HBs as well as 27 of 32 children (84.4 %) born to anti-HBs and anti-HBc positive mothers.

As expected, immunization at birth was efficacious even in children born to HBsAg carrier mothers : 10 of 11 were found anti-HBs positive at the 12th month control. One mother and her offspring were both HBsAg and anti-HBs positive. The child did not respond to the vaccine and became a carrier.

Efficacy - Children 3-24 months

During this controlled trial HB vaccine was compared to a placebo (DT-Polio triple vaccine) based on a village randomization. Results obtained in seronegative children after one and two years of follow-up are presented in Table III. HBsAg attack rate in the control group was found identical to the one described in base-line epidemiology studies performed in the same area prior to the introduction of the HB vaccine (15). 12.6 % of the children of the control group had became carrier at the end of the follow-up. On the contrary HBV attack rate was significantly reduced in children receiving the HB vaccine since only 3 % were HBsAg positive at the end of the follow-up, a protective efficacy rate of 76 %.

Table III - Efficacy of HB vaccine in seronegative children

	Seronegative who became HBsAg		
	one year follow-up		Two years follow-up
HB vaccine group	4/309 (1.3%)		3/101 (3.0%)
		$p < 0.01$	
Control group	17/252 (6.7%)		15/119 (12.6%)
Protective efficacy rate	81 %		76 %

12 children of the control group and 17 children of the HB vaccine group who were initially HBsAg positive were seen at the 24th month control. 9 of 12 (75 %) in the control group and 10 of 17 (58.8 %) in the vaccine group were still HBsAg carriers. The difference was not significant showing that once acquired HBV infection could not be modified by active immunization.

Global beneficial effect of the vaccination in a non-selective field trial can be appreciated by comparing HB vaccine and placebo populations at the end of the 24 months follow-up. 31 of the 164 children in the control group (18.9 %) were found HBsAg positive against 14 of 155 in the vaccine group (9.0 %) HBsAg carriers came from the group of children already HBsAg when immunized : 9 and 10 children in the control and vaccine group, respectively ; from seronegative children : 15 and 3, and from the children who were anti-HBc alone : 7 and 1.

Efficacy - Newborn infants

Although the vaccine trial did not include a control group, HBV status of immunized neonates could be confronted to epidemiology data being collected in Senegal. As part of a screening programme, 239 children aged 12 months have been tested for HBV markers : 18 (7.5 %) were found HBsAg positive and 23 (9.6 %) anti-HBs positive. By comparison, in the vaccine group only 1 of 60 infants (1.7 %) was HBsAg positive at the age of 12 months while 54 (90.0 %) were anti-HBs positive and thus protected from infection. These differences have even more meanings if one considers the prevalence of risk-mothers in the vaccine study : 11 HBsAg carrier mothers (18 %) and compares it to the general population of Senegal (13 % HBsAg carrier rate in women of child-bearing age).

Discussion

Anti-HBs response of Senegalese children to the Hepatitis B vaccine was excellent even in new born infants immunized within a few hours after birth. Seroconversion rate was similar to that observed in healthy Caucasian adults in France (14).

Data collected in long-term survey of adults showed that responsive vaccinees did not become chronic carriers after natural infections occuring when their anti-HBs had receeded. Although they are based upon a relatively short-term follow-up, results in children confirm this fact. For instance, 4 of the 309 seronegative children who received the HB vaccine as part of the controlled trial were found anti-HBs and anti-HBc positive at the 12th months end-point, indicating that abortive HBV infection had taken place. It is likely that in highly-exposed population such "natural booster" will occur frequently, thus sustaining the failing immunity in responsive vaccinees. This would suggest that in an endemic situation primovaccination could be sufficient without subsequent booster and that the 12th month booster which has been recommanded up to now might be skipped.

Efficacy of the HB vaccine in the prevention of child HBV infection was assessed after a two years follow-up. The prevalence of the HBsAg carrier stage was shown to be reduced by 76 % in susceptible children. Immunization of every child, irrespective of his or her HBV status prior to vaccination resulted in a more than 2-fold decrease of the prevalence of HBsAg carrier state : 14 of the 155 (9 %) in the HB vaccine group versus 31 of 164 (19 %) in the placebo group, $p < 0.02$. However, the study showed that once HBsAg had been acquired, vaccination did not alter the progression to a possible chronic carrier state.

Hepatitis B immune globulins (HBIG) given at birth are widely used to prevent perinatal transmission of HBV. Immediate protection against early chronic carrier state can be acheived in over 70 % of neonates ; most of them eventually developping an active immunity after natural infection. However, HBIG fail to protect 20-25 % of neonates and late chronic infections can be observed in infants when passive anti-HBs disappears (20). Immunization at birth is another alternative to prevent perinatal HBV infection. Our preliminary results in Senegalese neonates suggest that HB vaccine can indeed prevent most of perinatal and early infancy infections. A combination of passive and active immunization has been proposed to optimize the protection afforded to children born to carrier mothers (21). In an Asian population were over 40 % of carriers are also HBeAg positive serovaccination may be indicated. On the contrary, in an African population where the HBeAg rate is much lower (2 of 11 in our series), the advantage of HBIG added to the vaccine is less obvious. Further comparative trials in Africa and Asia will be necessary to provide cost-benefit analysis for the best choice amongst these various immunization strategies.

REFERENCES

1. Szmuness, W. (1978) Hepatocellular carcinoma and the Hepatitis B virus : Evidence for a causal association, Prog. Med. Virol. **24** : 40-69.
2. Maupas, P. Goudeau, A. Coursaget, P. Chiron, J.P. Drucker, J. Barin, F. Perrin, J. Denis, F. Diop Mar, I. and Summers, J. (1980) Hepatitis B virus infection and primary hepatocellular carcinoma : Epidemiological, and virological studies in Senegal from the perspective of prevention by active immunization in : Essex, M. Todarro, G. zur Hausen, H. (Ed.),"Viruses in naturally occuring cancers", Cold Spring Harbor Conference on cell proliferation, vol. 7, CSHL, New-York, 481-506.
3. Beasley, R.P. Hwang, L.Y. Lin, C.C. Chien, C.S. (1981) Hepatocellular carcinoma and hepatitis B virus. A prospective study of 22707 men in Taiwan, Lancet **ii** : 1129-1133, 1981.
4. Akagi, G. Furuya, K. and Otsuka, H. (1982) Hepatitis B antigen in the liver in hepatocellular carcinoma in Shikoku, Japon, Cancer, **49** : 678-682.
5. Goudeau, A. Maupas, P. Coursaget, P. Drucker, J. Chiron, J.P. Denis, F. and Diop-Mar, I. (1979) Hepatitis B virus antigen in human primary hepatocellular carcinoma tissues. Int. J. Cancer, **24** : 421-429.
6. Brechot, Ch. Hadchouel, M. Scotto, J. Fonck, M. Potet, F. Vyas, G.N. and Tiollais, P. (1981) State of hepatitis B virus DNA in hepatocytes of patients with hepatitis B surface antigen - positive and - negative liver diseases. Proc. Natl. Acad. Sci. USA **78** : 3906-3910, 1981.
7. Twist, E.M., Clark, H.F. Aden, D.P. Knowles, B.B. Plotkin, S.A. (1981) Integration pattern of hepatitis B virus DNA sequences in Human Hepatoma cell line. J. Virol. **37** : 239-243.
8. Lam, K.C. Tong, M.J. (1982) Analytical epidemiology of primary liver carcinoma in the Pacific basin. Natl. Cancer Inst. Monogr. **62** : 123-127.
9. Summers, J. Smolec, J.M. Snyder, R. (1979) A virus similar to human hepatitis B virus associated with hepatitis and hepatoma in woodchucks. Proc. Natl. Acad. Sci. USA, **75** ; 4533-4537.
10. Szmuness, W. Stevens, C.E. Harley, E.J. Zang, E.A. Olesko, W.R. William, D.C. Sadovsky, R. Morrison, J.M. Kellner, A. (1980) Hepatitis B vaccine. Demonstration of efficacy in a controlled trial in high-risk population in the United States. N. Engl. J. Med. **303** : 833-841.
11. Maupas, P. Chiron, J.P. Barin, F. Coursaget, P. Goudeau, A. Perrin, J. Denis, F. and Diop-Mar, I. (1981) Efficacy of Hepatitis B vaccine in prevention of early HBsAg carrier state in children. Lancet,**i** : 289-292
12. Crosnier, J. Jungers, P. Couroucé, A.M. Laplanche, A. Benhamou, E. Degos, F. Lacour, B. Prunet, P. Cerisier, Y. Guesry, P. (1981) Randomised placebo-controlled trial of hepatitis B surface antigen vaccine in French haemodialysis units : I - Medical staff. Lancet, **i** : 455-459
II - Hemodialysis patients. Lancet, **i** : 797-800.
13. WHO meeting on prevention of liver cancer (1983) WHO Techn. Report N° 691.
14. Goudeau, A. Coursaget, P. Barin, F. Dubois, F. Chiron, J.P. Denis, F. Diop Mar, I. (1982) Prevention of hepatitis B by active and passive-active immunization in : Szmuness, W. Alter, H.J. Maynard, J.E. (Ed.), Viral Hepatitis. The Franklin Institute Press, Philadelphia, pp 509-525.

15. Barin, F. Perrin, J. Chotard, J. Denis, F. N'Doye, R. Diop Mar, I. Chiron, J.P. Coursaget, P. Goudeau, A. Maupas, P. (1981) Cross-sectional and longitudinal epidemiology of hepatitis B in Senegal. Prog. Med. Virol. **27**, 148-162.

16. Goudeau, A., Yvonnet, B. Lesage, G. Barin, F. Denis, F. Coursaget, P. Chiron, J.P. Diop Mar, I. (1983) Lack of anti-HBc IgM in neonates with HBeAg carrier mothers argues against transplacental transmission of Hepatitis B virus infection. Lancet **II** : 1103-1104.

17. Maupas, P. Coursaget, P. Chiron, J.P. Goudeau, A. Barin, F. Perrin, J. Denis, F. Diop Mar, I. (1981) Active immunization against Hepatitis B in an aera of high endemicity. Part I : Field Design. Prog. Med. Virol. **27**, 168-184.

18. Yvonnet, B. Coursaget, P. Denis, F. Digoutte, P. Petat, E. Barin, F. Goudeau, A. Correa, P. Diop Mar, I. Chiron, J.P. (in press) Immune response to hepatitis B vaccine at birth. Pediatrics.

19. Barin, F. Yvonnet, B. Goudeau, A. Coursaget, P. Chiron, J.P. Denis, F. Diop Mar I. (1983) Hepatitis B vaccine : further studies in children with previously acquired Hepatitis B surface antigenemia. Inf. Imm. **41** : 83-87.

20. Beasley, R.P. Hwang, L.Y. Stevens, C.E. Lin, C.C. Hsieh, F.J. Wang, K.Y. Sun, T.S., Szmuness, W. (1983) Efficacy of Hepatitis B immune globulins (HBIG) for prevention of perinatal transmission of the HBV carrier state : Final report of a randomized double-blind placebo-controlled trial. Hepatology **3** : 135-141.

21. Hiroshi, I. Masahiko, Y. Jun, M. Toshi, F. Kiyoshi, B. Shuko, I. Shinobu, A. Fumio, T. Yuzo, M. Makoto, M. (1982) Combined passive and active immunization for preventing perinatal transmission of hepatitis B virus carrier state. Pediatrics **40** : 613-619.

The Role of Viruses in Human Cancer. Volume II
G. Giraldo and E. Beth editors

PREVALENCE AND SIGNIFICANCE OF ANTIBODY AGAINST AN ANTIGEN (HBV/T Ag) PRESENT IN A HUMAN HEPATOMA CELL LINE CARRYING INTEGRATED HEPATITIS B VIRUS DNA.

FRANCIS BARIN, GERARD LESAGE, JEAN-LOUP ROMET-LEMONNE AND ALAIN GOUDEAU.
Viral Oncology Study Group. Université François Rabelais.
Lab. Virology. CHR Bretonneau. 37044 Tours Cedex. France.

To date, numerous arguments strongly suggest that hepatitis B virus (HBV) is directly or indirectly involved in the etiology of primary hepatocellular carcinoma (PHC) (1). Viruses inducing cancers in animals usually promote the appearance of new cellular proteins whenever the transformation occurs. These proteins are known as T antigens (T for tumour) or neo antigens. Animals bearing virus-induced tumours develop antibodies specific of these antigens. We sought evidence for a similar antigen-antibody system in HBV-associated PHC.

DISCOVERY

The antigen was first detected by indirect immunofluorescence assay (IFA) in a liver necropsy obtained from a patient who died of HBs Ag positive PHC (2) using antisera from Senegalese patients suffering from PHC (Figure 1). Tissue samples used for the IFA contained integrated HBV sequences (3). The antigen was also detected in the PLC/PRF/5 hepatoma cell line (4) which is known to contain several integrated copies of HBV DNA (5) (Figure 1).

Antigen controls including normal liver tissue, non hepatic cancerous tissues, HBV DNA negative hepatoma cell lines {HepG2 (6), Malhavu (7)}, and antibody controls including sera from healthy Caucasian or African individuals and sera from patients with secondary liver cancers, assessed the specificity of this new antigen-antibody system. They also shown that the antigen was not related to known HBV structural antigens.

Because this antigen was detected only in PHC tissues containing integrated HBV DNA and in HBV DNA positive PLC/PRF/5 cell line, it was named HBV/T Ag for HBV tumour-associated antigen. The corresponding antibody, present in sera of PHC patients was named anti-HBV/T .

PREVALENCE AND SIGNIFICANCE OF ANTI-HBV/T

An enzyme-immunoassay (ELISA) using a PLC/PRF/5 cell lysate as HBV/T antigenic source was developped to study the prevalence and significance of anti-HBV/T in various populations.

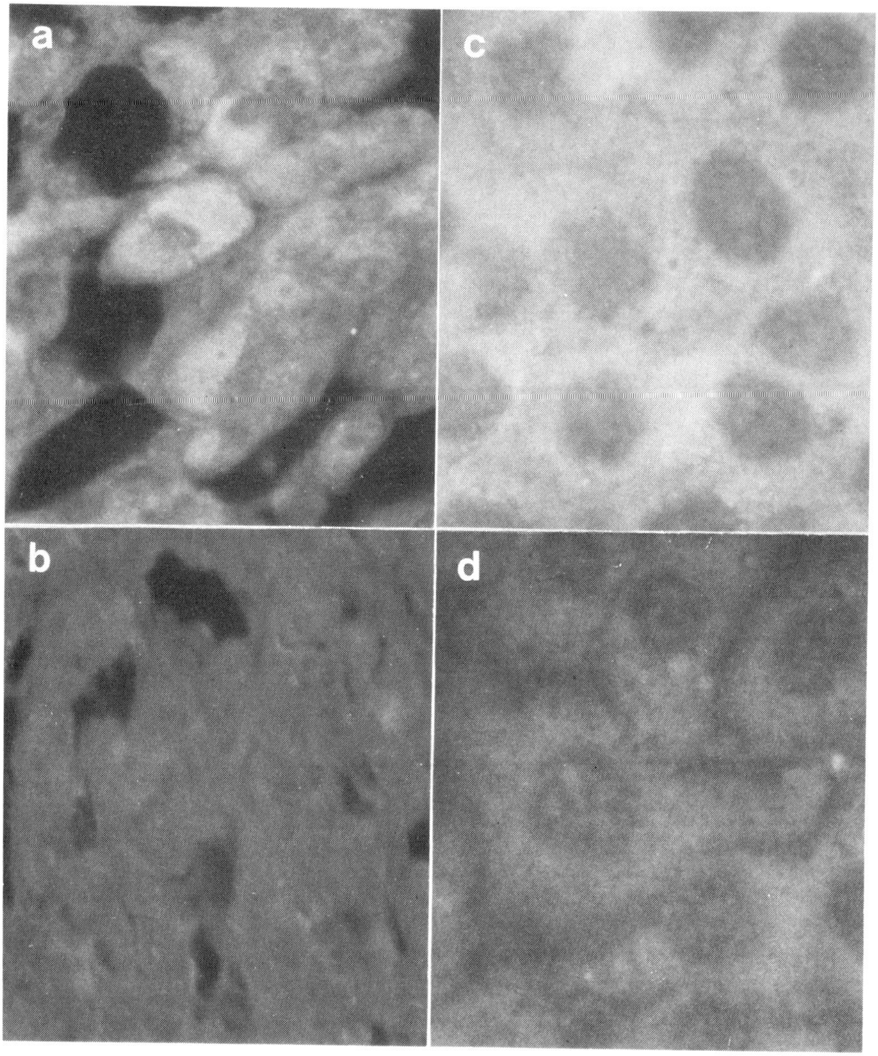

Fig. 1 : Indirect immunofluorescence assay for cytoplasmic HBV/T Ag
a) HBV DNA-positive PHC tissue stained with HBs Ag positive sera from a PHC patient.
b) HBV DNA-positive PHC tissue stained with a normal human sera.
c) PLC/PRF/5 cells stained with HBs Ag positive sera from a PHC patient.
d) PLC/PRF/5 cells stained with a normal human sera.

ELISA FOR ANTI-HBV/T

Source of antigen

HBV/T Ag used in the assay was prepared from a cell lysate of PLC/PRF/5 hepatoma cells. Subcellular fractionation was carried out according to a procedure described by Courtneidge et al. (9). Briefly, monolayers of cells were washed twice with phosphate-buffered saline (140 mM NaCl/1.8 mM KCl/8 mM Na_2HPO_4/0.7 mM $CaCl_2$/0.5 mM Mg Cl_2, pH 7.4), harvested by scraping into the same solution, and collected by centrifugation at 1000 x g for 5 min. The cells were suspended in 5 mM KCl/1 mM $MgCl_2$/20 mM Hepes, pH 7.1, at2x 10^7 cells/ml. After incubation at 4° C for 10 min., the cells were broken by 20 strokes in a tight-fitting Dounce homogenizer without apparent damage to nuclei. The lysate was centrifuged at 1000 x g for 5 min. to pellet the nuclear fraction. The supernatant was centrifuged at 100 000 x g for 30 min. The resulting supernatant fraction, corresponding to the cytosol, was stored at - 20° C until used as antigenic source.

Protein content of this cell lysate was 2 mg/ml by Lowry's (10). Concentration of HBS Ag was 5 ng/ml by Auszyme° (Abbott). The cell lysate did not contain α-foetoprotein (RIA from Abbott) nor carcinoembryonic antigen by radioimmunoassays (RIA from Commissariat à l'Energie Atomique - France).

Run of the test

Anti-HBV/T ELISA was set up according to the method of Voller et al. (11) with some modifications. 200 µl of PLC/PRF/5 cell lysate in 0.1 M bicarbonate buffer, pH 9.6, were coated on each well of microtiter plates (Dynatech micro-elisa M 129 B). Microplates were incubated overnight at room temperature in a moist chamber. Three washings were done with 250 µl of Phosphate-buffered saline (PBS) containing 0.1 % bovine serum albumin (BSA). 100 µl of each serum sample, diluted 10^{-2} in PBS containing 4 % Tween 20 and 1 % BSA, were set in duplicate. Plates were incubated for 1 hour at 37° C in a moist chamber. After 3 washings with 250 µl of PBS containing 1 % Tween 20, 50 µl of freshly diluted alkaline phosphatase rabbit anti-human IgG conjugate (Behring) were added to each well. A second incubation took place for 1 hour at 37° C and was followed by 3 washings as before. 100 µl of substrate solution (paranitrophenyl phosphate, Behring) were added and incubated for 30 min.. Enzymatic reaction was stopped with 50 µl of 2 N NaOH per well. Absorbance was read at λ = 405 nm within 1 hour. A substrate solution containing the appropriate amount of NaOH was used as a blank.

Standardisation

The threshold for absorbance was calculated from the mean absorbance obtained with 152 sera from healthy HBs Ag negative-blood donors majored by two standard deviations i.e., 0.110 + (2 x 0.070) = 0.250. These 152 sera were pooled and used as negative control in each run. For each sample, results were expressed as the S/N ratio : absorbance of sample/mean absorbance of 4 negative controls per run. Cut-off value for the S/N ratio was 2.3 (0.250/0.110).

High-titered anti-HBs sera containing more than 10^6 mIU/ml anti-HBs did not interfere in the test.

SERUM SAMPLES

Single serum samples from 502 subjects were studied (Table 1). 173 sera were collected in Senegal which is a high endemicity area of PHC. 82 Senegalese patients suffered from PHC and 20 from non alcoholic cirrhosis. 71 healthy Senegalese male adults (32 HBs Ag positive and 39 HBs Ag negative) were used as controls. 329 sera from patients with various liver diseases and controls were collected in France.

RESULTS

Results are shown in Table 1 and figure 2.

Anti-HBV/T antibody occured mainly in patients suffering from chronic liver diseases. The prevalence of anti-HBV/T was more than 90 % in patients suffering from PHC, cirrhosis and chronic hepatitis. In contrast the prevalence of anti-HBV/T was 0 % in HBs Ag negative-Senegalese healthy adults. Intermediate prevalences were found in the other populations under study. Further clinical evaluation of anti-HBV/T antibody was made by screening the sera with a S/N ratio equal to or greater than 5. The difference in the prevalence of high level anti-HBV/T was striking between patients with chronic liver diseases and patients with non chronic liver diseases or control populations (figure 2). None of patients with acute hepatitis A, none of HBs Ag positive healthy Senegalese adults and none of subjects in the recovery period from type B hepatitis had high positive anti-HBV/T antibody. 3 out of 15 patients with secondary liver cancers had high level anti-HBV/T ; 2 of them were HBs Ag chronic carriers suffering from colon cancer.

According to these results, it appears that anti-HBV/T antibody is produced at a low level during acute phase of some hepatitis B and disappears at the convalescent period. In contrast anti-HBV/T is common at a high level in chronic hepatitis and prevalence increases in cirrhosis and PHC. The sequence HBs Ag chronic carrier state-chronic hepatitis-cirrhosis and ultimately PHC

Table 1.
Prevalence of anti-HBV/T in various study populations in France and in Senegal.

GROUP	(N°)	No tested	S/N ≥ 2.3 No positive	(%)	S/N ≥ 5 No positive	(%)
PHC (Senegal)	(1)	82	81	(98.8)	67	(81.7)
Cirrhosis (Senegal)	(2)	20	19	(95.0)	18	(90.0)
Alcoholic-associated PHC (France)	(3)	40	36	(90.0)	19	(47.5)
Alcoholic cirrhosis (France)	(4)	54	52	(96.3)	36	(66.7)
Non-alcoholic cirrhosis (France)	(5)	45	43	(95.6)	34	(75.6)
Chronic hepatitis (France)	(6)	31	30	(96.8)	20	(64.5)
Acute type B hepatitis (France)	(7)	52	33	(63.5)	6	(11.5)
Acute type A hepatitis (France)	(8)	34	16	(47.1)	0	(-)
Fulminant hepatitis (France)	(9)	21	17	(81.0)	8	(38.1)
HBs Ag-positive healthy adults (Senegal)	(10)	32	7	(21.9)	0	(-)
HBs Ag-negative healthy adults (Senegal)	(11)	39	0	(-)	0	(-)
Secondary liver cancers (France)	(12)	15	7	(46.7)	3	(20.0)
Lymphoproliferative malignant diseases (France)	(13)	20	6	(30.0)	2	(10.0)
Convalescent phase of hepatitis B (France)	(14)	17	3	(17.6)	0	(-)

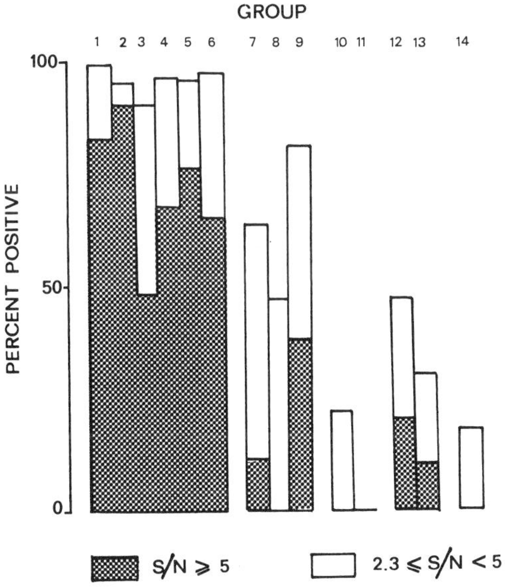

Fig. 2 : Prevalence of anti-HBV/T antibody in the different study populations.

is the long term natural course ofchronic HBV infection. Anti-HBV/T antibody could be a usefull indicator of this progression.

WESTERN BLOT ANALYSIS

A further study was undertaken to determine by western blot analysis the polypeptides which were recognized by sera from PHC patients (figure 3).

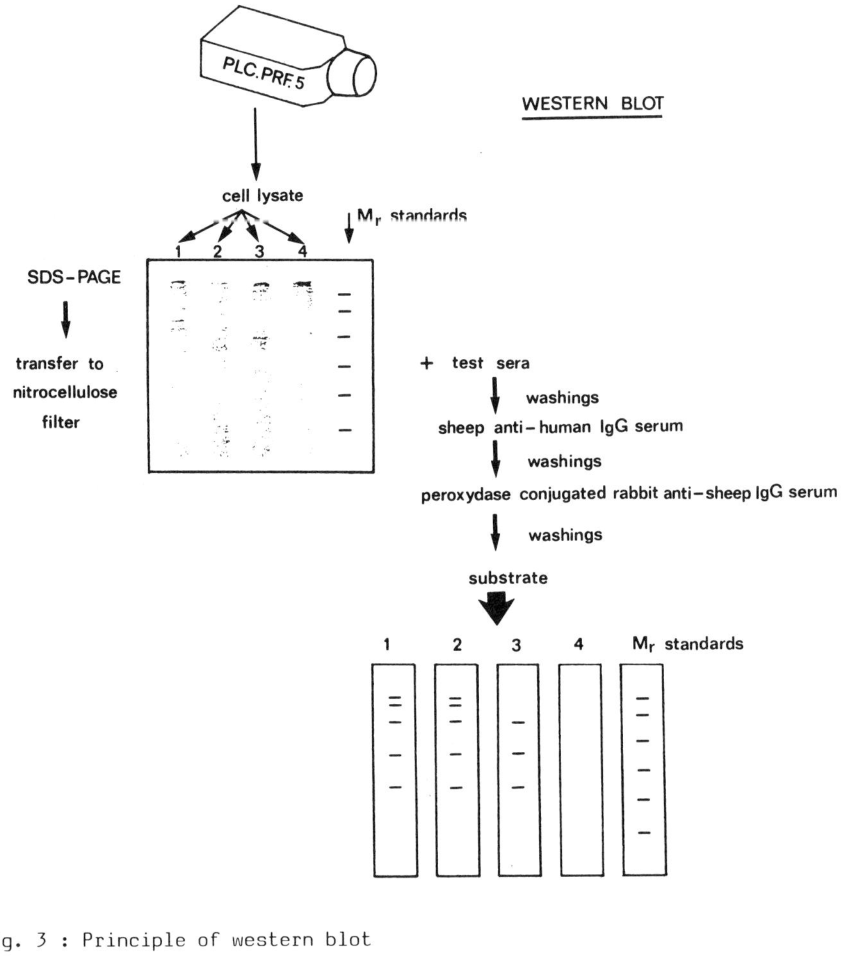

Fig. 3 : Principle of western blot

PLC/PRF/5 cell lysate, prepared as described above, was submitted to SDS-polyacrylamide gel electrophoresis (SDS-PAGE) in slab gels under denaturing conditions. Electrophoresis were performed in a 12.5 % separating gel with 5 % stacking gel using a discontinuous buffer system as described by Weber & Osborn (12). After electrophoresis, the gels were washed for 30 min. in three changes of transfer buffer (10 mM Tris-HCl, pH 7.0, 2 mM

EDTA, 50 mM NaCl, 0.1 mM dithiothreitol). Proteins were transfered passively from the gel to nitrocellulose filters (Schleicher and Schüll, BA 85) by incubating a nitrocellulose/gel "sandwich" overnight at room temperature. Then, the nitrocellulose sheets were soaked in phosphate buffer, pH 8.0, containing 3 % BSA for 1 hour at 40° C, followed by 3 washings in phosphate buffer, pH 8.0, containing 0.25 % BSA and 0.2 % Tween 20 (PBS-BT). Sheets were sliced in strips and immunological detection took place as shown in figure 3. Test sera and reactive immunosera were diluted 1:100 in PBS-BT before use. Incubations took place for 1 hour at 37° C. Peroxydase activity was detected colorimetrically by the addition of a fresh solution of diaminobenzidin as substrate (50 mg in 100 ml of 0.05 M Tris-HCl buffer, pH 7.6, supplemented with 0.01 % hydrogen peroxyde) for 30 seconds.

Molecular range (Mr) standards were run on all gels (low molecular weights calibration kit, Pharmacia). Mr of the polypeptides were estimated with a semi-logarithmic plot of the protein Mr versus their relative migration.

A polypeptide of Mr 42 K was strongly recognized by all 18 sera from PHC patients which were tested (figure 4). This polypeptide gave a weak band

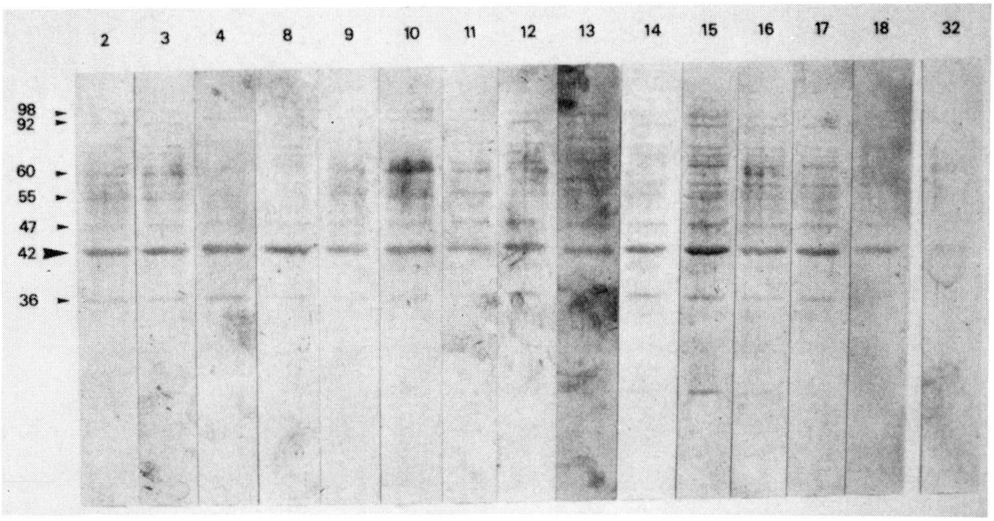

Fig. 4 : Western blot analysis of antigens present in PLC/PRF/5 hepatoma cell line recognized by antibodies of patients suffering from primary hepatocellular carcinoma.
N° 2 to 18 = sera from PHC patients
N° 32 = serum from a healthy adult

with some sera from healthy adults (8 tested) or with some sera from patients suffering from secondary liver cancer (9 tested). Two other polypeptides of Mr 36 K and 47 K were specifically and regulary recongnized by sera from PHC patients. Four other polypeptides of higher Mr were less regulary and less strongly recognized by sera from PHC patients. However those polypeptides reacted specifically with sera from PHC patients but not with sera from patients suffering from secondary liver cancers or with sera from healthy adults.

CONCLUSION

A new antigen, named HBV/T, has been found in the cytoplasm of cells transformed (or chronically infected) by HBV. The nature of HBV/T is unknown. Is HBV/T Ag a single or several of the polypeptides recongnized by sera from PHC patients when tested by Western blot analysis ? Is HBV/T Ag a product of a HBV gene, released as a consequence to HBV infection or HBV DNA integration ? Or, is HBV/T Ag a protein of the hepatocyte which is released or modified, so that it becomes immunogenic ?

Recently, Wen et al. identified by anticomplement immunofluorescence a nuclear antigen (HBNA) in PLC/PRF/5 hepatoma cell line (8). Antibody to HBNA was found in HBs Ag positive sera from patients with PHC or with chronic hepatitis. Thus, HBNA resembled more neoantigens associated with the oncogenesis of certain papoviruses or herpes viruses than did HBV/T Ag. This discrepancy could be due to different clones of PLC/PRF/5 cells used in the two studies.

A case control study was performed using an ELISA for anti-HBV/T antibody. It showed that anti-HBV/T is a relatively specific marker of chronic liver diseases, especially usefull for HBV-associated cirrhosis and PHC. As such the survey of anti-HBV/T in populations at a high risk to develop a chronic liver disease after HBV infection could be of particular interest for early diagnosis of cirrhosis and primary hepatocellular carcinoma.

ACKNOWLEDGEMENTS

This work was supported by INSTITUT NATIONAL DE LA SANTE ET DE LA RECHERCHE MEDICALE (INSERM grant ≠ 82 7003), by funds from UNIVERSITE FRANCOIS RABELAIS, TOURS and by dotation from the ASSOCIATION POUR LA RECHERCHE SUR LE CANCER.

Authors thank Drs I. Diop Mar, J.E. Bocande, B. Diop, A.M. Sow, F. Denis, S. M'Boup, T. Diop and B. Yvonnet for the Senegalese blood specimens, Drs E. Aron, J.P. Benhamou and P. Bougnoux for samples from French patients, Drs Chassaigne for the collection of blood donors sera.

Studies on tumour-associated antigens in PHC were initiated by Pr. Ph. Maupas.

REFERENCES

1. Maupas, P. and Melnick, J.L. (1981) Prog. Med. Virol., 27, 1.
2. Romet-Lemonne, J.L., Barin, F., Goudeau, A. and Maupas, P. (1982) C.R. Acad. Sc. Paris (série III), 294, 9.
3. Koshy, R., Maupas, P., Muller, R. and Hofschneider, P.H.(1981) J. Gen. Virol., 57, 95.
4. Alexander, J.J., Bey, E.M., Geddes, E.W. and Lecatsas, G. (1976) S. Afr. med. J., 50, 2124.
5. Marion, P.L., Salazar, F.H., Alexander, J.J. and Robinson, W.S. (1980) J. Virol., 33, 795.
6. Aden, D.P., Fogel, A., Plotkin, S., Damjanov, I. and Knowles, B.B. (1979) Nature (London), 282, 615.
7. Prozesky, O.W., Brits, C. and Grabon, W.O.K. (1973) In : S.J. Saunders and J. Terblanche (eds), Liver, Pitman Medical Publishing Co., London, p. 358.
8. Wen, Y.M., Mitamura, K., Merchant, B., Tang, Z.Y. and Purcell R.H. (1983) Infect. Immun., 39, 1361.
9. Courtneidge, S.A., Levinson, A.D. and Bischop, J.M. (1980) Proc. Natl. Acad. Sci. USA, 77, 3783.
10. Lowry, O.H., Rosebrough, N.J., Farr, A.L., Randall, R.J. (1951) J. biol. Chem., 193, 265.
11. Voller, A., Bidwell, D.E. and Bartlett, A. (1976) In : N.R. Rose and H. Friedman (eds), Manual of Clinical Microbiology, Am. Soc. Microbiol., Washington D.C., p. 506.
12. Weber, K. and Osborn, M. (1975) In : The proteins 1, Academic Press, New-York, p. 180.

© 1984 Elsevier Science Publishers B.V.
The Role of Viruses in Human Cancer. Volume II
G. Giraldo and E. Beth editors

EXPERIMENTAL ONCOGENICITY BY HUMAN PAPOVAVIRUSES AND POSSIBLE CORRELATIONS WITH HUMAN TUMORS

GIUSEPPE BARBANTI-BRODANO, ALFREDO CORALLINI AND MARIA PIA GROSSI

Institute of Microbiology, School of Medicine, University of Ferrara, Via Luigi Borsari, 46, I-44100 Ferrara, Italy

The human papovaviruses BK (BKV) and JC (JCV) are widespread in human populations. The viruses are worldwide in distribution and primary infections generally occur in childhood or adolescence.[1-5] They produce a persistent, latent, most often inapparent infection which is reactivated under conditions of impaired immunological response. Thus, BKV and JCV are rarely isolated from normal individuals, whereas they are found in the urine of organ transplant recipients and other immunosuppressed patients.[6] BKV and JCV persistent infection is also reactivated during pregnancy, although reports on transplacental transmission of the viruses were conflicting.[7-11] The kidney is the site of replication and persistence[12-14] and the urinary tract is the only known route of excretion. BKV is implicated as the etiologic agent of upper respiratory tract infections,[3,15-17] cases of Guillain-Barré syndrome[15] and acute infections of the urinary apparatus.[6,18,19] JCV is associated with progressive multifocal leukoencephalopathy,[20] a rare opportunistic viral infection of the central nervous system, occurring chiefly as a late complication in patients affected by neoplastic diseases of the lymphoid system. Both BKV and JCV induce neoplastic transformation of cells in culture and are oncogenic for laboratory animals. Such properties of human papovaviruses prompted investigations on their possible etiologic role in human tumors.

TRANSFORMATION AND ONCOGENICITY BY BKV

BKV or its complete DNA induced neoplastic transformation of hamster, mouse, rat, rabbit, monkey and human cells in culture.[21-30] In addition, hamster and human cells have been transformed by subgenomic fragments of BKV DNA containing the entire early region, whereas segments of the viral genome covering only the proximal part of the early region were unable to transform cells.[30,31] Transformation of permissive human cells by BKV or complete BKV DNA was difficult to achieve: in some cases transformation was abortive,[27,32] while in other experiments continuous cell lines were obtained.[28-30]

BKV oncogenicity was tested by inoculation of hamsters, mice and rats by different routes.[33-41] The incidence of tumors in hamsters was low after subcutaneous or intraperitoneal inoculation of the virus,[33-35] while a remarkable oncogenic effect was observed in animals inoculated intracerebrally or intravenously.[35-40] Tumor incidence was slightly lower in hamsters immunosuppressed with anti-lymphocyte serum (ALS), methyl-prednisolone acetate or methyl-prednisolone acetate plus gamma-radiation,[40] probably due to the toxic effect of the immunosuppressive treatment. A modest oncogenic activity was shown by BKV DNA, since few tumors were observed only in hamsters inoculated intracerebrally.[40]

A variety of tumors with different histologic appearance was observed in hamsters and several tumors of the same or different histotype often developed in a single animal.[38-42] Nevertheless, three types of tumors were clearly prevalent in hamsters: ependymomas (100% in animals inoculated intracerebrally, 68% in animals inoculated intravenously), tumors of pancreatic islets (16% in animals inoculated intravenously) and osteosarcomas (14% in animals inoculated intravenously). The high frequency of these tumors suggests either a specific tropism of BKV for ependymal, bone and islet tissues or a peculiar susceptibility of these tissues to the viral oncogenic activity. Mice inoculated intracerebrally or intravenously with BKV developed ependymomas and choroid plexus papillomas.[35,42] The latter type of tumor did not appear in hamsters and was observed only in mice. Immunosuppressed hamsters injected intravenously with BKV had a significantly higher incidence of lymphomas (11%),[40] especially after treatment with ALS, as compared to immunocompetent animals (1.5%).[38] It is likely that an efficient immunosuppressive treatment contributed to the induction of lymphomas by diminishing the immunological surveillance and/or by chronic stimulation of the immune system due to proteins of ALS. Moreover, 3 renal carcinomas and 2 carcinomas of the renal pelvis developed in immunosuppressed hamsters inoculated with BKV intravenously.[40] These tumors were not observed in BKV-injected immunocompetent hamsters, suggesting that immunosuppression may have activated the oncogenicity of BKV persisting in the kidneys of inoculated animals.

The morphological features of the most frequent tumors induced in hamsters by BKV inoculated intravenously are illustrated in Figures 1-6. Tumors of pancreatic islets contained secretion granules of four morphological types,[42] in agreement with the hypersecretion of insulin, C-peptide, glucagon and calcitonin found in tumor tissue.[38] Elevated concentrations of insulin and glucagon and altered levels of glucose were detected in blood of tumor-bearing animals.[38,40] Pancreatic carcinomas metastasized frequently to the liver and osteosarcomas to the lungs and peritoneum.[38,42] In general, tumors of all histotypes were more malignant, invasive and metastasizing in immunosuppressed than in immunocompetent animals.

BKV specificity of transformation and tumor induction was demonstrated by the presence of several BKV markers in transformed cells and tumors.[35,38,40] BKV tumor (T) antigen was detected by immunofluorescence and by immunoprecipitation in transformed cells. T antigen was also shown by immunofluorescence in tumor cells cultured in vitro or in tumor imprints and by complement fixation in

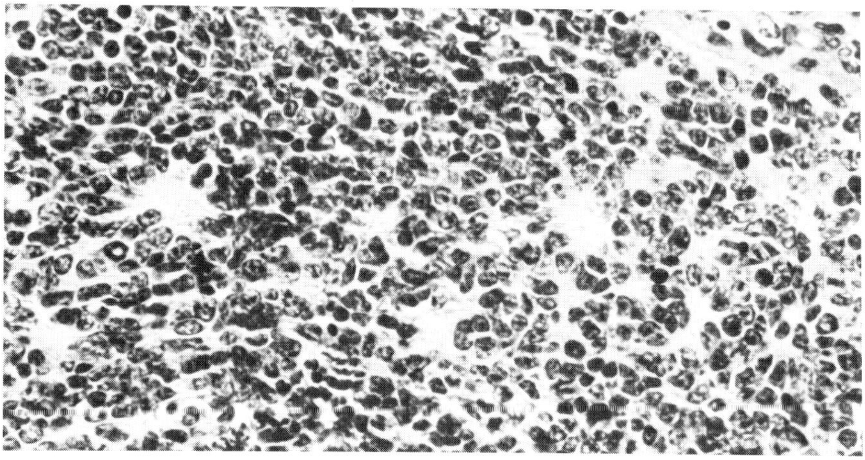

Figure 1. Ependymoma. Note perivascular rosettes, nuclear polymorphism and cellular atypia. (X 400)

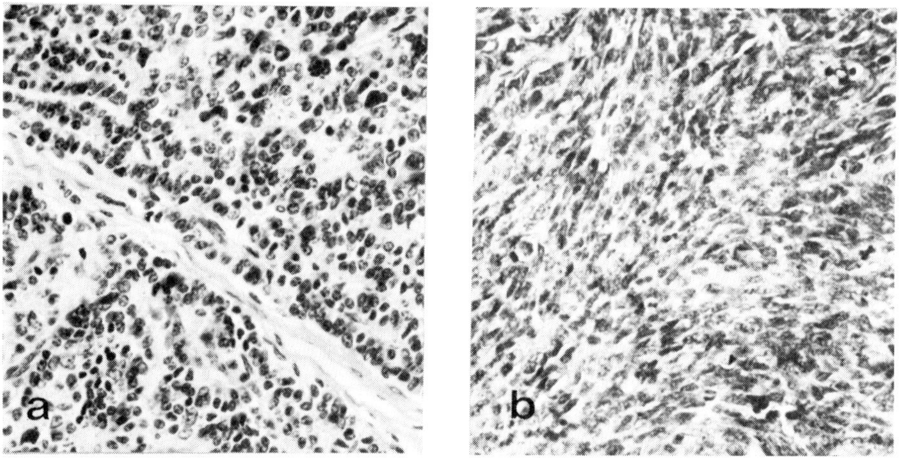

Figure 2 a and b. Tumors of pancreatic islets. Panel a: polygonal, monomorphic cells with modest anaplasia. Panel b: spindle-shaped, polymorphic cells with hyperchromatic nuclei. (Panels a and b, X 250)

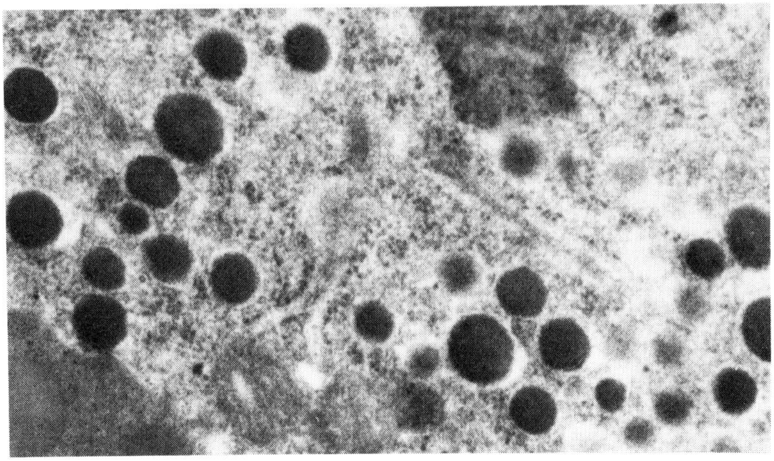

Figure 3. Tumor of pancreatic islets. Secretion granules of type I (130-150 nm) with a dense, sometimes eccentric, core separated from the limiting membrane by a clear halo. (X 19,000)

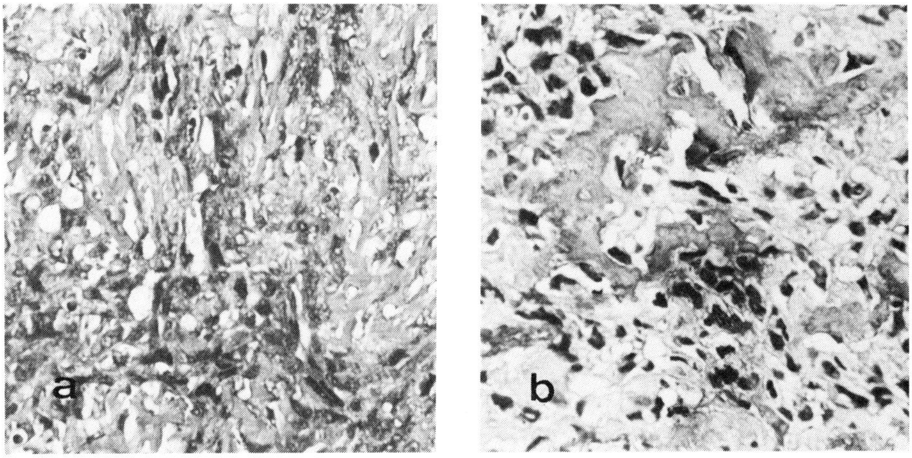

Figure 4 a and b. Osteosarcomas. Panel a: trabeculae of calcified bone tissue. Panel b: hypercellular sarcomatous zones with irregular deposition of osteoid substance. (Panels a and b, X 250)

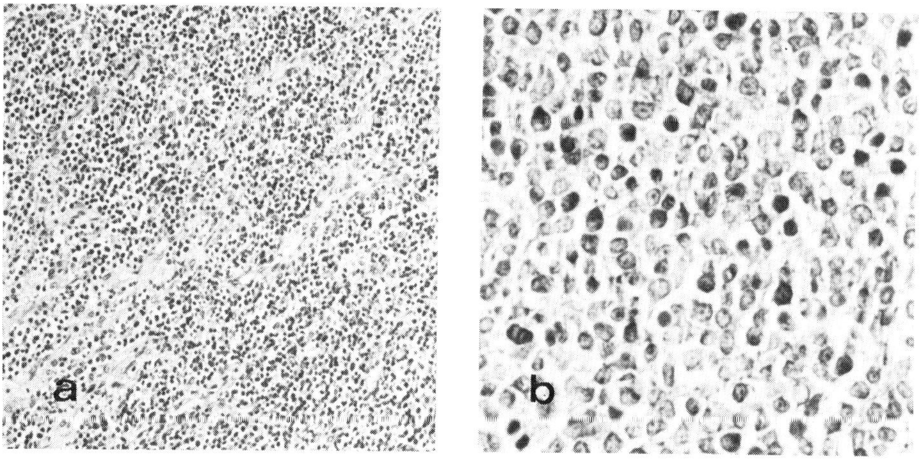

Figure 5 a and b. Lymphomas. Panels a and b: diffuse proliferation of lymphoid cells with scarce vascularization and stroma. (Panel a, X 100; Panel b, X 400)

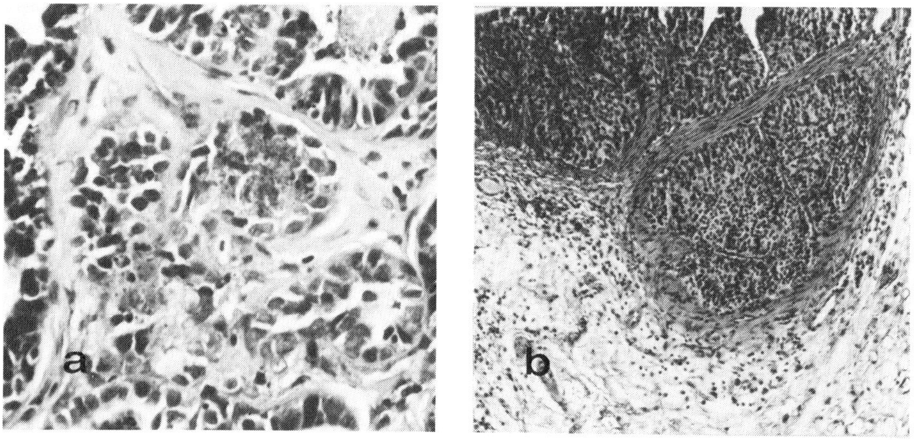

Figure 6 a and b. Panel a: renal carcinoma. Panel b: renal pelvis carcinoma. Both tumors show tubular or papillary structures of large, basophilic cells. (Panel a, X 250; Panel b, X 65)

tumor homogenates. Moreover, sera from tumor-bearing animals had antibodies to BKV T antigen detected by immunofluorescence and/or complement fixation. Infectious BKV was rescued by Sendai virus-mediated fusion of cultured tumor and transformed cells with permissive human or simian cells.

DNA-DNA blot-transfer hybridization was used to analyse the state and arrangement of viral sequences in BKV-induced hamster tumors and BKV-transformed cells. In tumors, viral DNA was present in several copies per cell. Most of the detected viral sequences were integrated into cellular DNA, often in a tandem head-to-tail array of oligomeric viral genomes, but some tumors also contained small amounts of free viral DNA.[40,43] The site of integration was not specific. On the contrary, integration patterns were different from each other and many different integration sites were available both on the cellular and on the viral DNA. Cell lines and single-cell clones derived from tumors had hybridization patterns considerably simpler than that of the original tumor, with lack of several integrated and all free viral genomes, suggesting a possible polyclonal origin of the tumors.[43] BKV-transformed hamster, mouse and rabbit cells gave similar results, with most of the viral sequences integrated into cellular DNA and some free viral genomes that were more numerous in mouse and rabbit than in hamster cells.[31, 44-46] On the other hand, a completely different situation was observed in BKV-transformed human cells. In these cells no integrated viral genomes could be detected and only free viral DNA at a high copy number per cell (from 60 to 550 genome equivalents) was present.[28,30] These results suggest that in human cells transformed by BKV, as in mouse cells transformed and in animal tumors induced by bovine papilloma virus,[47] the transformed phenotype is maintained by a resident viral genome in a free, episomal state.

TRANSFORMATION AND ONCOGENICITY BY JCV

JCV transformed hamster as well as human amnion and vascular endothelial cells in culture[48-50] and induced a variety of tumors with high frequency in newborn hamsters inoculated intracerebrally.[51] Medulloblastomas of the cerebellum were the most frequent tumors, followed by ependymomas, neuroblastomas, astrocytomas, glioblastomas and retinoblastomas.[51,52] Astrocytomas were also induced by intracerebral inoculation of owl and squirrel monkeys with JCV.[53,54] JCV T antigen was detected in hamster tumors and in one of four owl monkey tumor cell lines. Tumor-bearing animals developed antibodies to JCV T antigen and JCV DNA was found integrated into cellular DNA of owl monkey brain tumors.

POSSIBLE ROLE OF HUMAN PAPOVAVIRUSES IN HUMAN ONCOGENESIS

Studies on the possible association of human papovaviruses with human tumors yielded conflicting results. In an early serological investigation, 952 sera from tumor patients were analysed for the presence of antibodies to BKV T antigen. A small percentage (1.15%) of sera from tumor patients contained T antibodies, but sera from normal persons (0.8%) were also positive.[55] The difference is not significant and this result may depend on production of T antibodies during a lytic infection by BKV in tumor patients and normal people. In a second study by the same authors[56] no antibodies to BKV T antigen were detected in sera and cerebro-spinal fluids from 142 patients with brain and urinary tract tumors. Similar results were reported by Costa et al.[57] who detected antibodies to BKV T antigen by indirect immunofluorescence at a low serum dilution (1:2) in 12 of 113 sera from tumor patients, but the sera positive by indirect immunofluorescence gave negative results when tested by anticomplement immunofluorescence at higher dilutions.

Another approach to the relationship of papovaviruses with human tumors was the search for T antigen in tumor tissues. Tabuchi et al.[58] analysed 39 human brain tumors and reported the presence of T antigen in two tumors of ependymal origin (a malignant ependymoma and a choroid plexus papilloma) using immunoperoxidase methods and a serum to SV40 T antigen. Since SV40 T antigen strongly cross-reacts with BKV and JCV T antigens, this serum would have detected also T antigen of human papovaviruses. On the contrary, a series of 80 human brain tumors, comprising 6 ependymomas, was found to be negative for papovavirus T antigen by immunofluorescence.[59] Likewise, Shah et al.[60] did not detect papovavirus T antigen by immunofluorescence in 123 tumors of the urogenital tract. Similar results were reported by Corallini et al.[55] in a series of 17 meningiomas and by Grossi et al.[56] in 142 brain and urinary tract tumors.

DNA-DNA reassociation kinetics and DNA-DNA blot-transfer hybridization were employed to detect BKV DNA in human tumors. Fiori and Di Mayorca[61] found BKV DNA sequences in 5 of 12 human tumors and 3 of 4 human tumor cell lines. These authors recently confirmed their results,[62] whereas three other reports[56,63,64] failed to support these findings.

While these studies had analysed a wide range of the most common types of human tumors, a reasonable approach was to focus attention on those rare types of human tumors most frequently induced by BKV in experimental animals, that is ependymomas, tumors of pancreatic islets and osteosarcomas. Indeed, BKV DNA was recently detected in a human pancreatic insulinoma.[65] BKV DNA was free in tumor cells and no evidence was found of viral sequences integrated into cellular DNA. On the other hand, the amount of viral DNA was low (0.2 genome equivalents per cell) arguing against a lytic or persistent infection. A virus was rescued by transfection of human embryonic fibroblasts with tumor DNA. This virus (BKV-IR) had biological and antigenic characteristics identical to BKV, but its DNA was different from that of wild type BKV by restriction endonuclease mapping. The genome of BKV-IR is 235 base pairs shorter than the genome of wild type BKV. This alteration originates from a deletion of approximately 300 base pairs involving HindIII fragments B and D, and an insertion of 70 base pairs in the region of HindIII fragment C (Figure 7).

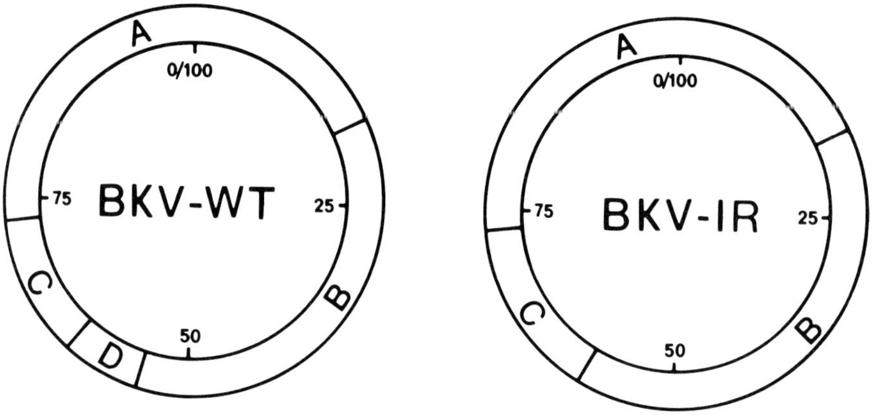

Figure 7. Genome structure of BKV wild type (BKV-WT) and BKV-IR digested with the restriction endonuclease HindIII. Numbers within the first circle represent map units. The single cut by the restriction endonuclease EcoRI is taken as 0/100 map units. A, B, C and D label the four HindIII fragments. BKV-IR has only three HindIII fragments because a deletion at 54 map units eliminated the HindIII site between fragments B and D. In addition, HindIII fragment C is larger in BKV-IR than in BKV-WT, due to an insertion of 70 base pairs.

The genome organization of BKV-IR closely resembles that of MMV, another BKV variant isolated from a brain reticulum cell sarcoma.[66] In view of these findings, one could speculate that the peculiar genomic structure of BKV-IR and MMV allows these viruses to display their oncogenic potential for human cells in vivo better than wild type BKV. Transformation of hamster kidney cells was obtained with total tumor DNA as well as with BKV-IR and BKV-IR DNA.

BKV-IR induced malignant ependymomas after intracerebral inoculation of newborn hamsters. Antibodies to BKV T antigen were not detected in the patient's serum by immunofluorescence. However, two bands, corresponding to the 94,000 daltons large T antigen and to the 56,000 daltons T antigen-associated cellular protein, were observed in immunoprecipitates of BKV-transformed hamster cell homogenate with the patient's serum. Although more of these tumors must be analysed to draw any conclusions, it is suggestive that free, episomal BKV DNA, which had been shown to maintain transformation in cultured human cells,[28,30] was found in a human tumor of pancreatic islets, corresponding histologically to one of the most frequent types of tumors induced by BKV in laboratory animals.[35-42]

The correlation of JCV with human neoplasia has not yet been studied. Nevertheless, it has been noted that the majority of brain tumor types induced by JCV in hamsters have their counterpart in human tumors of childhood[52] which therefore would be good candidates for a possible JCV etiology. Moreover, JCV oncogenicity in laboratory animals yielded two experimental models, medulloblastoma in hamsters[51,52] and astrocytoma in owl and squirrel monkeys,[53,54] which should be useful to study existing and new diagnostic techniques or to evaluate different therapeutic treatments.

CONCLUSION

Much work was performed in the past years on the experimental oncogenicity of human papovaviruses. The outcome of such work has been an extensive effort devoted to study the possible involvement of human papovaviruses in human oncogenesis. In spite of contrasting data, it is reasonable to conclude that BKV does not seem to be related to the most common types of human tumors. However, the recent finding of BKV DNA in a human tumor of pancreatic islets suggests it would be advisable to persist in testing those rare types of human tumors that are induced by BKV with high frequency in experimental animals. As to JCV, we can be confident that reports on its possible correlation with human tumors will certainly appear in the near future.

ACKNOWLEDGMENTS

Authors' work described in this article was supported by Consiglio Nazionale delle Ricerche "Progetto Finalizzato Virus" grants 78.00338.84, 79.00367.84, 80.00577.84 and "Progetto Finalizzato Controllo della Crescita Neoplastica" grant 82.00224.96. The work was also supported in part by the NATO Research Grant 284.81.

REFERENCES

1. Gardner, S.D.: Br. Med. J. 1:77, 1973.
2. Shah, K.V., Daniel, R.W., and Warszawski, R.M.: J. Infect. Dis. 128:284, 1973.
3. Mantyjarvi, R.A., Meurman, O.H., Vihma, L., and Berglund, B.: Ann. Clin. Res. 5:283, 1973.
4. Portolani, M., Marzocchi, A., Barbanti-Brodano, G., and La Placa, M.: J. Med. Microbiol. 7:543, 1974.
5. Brown, P., Tsai, T., and Gajdusek, D.C.: Am. J. Epidemiol. 102: 331, 1975.
6. Coleman, D.V.: In: Recent Advances in Clinical Virology, A.P. Waterson (ed.), Churchill Livingstone, London, 1980, p. 89.
7. Taguchi, F., Nagaki, D., Saito, M., Haruyama, C., Iwasaki, K., and Suzuki, T.: Japan. J. Microbiol. 19:395, 1975.
8. Rziha, H.J., Bornkamm, G.W., and zur Hausen, H.: Med. Microbiol. Immunol. 165:73, 1978.
9. Borgatti, M., Costanzo, F., Portolani, M., Vullo, C., Osti, L., Masi, M., and Barbanti-Brodano, G.: Microbiologica 2:173, 1979.
10. Shah, K.V., Daniel, R., Madden, D., and Stagno, S.: Infect. Immun. 30:29, 1980.
11. Coleman, D.V., Wolfendale, M.R., Daniel, R.A., Dhanjal, N.K., Gardner, S.D., Gibson, P.E., and Field, A.M.: J. Infect. Dis. 142:1, 1980.
12. Heritage, J., Chesters, P.M., and McCance, D.J.: J. Med. Virol. 8:143, 1981.

13. McCance, D.J.: In: Polyomaviruses and Human Neurological Disease, J.L. Sever and D.L. Madden (eds.), Alan R. Liss, New York, 1983, p. 343.

14. Chesters, P.M., Heritage, J., and McCance, D.J.: J. Infect. Dis. 147:676, 1983.

15. van der Noordaa, J. and Wertheim-van Dillen, P.: Br. Med. J. 1:1471, 1977.

16. Goudsmit,J., Baak, M.L., Slaterus, K.W., and van der Noordaa, J.: Br. Med. J. 283:1363, 1981.

17. Goudsmit, J., Wertheim-van Dillen, P., van Strien, A., and van der Noordaa, J.: J. Med. Virol. 10:91, 1982.

18. Padgett, B.L., Walker, D.L., Desquitado, M.M., and Kim, D.V.: Lancet 1:770, 1983.

19. Rosen, S., Harmon, W., Krensky, A.M., Edelson, P.J., Padgett, B.L., Grinnell, B.W., Rubino, M.J., and Walker, D.L.: N. Engl. J. Med. 308:1192, 1983.

20. Zu Rhein, G.M.: Progr. Med. Virol. 11:185, 1969.

21. Major, E.O. and Di Mayorca, G.: Proc. Natl. Acad. Sci. USA 70: 3210, 1973.

22. Portolani, M., Barbanti-Brodano, G., and La Placa, M.: J. Virol. 15:420, 1975.

23. van der Noordaa, J.: J. Gen. Virol. 30:371, 1976.

24. Tanaka, R., Koprowski, H., and Iwasaki, Y.: J. Natl. Cancer. Inst. 56:671, 1976.

25. Takemoto, K.K. and Martin, M.A.: J. Virol. 17:247, 1976.

26. Mason, D.H. and Takemoto, K.K.: Int. J. Cancer 19:391, 1977.

27. Portolani, M., Borgatti, M., Corallini, A., Cassai E., Grossi, M.P., Barbanti-Brodano, G., and Possati, L.: J. Gen. Virol. 38: 369, 1978.

28. Purchio, A.F. and Fareed, G.C.: J. Virol. 29:763, 1979.

29. Takemoto, K.K., Linke, H., Miyamura, T., and Fareed, G.C.: J. Virol. 29:177, 1979.

30. Grossi, M.P., Caputo, A., Meneguzzi, G., Corallini, A., Carrà, L., Portolani, M., Borgatti, M., Milanesi, G., and Barbanti-Brodano, G.: J. Gen. Virol. 63: 393, 1982.

31. Grossi, M.P., Corallini, A., Valieri, A., Balboni, P.G., Poli, F., Caputo, A., Milanesi, G., and Barbanti-Brodano, G.: J. Virol. 41:319, 1982.

32. Shah, K.V., Hudson, C., Valis, J., and Strandberg, J.D.: Proc. Soc. Exp. Biol. Med. 153:180, 1976.

33. Nase, L.M., Karkkaiven, M., and Mantyjarvi, R.A.: Acta Pathol. Microbiol. Scand. 83:347, 1975.

34. Shah, K.V., Daniel, R.W., and Strandberg, J.: J. Natl. Cancer Inst. 54:945, 1975.

35. Corallini, A., Barbanti-Brodano, G., Bortoloni, W., Nenci, I., Cassai, E., Tampieri, M., Portolani, M., and Borgatti, M.: J. Natl. Cancer, Inst. 59:1561, 1977.

36. Costa, T., Yee, C., Tralka, T.S., and Rabson, A.S.: J. Natl. Cancer Inst. 56:863, 1976.

37. Uchida, S., Watanabe, S., Aizawa, T., Kato, K., Furuno, A., and Muto, T.: Gann 67:857, 1976.

38. Corallini, A., Altavilla, G., Cecchetti, M.G., Fabris, G., Grossi, M.P., Balboni, P.G., Lanza, G., and Barbanti- Brodano, G.: J. Natl. Cancer Inst. 61:875, 1978.

39. Uchida, S., Watanabe, S., Aizawa, T., Furuno, A., and Muto, T.: J. Natl. Cancer. Inst. 63:119, 1979.

40. Corallini, A., Altavilla, G., Carrà, L., Grossi, M.P., Federspil, G., Caputo, A., Negrini, M., and Barbanti-Brodano, G.: Arch. Virol. 73:243, 1982.

41. Noss, G., Stauch, G., Mehraein, P., and Georgii, A.: Arch. Virol. 69:239, 1981.

42. Altavilla, G., Carrà, L., Alberti, S., Corallini, A., Cavazzini, L., Fabris, G., Aleotti, A., and Barbanti-Brodano, G.: Oncology 1983 (in press.)

43. Chenciner, N., Meneguzzi, G., Corallini, A., Grossi, M.P., Grassi, P., Barbanti-Brodano, G., and Milanesi, G.: Proc. Natl. Acad. Sci. USA 77:975, 1980.

44. Chenciner, N., Grossi, M.P., Meneguzzi, G., Corallini, A., Manservigi, R., Barbanti-Brodano, G., and Milanesi, G.: Virology 103:138, 1980.

45. Meneguzzi, G., Chenciner, N., Corallini, A., Grossi, M.P., Barbanti-Brodano, G., and Milanesi, G.: Virology 111:139, 1981.

46. Grossi, M.P., Corallini, A., Meneguzzi, G., Chenciner, N., Barbanti-Brodano, G., and Milanesi, G.: Virology 120:500, 1982.

47. Law, M.F., Lowy, D.R., Dvoretzky, I., and Howley, P.M.: Proc. Natl. Acad. Sci. USA 78:2727, 1981.

48. Frisque, R.J., Rifkin, D.B., and Walker, D.L.: J. Virol. 35: 265, 1980.

49. Howley, P.M., Rentier-Delrue, F., Heilman, C.A., Law, M.F., Chowdhury, K., Israel, M.A., and Takemoto, K.K.: J. Virol. 36: 878, 1980.

50. Fareed, G.C., Takemoto, K.K., and Gimbrone, M.: In: Microbiology 1978, D. Schlessinger (ed.), American Society for Microbiology, Washington, D.C., 1978, p. 427.

51. Walker, D.L., Padgett, B.L., Zu Rhein, G.M., Albert, A.E., and Marsh, R.F.: Science 181:674, 1973.

52. Zu Rhein, G.M.: In: Polyomaviruses and Human Neurological Disease, J.L. Sever and D.L. Madden (eds.), Alan R. Liss, New York, 1983, p. 205.

53. London, W.T., Houff, S.A., Madden, D.L., Fuccillo, D.A., Gravell, M., Wallen, W.C., Palmer, A.E., Sever, J.C., Padgett, B.L., Walker, D.L., Zu Rhein, G.M., and Ohashi, T.: Science 201:1246, 1978.

54. Houff, S.A., London, W.T., Zu Rhein, G.M., Padgett, B.L., Walker, D.L., and Sever, J.L.: In: Polyomaviruses and Human Neurological Disease, J.L., Sever, and D.L. Madden (eds.), Alan R. Liss, New York, 1983, p. 223.

55. Corallini, A., Barbanti-Brodano, G., Portolani, M., Balboni, P.G., and Grossi, M.P.: Infect. Immun. 13:1684, 1976.

56. Grossi, M.P., Meneguzzi, G., Chenciner, N., Corallini, A., Poli, F., Altavilla, G., Alberti, S., Milanesi, G., and Barbanti-Brodano, G.: Intervirology 15:10, 1981.

57. Costa, J., Yee, C., and Rabson, A.S.: Lancet 2:709, 1977.

58. Tabuchi, K., Kirsch, W.M., Law, M., Gaskin, D., van Buskirk, J., and Maa, S.: Int. J. Cancer 21:12, 1978.

59. Greenlee, J.E., Becker, L.E. Narayan, O., and Johnson, R.T.: Ann. Neurol. 3:479, 1978.

60. Shah, K.V., Daniel, R.W., Stone, K.R., and Elliot, A.Y.: J. Natl. Cancer, Inst. 60:579, 1978.

61. Fiori, M. and Di Mayorca, G.: Proc. Natl. Acad. Sci. USA 73: 4662, 1976.

62. Pater, M.M., Pater, A., Fiori, M., Slota, J., and Di Mayorca, G.: In: Viruses in Naturally Occurring Cancers, M. Essex, G. Todaro and H. zur Hausen (eds.), Cold Spring Harbor Laboratory, Cold Spring Harbor, New York, 1980, p. 329.

63. Israel, M.A., Martin, M.A., Takemoto, K.K., Howley, P.M., Aaronson, S.A., Solomon, D., and Khoury, G.: Virology 90:187, 1978.

64. Wold, W.S.M., Mackey, J.K., Brackman, K.H., Takemori, N., Rigden, P., and Green, M.: Proc. Natl. Acad. Sci. USA 75:454, 1978.

65. Caputo, A. Corallini, A., Grossi, M.P., Carrà, L., Balboni, P.G., Negrini, M., Milanesi, G., Federspil, G., and Barbanti-Brodano, G.: J. Med. Virol. 12:37, 1983.

66. Takemoto, K.K., Rabson, A.S., Mullarkey, M.F., Blaese, R.M., Garon, C.F., and Nelson, D.: J. Natl. Cancer, Inst. 53:1205, 1974.

The Role of Viruses in Human Cancer. Volume II
G. Giraldo and E. Beth editors

SV40 IN HUMAN BRAIN TUMORS: RISK FACTOR OR PASSENGER ?

ERHARD GEISSLER[1], SIEGFRIED SCHERNECK[1], HELMUT PROKOPH[1], WOLFGANG ZIMMERMANN[1] AND WOLFHARD STANECZEK[2]
Akademie der Wissenschaften der DDR, [1]Zentralinstitut für Molekularbiologie, Abteilung Virologie, Robert-Rössle-Str. 10, DDR-1115 Berlin (German Democratic Republic) and [2]Zentralinstitut für Krebsforschung, Bereich Nationales Krebsregister, DDR-1197 Berlin Sterndamm 12 (German Democratic Republic)

INTRODUCTION

Conflicting results regarding the possible involvement of SV40 in the induction of human cancer have been published by several groups and have recently been summarized by Takemoto (1). The most widely accepted view so far is that SV40 "apparently ... is not oncogenic in humans" (2). During the last years, however, at least in Sauer's laboratory (3) and ours (4-6) some evidence has been obtained indicating a possible involvement of SV40 as a risk factor in the induction and/or development of human brain tumors and other intracranial malignancies. Here we summarize our data including some recent results.

FOOTPRINTS OF SV40-LIKE VIRUSES IN HUMAN INTRACRANIAL TUMORS

One hundred brain tumors and other intracranial malignancies were studied for the presence of viral antigens and/or of viral DNA.

Presence of SV40-like T-antigen

68 of these tumors were tested for the presence of SV40-like T-antigen by indirect immunofluorescence using standard methods (7). The results are summarized in Table 1: eighteen tumors reacted positively.

These results are in good agreement with data published by Weiss et al. (8) and by Zang et al. (9) who found SV40-like T-antigen in 2 out of 7 and 11 out of 27 meningiomas, respectively.

SV40-like T-antigen could not be detected in normal brain tissue of 54 patients whose intracranial tumors were tested. However, in one case in which the control tissue was taken from an area in very close proximity to the tumor, a meningioma, T-antigen was detected.

TABLE 1: FOOTPRINTS OF SV40-LIKE VIRUSES IN INTRACRANIAL TUMORS

Tumor type	No. Tumors studied	T-antigen	SV40-like DNA Spot test	SV40-like DNA Southern blot	Remarks
Meningiomas	6		+	+	Plasmids
	1	-	+		DNA but no T!
	15	+			1 patient 6 year old
	3	-	-		
	10		-	-	
	32	-			
Glioblastomas	1	+			Rescue of GBM
	2	+	+		Plasmids
	1	-	-		
	3		-	-	
	1		-		
	7	-			
Astrocytomas	2		+	+	Plasmids
	1	-	-	-	
	5		-	-	1 patient 14 year old
	1	-			
Oligodendrogliomas	1		+	+	Plasmid
	2		-	-	
other i. c. tumors	6	-	-	-	each 1 pt. 11 & 15 y.old
Total	100	18/68	12/41	9/32	

Although the presence of SV40-like T-antigen seems to be highly specific these results are not completely unambiguous because monoclonal antibodies reacting exclusively with SV40 T-antigen could not be used. Therefore the material detected by indirect immunofluorescence might be either SV40 T-antigen or T-antigen coded for by one of the human papovaviruses, BKV or JCV, and/or the 55k protein often found in cells transformed by viruses or chemicals or even spontaneously. The 55k protein forms complexes with (SV40-)T-antigen and is therefore thought to react with antibodies directed against SV40 induced hamster tumors. 55k can be detected even in SV40 infected F9 teratocarcinoma mouse cells expressing no SV40 T-antigen (10).

We looked therefore for other footprints of SV40-like viruses.

Presence of viral DNA in tumor cells

In order to detect viral DNA by hybridization of DNA extracted from tumor or control tissues with ^{32}P labeled SV40 DNA, two methods were used (H. Prokoph et al., in preparation).

The spot hybridization technique originally developed by Brandsma and Miller (11) to detect Epstein-Barr Virus DNA is an easy and convenient method to screen for viral DNA sequences. The method is highly sensitive and allows the demonstration of a few picograms of DNA, as illustrated in Figure 2. In Table 1 the results of spot testing 41 tumors are summarized: 12 tumors gave positive reactions some of which are shown in Figure 2. As the spot hybridization was done under stringent conditions at least 85% of the sequences detected are homologous to the ^{32}P labeled SV40 DNA. A hybridization with BKV DNA cannot be excluded, however, even under these conditions.

By Southern blotting a reaffirmation of the results obtained with the spot test was possible as well as a characterization of the viral DNA found. Using DNA extracted from the cells of 32 tumors and cut with restriction endonuclease Eco RI, a positive reaction was obtained in 9 cases, as demonstrated in Figure 1 and summarized in Table 1. The position of the hybridizing sequences in the electropherograms corresponded in each case with the position of linear SV40 DNA. This suggests that most if not all SV40-like DNA does exist in these tumors in an extrachromosomal plasmid-like state. These data correspond closely to the results obtained by Sauer et al. (3) both with respect to the fraction of

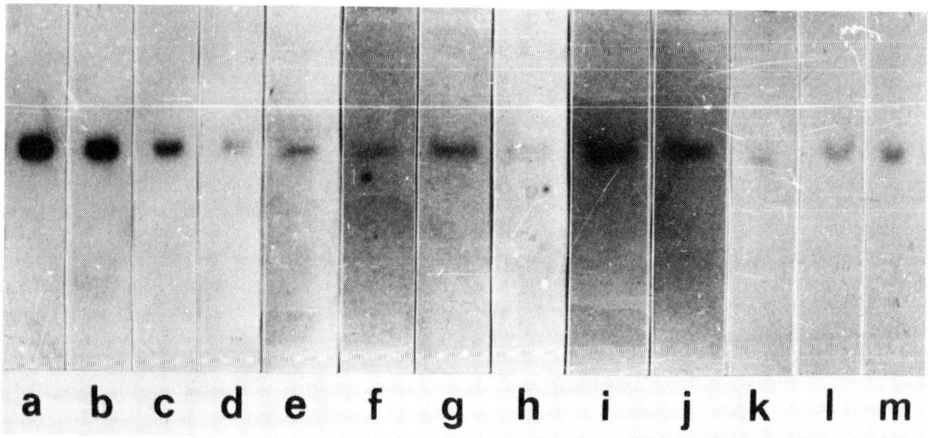

Fig. 1. Compilation of SV40-like DNA sequences detected by Southern blotting. Lanes a - d: SV40 form-III DNA (a: 20 pg, b: 10 pg, c: 5 pg, d: 1 pg); Meningiomas: ME3 (e), ME8 (f), ME11 (g), ME13 (h), ME14 (i), ME20 (m); Oligodendroglioma OG1 (j); Astrocytoma AZ6 (k)

positive tumors (about 1/3) and to the extrachromosomal location of the SV40-like DNA.

In order to exclude the possibility that BKV sequences were detected by spot hybridization and by Southern blotting, the DNAs of five positive tumors were cut with two other restriction endonucleases, BamHI and KpnI. The DNAs of SV40 and of BKV differ markedly in the distribution of their recognition sites for these two enzymes thus giving rise to totally different cleavage patterns. Four of the tumor DNAs were cut into fragments corresponding to fragments obtained after cleavage of SV40 DNA with the very same two enzymes. These results suggest that at least these four tumors contain SV40 DNA.

The cleavage pattern of the fifth tumor DNA could not be analyzed because the resulting bands were of too low density.

Among the 12 tumors containing SV40-like DNA, 11 also expressed SV40-like T-antigen while one viral DNA containing meningioma did not express T-antigen. These results differ from those obtained in Sauer's laboratory where viral DNA containing intracranial tumors did not express T-antigen at all (3).

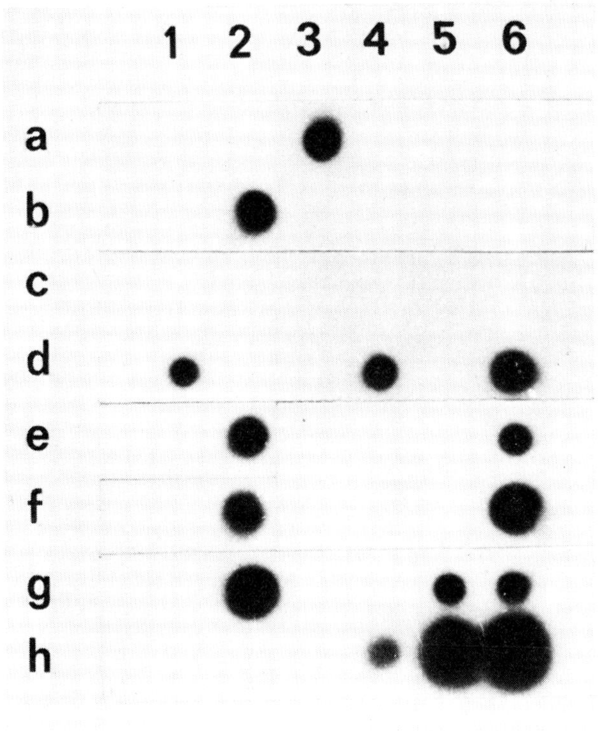

Fig. 2. Compilation of autoradiographs with SV40 positive tumor DNA spots. Meningiomas: ME13 (a3), ME14 (b2), ME8 (d1), ME11 (d6), ME3 (e2), ME20 (f2), ME5 (g6); Oligodendroglioma OG1 (d4); Astrocytomas: AZ6 (e6), AZ8 (f6); Glioblastomas: GB3 (g2), GB4 (g5); Controls: Salmon sperm DNA (h1); SV40 DNA 2pg (h4), 10 pg (h5), 20 pg (h6). All other positions: tumor DNAs without demonstrable SV40-like sequences.

Among a further 6 tumors in which no viral DNA could be detected, none was found to express T-antigen.

SV40 DNA could not be detected by Southern blotting in normal brain tissue of 4 patients, whose tumors contained SV40 DNA (1 astrocytoma, 3 meningiomas), or in normal brain tissue of 2 patients with SV40-free tumors. In another 14 patients without intracranial tumors no SV40 DNA could be detected by this method in brain tissue.

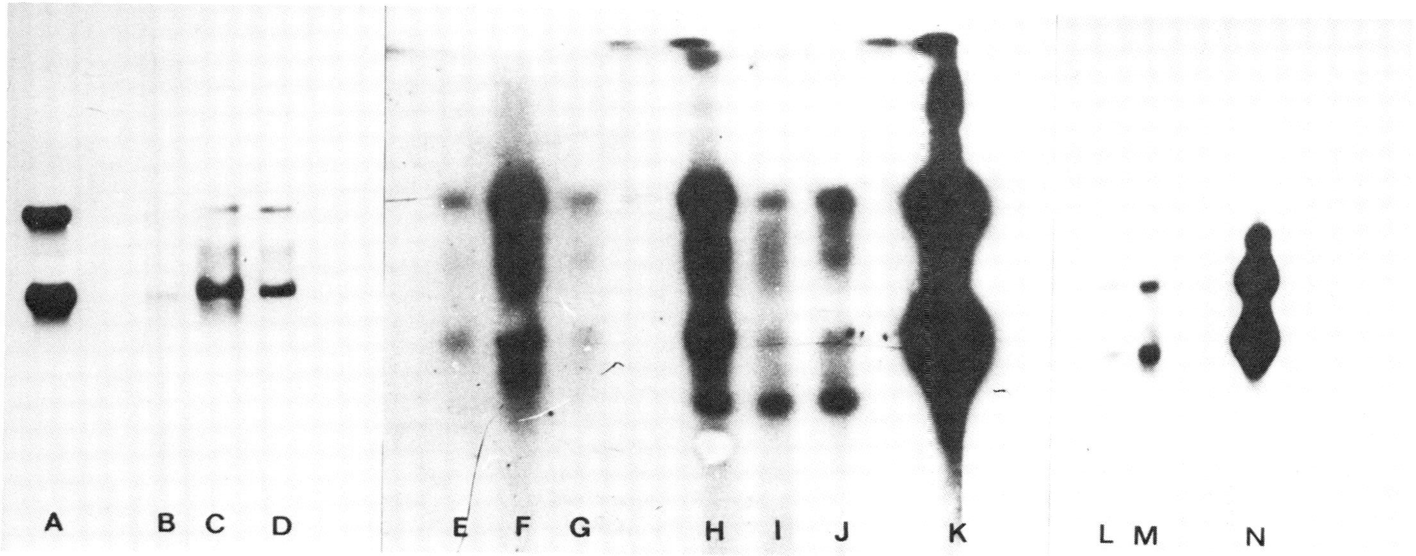

Fig. 3. Autoradiographies of electrophoretic separated Hirt supernatants of human cell lines 10 days after infection with SV40 wildtype, SV40 GBM and SV40 DAR virions, respectively. The gels were blotted on to a nitrocellulose filter and hybridized under stringent conditions (Tm-28°C) with ^{32}P labeled uncloned SV40 WT DNA. In control experiments using either ^{32}P labeled GBM or DAR DNA the same results were obtained. Lanes A,K,N: SV40 WT DNA, lanes B-D: SV40 (B), GBM (C) and DAR virus (D) DNA replication in cells obtained from a human mammary tumor. Lanes E-G: SV40 (E), GBM (F) and DAR virus (G) DNA replication in human amnion cells. Lanes H-J: SV40 (H), GBM (I) and DAR virus (J) DNA replication in A172 cells. Lanes L-M: SV40 (L) and GBM (M) DNA replication in cells a isolated from a human rhabdomyosarcoma.

Rescue of an SV40-like virus, SV40 GBM

As already described an SV40-like virus designated SV40 GBM could be rescued by fusion of cells of a glioblastoma multiforme expressing SV40-like T-antigen with permissive CVI monkey cells (12). This tumor must be regarded therefore as No. 13 of the SV40-like DNA-containing malignancies.

SV40 GBM is similar to SV40 wildtype (WT) (5) with respect to morphology, size, DNA characteristics including physical map and antigenicity. Like SV40 WT it does not agglutinate human 0 erythrocytes. Thus, it is different from the human papovaviruses, BKV and JCV.

SV40 GBM differs however from SV40 WT in several respects:

i) Being highly unstable it gives rise to two major fractions of DNA molecules already after only three passages at low multiplicities (5). Both DNA fractions have been cloned (13) and characterized by electron microscopy, cleavage with restriction endonucleases, and heteroduplex analysis (14). One of the DNA fractions, GBM-H, is indistinguishable from the DNA of the original GBM isolate. The molecules of the smaller variant, GBM-L, are about 19% shorter and have lost the EcoRI, HpaII and PvuII cleavage sites which are located in the late region of genome of SV40.

ii) It enhances the transforming activity of SV40 in mixed infections of primary Chinese hamster fibroblasts four to five-fold (4).

iii) It replicates better than SV40 and SV40 DAR in human amnion cells, as well as in cells isolated from a human mammary tumor and a human rhabdomyosarcoma (Figure 3). Similar results have been obtained earlier with primary human fibroblasts and with primary glial cells (5). Only the human glioblastoma cell line A172 established by O'Neil (15) supports productive infection by SV40 much better than by SV40 GBM and SV40 DAR, respectively. Similar results were obtained when the abilities to form plaques and to synthesize T-antigen in the different human cell lines were compared.

With the methods available to us so far it was not possible to determine whether SV40 GBM is already present in the glioblastoma multiforme cells from which it was rescued or whether this derivative of SV40 evolved during the processes of cell fusion leading to its rescue. The latter speculation is supported by our observation that the original tumor cells neither expressed V antigen nor did they produce virus particles. In contrast SV40 GBM does replicate in human cells. Moreover, Botchan et al. (16) could demonstrate that rescue of SV40 does lead to rearrangements of most of the viral chromosomes involved.

Age and sex distribution

The sex distribution of patients with intracranial tumors containing SV40-footprints (15 ♀, 13 ♂) does not differ from the sex distribution of patients with SV40-free tumors (39 ♀, 33 ♂). The mean age of persons in whose tumors footprints could be found was 52 compared to 46.6 in patients without footprints ($0.05<p<0.1$). The mean age of meningioma patients was 53.2 compared to 47.1 with a p of approximately 0.05. The meaning of this apparent increase in mean age in patients with SV40-like footprints is at present not clear.

IS THERE A LINK BETWEEN THE FOOTPRINTS OF SV40 AND POLIO VACCINE?

At first sight one could speculate on a possible relationship between usage of vaccines contaminated with SV40 - one of which was even in use in 1980 (!) (17) - and the demonstration of SV40 and its footprints among human beings.

Hundreds of millions of people between 1955 and 1962 were treated with polio vaccine inadvertently containing SV40. The vaccine was prepared using poliovirus grown on rhesus kidney cell cultures in which SV40 persists latently without causing any symptoms (18).

In the GDR vaccination against polio was begun in 1960 using Sabin-Tschumakov live vaccine (4). More than 86% of the 885,783 children born between 1959 and 1961 as well as an unknown number of older people were treated with vaccine that presumably contained SV40 in various degrees. The use of SV40-free vaccine as recommended by WHO started at the beginning of 1963.

Since all cancer cases are centrally registered in the GDR we can use the data of our National Cancer Registry to compare the incidence of cancer in children born between 1959 and 1961 (who were treated with vaccine presumably containing SV40 during their first year of life) with the cancer morbidity in the 893,033 children born between 1962 and 1964 who received SV40 free vacci-

TABLE 2

INCIDENCE OF MALIGNANCIES IN CHILDREN TREATED WITH POLIOVACCINE PRESUMABLY CONTAMINATED WITH SV40 OR FREE OF SV40

Type of tumors	1959 - 1961 born (885,783)[a] treated with vaccine SV40-contaminated	1962 - 1964 born (893,033)[a] treated with vaccine not contaminated
Astrocytomas	44	49
Ependymomas	33	51
Gliomas & Glioblastomas	39	29
Medulloblastomas	71	57
Meningiomas	7	11
Neurinomas	16	19
Oligodendrogliomas	8	6
Plexuspapillomas	5	4
Retinoblastomas	32	38
Spongioblastomas	84	69
Other intracranial tumors	16	30
Hemoblastoses	515	555
Melanomas	20	21
Sarcomas	292	291
Wilms tumors	60	59
Other malignancies	285	332
Preneoplasias	46	43
Total	1,573	1,664
Rate per 10,000	17.8	18.6

[a]Number of newborn

ne. Preliminary data concerning the incidence of cancer in these groups have already been published (4-6). Table 2 presents the most recent figures representing 16 years of follow-up after vaccination.

The data demonstrate that persons treated with polio vaccine presumably contaminated with SV40 did not develop more tumors within 16 years after vaccination than did those who had received SV40-free vaccine. These results correspond completely with data recently published by Mortimer et al. (21).

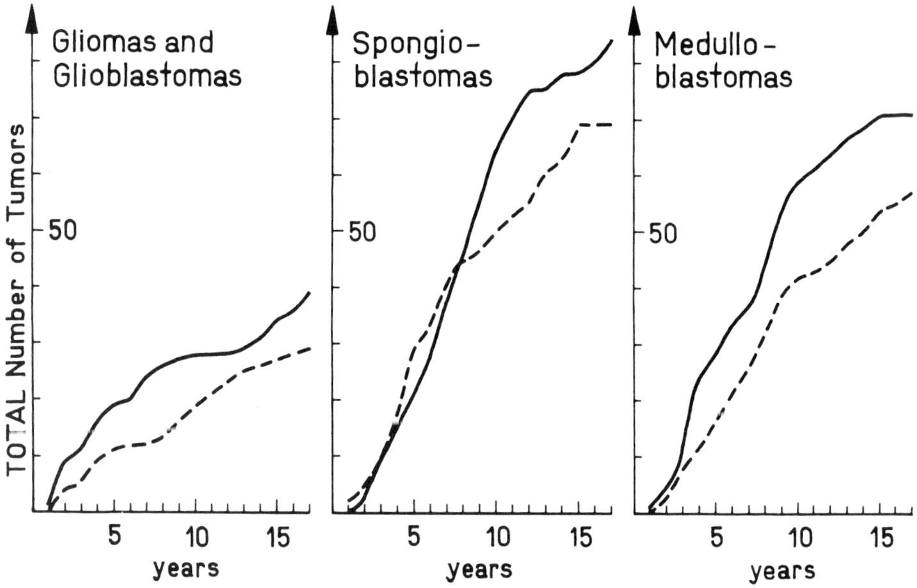

Fig. 4. Occurrence of gliomas and glioblastomas, spongioblastomas, and medulloblastomas in persons vaccinated in their first year of life with polio vaccine presumably containing SV40 (——) or free of SV40 (- - -).

Children born in 1962 and 1963 represent a mixed population, however, since it is possible that their mothers were treated with contaminated vaccine during pregnancy. This is suggested by the results obtained by Heinonen et al. (20, 21) demonstrating more tumors among 4 year-old children of mothers treated with killed polio vaccine during pregnancy (14 cases in 18,342 children) in comparison to children whose mothers were not vaccinated (10 of 32,555). Similarly, Farwell et al. (22) observed that children whose mothers had received polio vaccine contaminated with SV40 developed more medulloblastomas than a comparison group. In order to have a group without any exposure, therefore we are now determining the incidence of tumors among children born between 1964 and 1966.

Our statistical evidence does not rule out some relationship between polio vaccination and the development of brain tumors, however. More gliomas and glioblastomas, medulloblastomas, and spongioblastomas, respectively, seem to develop in the first 17 years of age among persons treated with presumably contaminated vaccine in comparison to persons who received SV40-free vaccine (Figure 4).

One should take into account in this connection that the mean age of those patients in whose intracranial tumors footprints of SV40 could be detected, was 52 years. With the exception of a 6 year old girl, all patients were over age 40. In contrast, our follow-up statistics now cover only the first 17 years of life. These studies will therefore be continued.

DISCUSSION

A substantial fraction of the human intracranial tumors which we have studied express SV40-like T-antigen (18/68) and contain SV40-like DNA (12+1/41+1). Among the latter, one tumor (glioblastoma multiforme) was found to contain a complete SV40 genome or its close variant SV40 GBM - which, on the other hand, might have been arisen during cell fusion. Four additional tumors (1 astrocytoma, 2 meningiomas, 1 oligodendroglioma) contained viral DNA with a cleavage pattern identical to SV40 but different from other human papovaviruses. With the exception of the missing expression of T-antigen by viral DNA containing cells, similar results have been obtained by Sauer et al. (3). Moreover, other authors, as previously cited (6), reported on the occasional presence of SV40 footprints. These results indicate that a certain relationship does exist between human intracranial tumors and SV40 and SV40-like viruses.

This very fact does pose two questions: Where do the viruses in question come from and how are they related to the tumors in which they are found?

Speculations about the source of footprinting SV40's

The SV40 viruses whose footprints have been detected seem not to be directly correlated with exposure to polio vaccine presumably contaminated with SV40 some 20 years ago for the following reasons:

i) Footprints have also occasionally been found in persons who did not receive SV40-contaminated vaccine. Our material includes a meningioma of a 6-year old girl which expresses SV40-like T-antigen (Table 1). Krieg et al. (3) found SV40 in an astrocytoma of a 10-year old patient as well as in an oligodendroglioma of a 54-year old patient who did not receive polio vaccine. Similarly Tabuchi et al. (23) described an ependymoma expressing SV40-like T antigen which developed in a 9-year old girl.

ii) When the sera of several groups of persons were tested for the presence of anti-SV40 IgG antibodies (24), no obvious correlation between the presence of antibodies and the presence or absence of malignancies could be found (6). On the other hand anti-SV40 antibodies could also be detected in the sera of 13% of those persons who were born in 1962 or later and who therefore did not receive SV40 contaminated polio vaccine (6). However, as discussed above, some of these persons might have had inter utero exposures.

Nevertheless the incidence of gliomas and glioblastomas, medulloblastomas, and spongioblastomas, respectively (Figure 4) is somewhat higher among persons who received polio vaccine presumably contaminated with SV40. Further conclusions regarding possible involvement of SV40 contaminated polio vaccine some 20 years ago must be withheld until the vaccinated persons reach an age where intracranial tumors usually develop.

Speculations about the role of footprinting SV40's

The presence of SV40 material might have no connection whatsoever to the induction and/or development of human tumors. The tumor tissue might attract papovaviral passengers which are present in the organism by chance.

This presumption is not in agreement with the following two observations, however:

i) Viral DNA and/or early proteins were almost completely confined to tumor tissue.

ii) Brain tumors of growing rats induced in utero by treatment of their pregnant mothers with the carcinogen, N-Methyl-nitrosourea, did not express SV40 T-antigen even after the animals were infected with SV40 four and eight weeks after birth, respectively (S. Scherneck et al., in preparation). We are now studying whether these tumors likewise are free of viral DNA.

Alternatively SV40 might be one of the risk factors involved in the development of human (intracranial) tumors. It could be even jointly responsible for all such tumors by a "hit-and-run" mechanism currently under discussion with respect to the oncogenic activity of Herpes simplex (25) and other viruses. Such speculation is justified since SV40 is clearly mutagenic (26) as is its brain tumor-derived variant SV40 GBM (5).

One could speculate then that SV40 might transform an (intracranial) host cell by mutating an oncogene through induction of either base substitution and/or chromosomal rearrangement. Thereafter the viral genome might be eliminated totally or in part so that viral footprints can be found only occasionally in those tumors which by chance retained and expressed viral genetic material.

This speculation is supported by the following facts:

i) SV40-induced gene mutations, at least those studied by us (27), do not arise by insertion of viral DNA into the affected gene, i.e. not by "gene splitting". Nevertheless they are stable.

ii) In preliminary studies three SV40-induced mutants were shown to contain SV40 DNA and two not (28). Three of 12 mutants did not express SV40 T-antigen (29). We are presently studying additional mutants for presence and status of SV40 DNA.

iii) Several results indicate that mutagenic and transforming activities of SV40 might have at least some steps in common (26, 30).

iv) Besides gene mutations, SV40, like many other viruses, does induce chromosomal aberrations.

v) In all cases studied by Sauer (3) and by our group SV40-like DNA was localized extrachromosomally. A plasmid-like state of viral DNA should facilitate its random loss during cell division.

vi) Indeed, cells derived from viral-induced tumors may lose viral genetic material and/or its capacity for expression: O. de Lapeyroere, B. Hayot, M. Courcoul, D. Arnaud and F. Birg (30) found that none of the cell lines derived from tumors grown out in syngeneic immunocompetent Fisher rats which had been inoculated with SV40 transformed FR 3T3 rat cells produced SV40 T-antigen. Some of these cell lines still contained SV40 DNA (similar to the meningioma studied by us which contains SV40 DNA but does not synthesize T-antigen), while in others no viral DNA could be detected. These data indicate that the host organism exerts a strong immunological pressure against the expression of viral antigens even to the point of eliminating the viral genetic material. This clearly might also happen in human beings. If that were the case convincing answers regarding a possible causal relationship between SV40 and human tumors could result only from combining epidemiological studies with cell and molecular biological, genetic and immunologic approaches. The present work is a step in this direction.

ACKNOWLEDGEMENTS

Mrs. Helga Zeidler provided expert technical assistance. Dr. J. F. Haas assisted us in editing the manuscript. We thank the staff of the Städtisches Klinikum Berlin-Buch, Neurochirurgische Klinik (Chefarzt: W.-D. Siedschlag) and the Humboldt-Universität zu Berlin, Bereich Medizin, Charité, Chirurgische Klinik (Direktor H. Wolf) for providing us with tumor material.

REFERENCES

1. Takemoto, K. (1980) in: Essex, M., Todaro, G. and zur Hausen, H. (Eds.), Viruses in Naturally Occurring Cancers, 7. Cold Spring Harbor Conference on Cell Proliferation, Cold Spring Harbor Laboratory, Cold Spring Harbor, N.Y., p. 311.
2. Topp, W.C., Lane, D. and R. Pollack (1981) in: Tooze, J. (Ed.), DNA Tumor Viruses, Cold Spring Harbor Laboratory, Cold Spring Harbor, N.Y., p. 207.
3. Krieg, Pl, Amtmann, E., Jonas, D., Fischer, H., Zang, K. and Sauer, G. (1981) Proc. Natl. Acad. Sci. USA 78, 6446.
4. Geissler, E., Scherneck, S., Waehlte, H., Zimmermann, W., Luebbe, L., Krause, H., Theile, M., Herold, H.J., Rudolph, M., Weickmann, F. and Nisch, G. (1980) in: Essex, M., Todaro, G. and zur Hausen, H. (Eds.), Viruses in Naturally Occurring Cancers, 7. Cold Spring Harbor Conference on Cell Proliferation, Cold Spring Harbor Laboratory, Cold Spring Harbor, N.Y., p. 343.
5. Geissler, E., Scherneck, S., Theile, M., Herold, H.J., Staneczek, W., Zimmermann, W., Krause, H., Prokoph, H., Vogel, F. and Platzer, H. (1980) in: Bachmann, P.A. (Ed.), Leukaemias, Lymphomas and Papillomas, Taylor and Francis Ltd., London, p. 43.
6. Geissler, E. (1983) Arch. Geschwulstforsch. 53, 217.
7. Scherneck, S., Luebbe, L., Geissler, E., Nisch, G., Rudolph, M., Waehlte, H., Weickmann, F. and Zimmermann, W. (1979) Zbl. Neurochir. 40, 121.

8. Weiss, A.F., Portmann, R., Fischer, H., Simon, J. and Zang, K.D. (1975) Proc. Natl. Acad. Sci. USA 72, 609.

9. Zang, K.D., May, G. and Fischer, H. (1979) 66, 59.

10. Linzer, D.I.H., Maltzman, W. and Levine, A.J. (1980) Cold Spring Harbor Symp. Quant. Biol. 44, 215.

11. Brandsma, J. and Miller, G. (1980) Proc. Natl. Acad. Sci. USA 77, 6851.

12. Scherneck, S., Rudolph, M., Geissler, E., Vogel, F., Luebbe, L., Waehlte, H., Nisch, G., Weickmann, F. and Zimmermann, W. (1979) Int. J. Cancer 24, 523.

13. Krause, H., Garajev, M.M., Geissler, E., Platzer, M., Scherneck, S. and Zimmermann, W. (in press) Gene.

14. Zimmermann, W., Platzer, M., Scherneck, S., Vogel, F., Krause, H., Prokoph, H. and Geissler, E. (1983) J. Gen. Virol. 64, 733.

15. O'Neill, F.J. and Carroll, D. (1978) Virol. 87, 109.

16. Botchan, M., Topp, W. and Sambrook, J. (1979) Cold Spring Harbor Symp. Quant. Biol. 43, 709.

17. Robbins, F.C. (1981) in: Pollard, M. (Ed.), Perspectives in Virology XI, Alan R. Liss, Inc., New York, p. 247.

18. Shah, K.V. and Nathanson, N. (1976) Am. J. Epidemiol. 103, 1

19. Mortimer, E.A., Lepow, M.L., Gold, E., Robbins, F.C., Burton, G.J. and Fraumeni, J.F. (1981) New England J. Med. 305, 1517.

20. Heinonen, O.P., Shapiro, S., Monson, R.R., Hartz, S.C., Rosenberg, L. and Slone, D. (1973) Int. J. Epidemiol. 2, 229.

21. Madden, D.L., Iltis, J., Tzan, N. and Sever, J.L. (1983) in: Polyomaviruses and Human Neurological Diseases, Alan R. Liss, Inc., New York, p. 149.

22. Farwell, J.R., Dohrmann, G.C., Marrett, L.D. and Meigs, J.W. (1979) Ann. Neurol. 6, 166.

23. Tabuchi, K., Kirsch, W.M., Low, M., Gaskin, D., van Buskirk, J. and Maa, S. (1978) Int. J. Cancer 21, 12.

24. Zimmermann, W., Scherneck, S. and Geissler, E. (1983) Zbl. Bakt. Hyg., I. Abt. Orig. A 254, 187.

25. Galloway, D.A. and McDougall, J.K. (1983) Nature 301, 21

26. Theile, M. (1982) Biol. Zbl. 101, 321.

27. Geissler, E. and Theile, M. (1983) Human Genet. 63, 1

28. Theile, M., Scherneck, S., Waehlte, H. and Geissler, E. (1983) Arch. Geschwulstforsch. 53, 227.

29. Theile, M., Scherneck, S. and Geissler, E. (1980) Arch. Virol. 55, 293.

30. Theile, M., Strauss, M., Luebbe, L., Scherneck, S., Krause, H. and Geissler, E. (1980) Cold Spring Harbor Symp. Quant. Biol. 44, 377.

31. Birg, F. (1983) personal communication.

© 1984 Elsevier Science Publishers B.V.
The Role of Viruses in Human Cancer. Volume II
G. Giraldo and E. Beth editors

PLASMA MEMBRANE-BOUND M_r 94 000 SIMIAN VIRUS 40 AND BK VIRUS TUMOR ANTIGENS ACT AS CROSS-REACTING TUMOR-SPECIFIC TRANSPLANTATION ANTIGENS

RUPERT SCHMIDT-ULLRICH[1], ELKE BETH[2] AND GAETANO GIRALDO[2]
[1]Tufts-New England Medical Center Hospital, Boston, MA 02111 (U.S.A.) and
[2]National Cancer Institute, "Fondazione Pascale", Naples 80131 (Italy)

The purpose of this overview is to summarize the information on the molecular nature of tumor-specific transplantation antigens (TSTA) and tumor-specific surface antigens (TSSA) that are induced in human and rodent cells neoplastically transformed by Simian virus 40 (SV40; 1-5) and the human BK virus (BKV; 6-8), two closely related oncogenic papova viruses. TSTA and TSSA are presumably identical antigens assayed by different test systems (4,5,9,10). TSTA is characterized as an antigen that represents the target for a specific cell-mediated *in vivo* rejection of SV40 transformed cells in hosts immunized with SV40, SV40-transformed cells, membrane isolated thereof, or solubilized membrane antigens (7,11,12). Similar assays have been reported for BKV and BKV-transformed cells (7,8,13). SV40 TSTA can also be documented by a cell mediated cytotoxicity assay *in vitro* (9). TSSA mediates an antibody-dependent cytolytic reaction against SV40-transformed cells (4,5). A limited number of SV40/BKV virally-induced antigens can be expected because the early region of the viral genome, required for maintenance of the transformed state (14,15), can only encode for a protein of a M_r near 100 000. As cells transformed by SV40 or BKV express a M_r 94 000 nuclear T antigen (T_N; 16,17), other virally-induced proteins, i.e. the M_r 17 000 t-antigen (18-21) and TSTA/TSSA (1-5) must share polypeptide sequences with the T_N antigen.

Recent findings indicate that SV40-/BKV-transformed cells also express a M_r 94 000, T antigen-related molecule in their plasma membranes, here called T_M (22). The identification of T_M, the association of TSTA/TSSA with the cell surface and TSTA activity of the M_r 94 000 T antigen lead us to examine whether T_M represents TSTA, whether T_M and T_N differ in their activity in inducing a TSTA response and to determine the degree of cross-reactivity between T_M antigens induced in cells transformed by SV40 and BKV. The SV40 TSTA has been identified in lytically-infected and transformed cells whereas the equivalent antigen has as yet only been described in BKV-transformed cells (e.g. 8). The substantial homology between the DNA of SV40 and BKV and the cross-reactivity between their non-structural virus-specific T antigens link studies on these two viruses of which BKV has been found to be associated with certain human tumors (23,24).

HOMOLOGY BETWEEN SV40 AND BKV

Cells infected with or transformed by SV40 or BKV produce two forms of virus-specific antigens, the M_r 94 000 T- and the M_r 17 000 t-antigen (25-27). They are encoded by two different early mRNA that are transcribed counter-clockwise from the map position 0.65-0.67 to 0.17. The mRNAs of both viruses vary similarly in their size and splicing pattern (e.g. 28,29). The DNA sequence homology (30) within the early region of the genome varies greatly depending on the DNA-DNA hybridization technique employed. Under stringent hybridization conditions an overall nucleotide sequence homology of 11 to 20% has been reported (e.g. 31).

Under a range of non-stringent conditions extensive nucleotide homology has been found with an as low as 20 to 30% base mismatch. This translates into a predicted 71% amino acid homology between the T antigens induced by SV40 and BKV (26). The highest degree of amino acid homology in the early region appears to be near the N-terminus (26).

Functionally and antigenically the T antigens induced by both viruses are closely related as (i) BKV can complement early temperature-sensitive SV40 mutants (32), (ii) there is immunological cross-protection between SV40- and BKV transformed cells (e.g. 3) and (iii) cross-reacting virus-specific surface (33) and nuclear (34) antigens are expressed by cells of a given host cell species.

RELATIONSHIP BETWEEN THE T_M AND TSTA IN CELLS TRANSFORMED BY SV40 AND BKV

Serological evidence for the existence of T_M. The antigenic characteristics of the surface membrane-associated SV40 T_M antigen have been examined using sera raised in syngeneic hosts against SV40-transformed mouse (VLM) and hamster (H-50) cells, membrane/nuclear fractions isolated thereof, and SV40 T antigen purified by indirect immune precipitation (IIP) and sodium dodecyl-sulfate polyacrylamide gel electrophoresis (SDS-PAGE). The sera produced against membranes react exclusively with the surface membranes of suspended, live SV40-transformed cells as determined by IIF, or by a ^{125}I-*Staphylococcus aureus* (*S. aureus*) protein A and double-antibody radioimmune assays for the surface antigen (35). In contrast, sera that are raised against whole SV40-transformed cells or nuclei isolated therefrom react with both the surface membrane and the nucleus (35). All antisera described immune precipitate the detergent-solubilized M_r 96 000 SV40 T antigen from both nuclei and membranes.

However, a hamster serum against membranes of syngeneic H-50 cells precipitates T antigen from this fraction much more effectively than from nuclei. These data suggest that T antigen exhibits different antigenic reactivity in the nucleus and the membrane and that T_M may express unique antigenic determinants (35).

The presence of a SV40 T antigen-like molecule at the surface of SV40-transformed SV3T3 and VLM mouse cells has been also concluded based on the reactivity of rabbit antisera against SDS-PAGE-purified T antigen (36). Live monolayer cells, when reacted with this antibody, exhibited no surface reactivity by indirect immune fluorescence (IIF). However, glutaraldehyde fixation and heating of the cells (to eliminate the perinuclear T antigen reactivity) allowed to detect antigen in both the surface membrane and the nucleus (36). The binding of S. aureus bacteria to antibodies against the membrane T antigen of transformed cells was demonstrated by scanning electron microscopy (36). While the data support the evidence for a plasma membrane-associated T_M no conclusions can be drawn about the surface exposure of T_M. The same reservations apply to the claim that the C-terminus of T_M is exposed at the cell surface based on the binding of a monoclonal antibody (clone 412), specific for C-terminal polypeptide sequences of T antigen, to the cell surface of glutaraldehyde-fixed HeLa cells infected with Adenovirus 2 (Ad 2)-SV40 hybrids (37).

We have produced antisera in guinea pigs against highly purified plasma membranes (38), free of nuclear contaminants, from SV40-transformed hamster lymphoid cells (GD248) that reacted with an isoelectric point (pI) 4.7/94 000 component of Triton X-100-solubilized plasma membranes using bidimensional isoelectric focussing-crossed immune electrophoresis (IEF-CIE; 39,40). This component has been identified as SV40 T_M using sera of tumor bearing hamsters and rabbit sera against SDS-PAGE purified SV40 T antigen (41). Despite strong reactivity of the guinea pig sera with SV40 T_M of GD248 and SV3T3 cells they did not yield an IIF reaction with T_N of either cell line.

The existence of an antigenically unique T_M has been also suggested for BKV-transformed hamster brain cells (LSH-BR-BK; 13). LSH-BR-BK cells produce tumors in the syngeneic LSH hamsters and induce in more than 60% of the animals a strong humoral immune response against BKV T_N and, in 21% of the hamsters, also against a BKV-specific surface (S) antigen (13). Rabbit antisera against plasma membranes of LSH-BR-BK cells, after extensive absorption with live non-transformed LSH-BR cells, react with the surface of live and methanol-acetone-fixed LSH-BR-BK cells. While there is a strong BKV-specific surface IIF reaction with live LSH-BR-BK cells no nuclear IIF reaction is seen after fixation (13). $F(ab)_2$ fragments of the rabbit anti-membrane sera inhibit the binding of surface-reactive antibodies from sera of tumor-bearing LSH hamsters suggesting that both sera reacted with the same antigenic determinants (13). Both the hamster and absorbed rabbit sera immune precipitated a M_r 94 000 protein from plasma membranes of LSH-BR-BK and SV3T3 cells representing the virus-specific T_M (Figure 1). Minor antigenic differences between the SV40 and BKV T_M, already described for T_M in cells transformed by different papova viruses (34), are documented by the sequential IIP of T_M with sera against T_M induced by the homologous and heterologous viral strain (Figure 1).

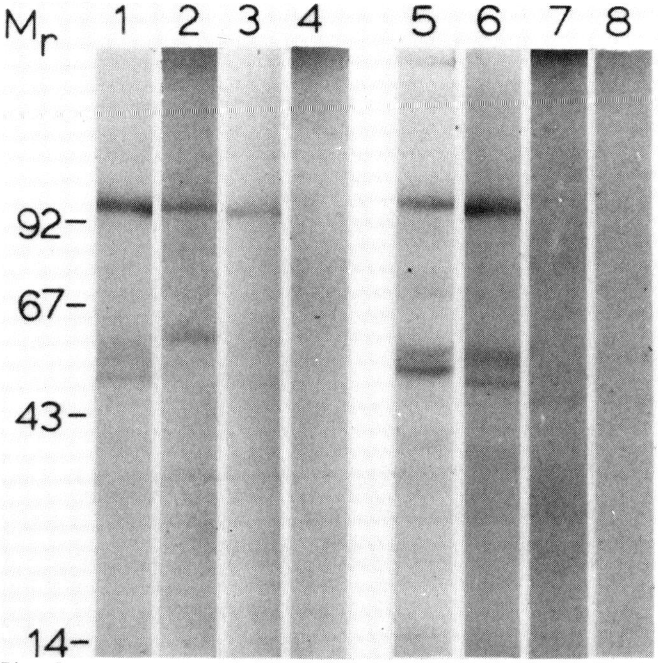

Fig. 1. Immunological cross-reactivity of T_M antigens in plasma membranes of LSH-BR-BK hamster and SV3T3 mouse cells. Assessment by sequential immune precipitation. Identical quantities of Triton X-100-solubilized plasma membranes (PM) from [^{35}S]-methionine-labeled cells were reacted at conditions at which maximum immune precipitation of T_M occurred. The antigen was first reacted with the heterologous anti-T_M, serum and then with the homologous one. Antigen-antibody complexes were deposited by protein A-Sepharose. [^{35}S]-methionine autoradiograms. 1. SV3T3 PM, rabbit-anti-SV40 T_M serum; 2. SV3T3 PM, rabbit anti-LSH-BR-BK PM serum; 3. SV3T3 PM,rabbit anti-LSH-BR-BK PM serum followed by rabbit-anti-SV40 T_M serum; 4. SV3T3 PM, normal rabbit serum; 5. LSH-BR-BK, rabbit-anti-LSH-BR-BK serum; 6. LSH-BR-BK PM, rabbit anti-SV40 T_M serum; 7. LSH-BR-BK PM, rabbit-anti-SV40 T_M serum followed by rabbit anti-LSH-BR-BK PM serum; 8. LSH-BR-BK PM, normal rabbit serum.

Biochemical characterization of T_M and TSTA. The assay for TSSA on SV40-transformed cells and the demonstration of TSTA activity in membranes from SV40- and BKV-transformed cells localize these virus-specific antigens into the surface membrane (1-5,13,33). There is evidence for a close structural and antigenic relationship between SV40 TSTA and the M_r 94 000 T antigen. T antigen, purified to homogeneity by SDS-PAGE exhibits a potent TSTA activity *in vivo* (42,43) and the biochemical purification of SV40 T antigen by $(NH_4)_2SO_4$ precipitation and DEAE-cellulose chromatography indicates that TSTA copurifies with T antigen (10,44). Further chromatographic purification of SV40 T antigen by phosphocellulose chromatography reveals proteins with M_rs between 45 000 and 56 000 that exhibit TSTA/TSSA activity (44). Studies on various non-defective Ad 2-SV40 hybrid viruses show that the synthesis of SV40 TSTA is associated with the DNA sequence within the distal (3'-terminal) portion of the early region of the SV40 genome (45,46). The mRNA of Ad 2 ND_2 mediates the synthesis of a M_r 56 000 and a M_r 42 000 SV40-specific protein (47). Expression of these proteins, suggestively residing in the surface membrane of HeLa cells, coincides with expression of TSTA (47).

Using highly purified plasma membranes of SV40-transformed human (SV80) and mouse (SV3T3) cells we have identified a M_r 94 000 protein unique for membranes of transformed cells. This protein reacts specifically with sera of hamsters bearing SV40-induced sarcomas (SK, line B; 48) and represents SV40 T_M (41). It differs from the T_N by the lower pI (pI 4.7 instead of pI 5.2) and is glycosylated based on ^{14}C-glucosamine incorporation (39,40). The M_r 94 000 T_N and T_M antigens, isolated by IIP and SDS-PAGE from nuclei and plasma membranes of SV80 and SV3T3 cells, respectively, were labeled with ^{125}I using the chloramine T technique and compared by thin layer chromatography (TLC) tryptic peptide mapping (41). T_N and T_M differed in 5 of 27 peptides. Three of these five peptides were glycopeptides that were also exposed at the cell surface as determined by tryptic peptide analysis of T_M after lactoperoxidase-catalyzed surface radioiodination of intact cells prior to isolation of plasma membranes and purification of T_M (41).

These studies have been extended to comparative analyses of T_M in SV3T3 and LSH-BR-BK cells. Purified plasma membranes from both cell lines contain only the M_r 94 000 T_M. Tryptic peptides of immune precipitated, SDS-PAGE purified SV40 and BKV T_M and T_N were analyzed by high pressure liquid chromatography (HPLC) and revealed at least 20 well defined ^{125}I-peptides (Figure 2). Reelectrophoresis-chromatography of HPLC-peptides on TLC plates confirmed that each peak represented only one peptide. Analysis of T_M from SV40 transformed cells,

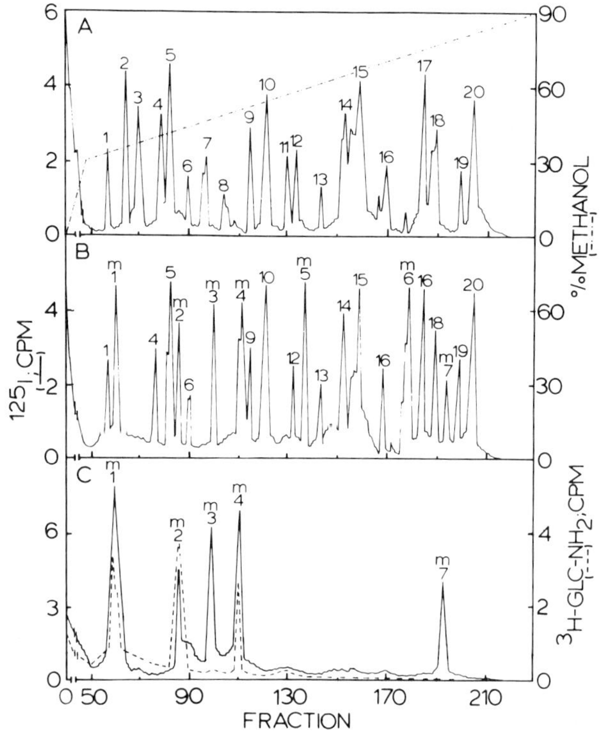

Fig. 2. High pressure liquid chromatography (HPLC) tryptic peptide maps of SV40 nuclear (T_N) and plasma membrane (T_M) T antigens isolated from SV3T3 cells. The isolation of T_N and T_M, their radioiodination using the chloramine T technique, and tryptic cleavage was according to established techniques (41). T was isolated from membranes of cells subjected to lactoperoxidase-catalyzed surface radioiodination or to metabolic labeling with $[^3H]$- glucosamine. The antigens were subject to the chloramine procedure in the absence of ^{125}I. HPLC peptide mapping was performed using an Altex Ultrasphere T_M-ODS reverse phase column attached to a programmable Beckman, Model 110A, chromatograph interphased with a LKB Ultravac fraction collector. The column solvent was formic acid/H_2O (4.5/95.5, v/v). The elution of peptides was by a linear methanol gradient between concentrations of 30-90% (v/v) at flow rate of 2ml/min over 90 min.

Ordinates: ^{125}I-radioactivity in cpm x 10^{-3} (panel A and B) and ^{125}I-radioactivity in cpm · 10^{-2} (panel C); $[^3H]$-glucosamine (glc-NH2) activity in cpm · 10^{-2} (panel C); methanol concentration (-·-·-) in % (v/v). Abscissa: Fractions collected (0.5 ml volume). Panel A: $[^{125}I]$-peptides of SV40 T_N. Panel B: $[^{125}I]$-peptides of SV40 T_M. Panel C: $[^{125}I]$-peptides of T_M from surface-radioiodinated cells (——) and $[^3H]$-glucosamine-labeled glycopeptides of T_M (---). The peptides are numbered in the sequence of their elution from the HPLC column. Peptides unique to T_M are labeled with m, in addition (Panels B and C).

labeled with [^{3}H]-glucosamine, reveals selective labeling of three peptides. Glycopeptides m1, m2 and m4 and two additional peptides, m3 and m7, are exposed at the cell surface as determined by lactoperoxidase-catalyzed radioiodination (Figure 2). The surface-exposed peptides/glycopeptides and peptides m5 and m6 (Figure 2) represent domains of the M_r 94 000 SV40 molecules that are unique to T_M.

Similar studies on T_N and T_M from BKV-transformed LSH-BR cells reveal analogous results (Figure 3.). About 50% of ^{125}I-peptides are shared between BKV T_N and SV40 T_N (Figures 2 and 3). This is in reasonable agreement with previous analyses of a 30% methionine tryptic peptide homology between the two antigen (18). BKV T_M also represents a glycoprotein as revealed by the identification of three glycopeptides, m1, m2, and m3 (Figure 3). All glycopeptides and two additional peptides, 5 and m5, are labeled with lactoperoxidase-catalyzed surface radioiodination (Figure 3). The data indicates that both SV40 and BKV induce highly homologous T_M molecules that are subjected to host cell processing and appear to exhibit a similar membrane assembly.

In addition to glycosylation, T_M may also differ from T_N in its amino acid composition (peptides m5 and m6 of SV40 T_M and peptide m4 of BKV T_M). Such a molecule, as a result of a third splice in the SV40 mRNA, has been predicted to be rich in hydrophobic amino acids near its C-terminus and would qualify as a membrane protein (49). The presence of a M_r 94 000 T antigen in membranes of SV40-/BKV-transformed cells, that is host-cell modified by glycosylation, can also explain the apparently discordant findings of a broad immunological cross-reactivity between the antigens induced by SV40 or BKV and the presence of virus-/host cell-specific antigenic determinants described for TSTA (50).

TSTA activity of T_M. Based on the structural and antigenic similarity between TSTA and T antigen of different strains of papova viruses, the antigenic cross reactivity between the TSTA molecules induced by SV40 and BKV has been examined. Isolated membranes from BKV-transformed hamster cells were used for immunization of Balb/c mice and Syrian hamsters. The animals were challenged with syngeneic SV40-transformed mKSA murine fibroblasts and SV34 hamster cells (7). The mice were equally well protected against cells transformed by the homologous and heterologous virus. These results and the data that membranes of cells transformed by SV40 and BKV express a highly homologous T_M antigen that represents a host-cell modified T antigen lead us to examine the activity of the SV40 and BKV T_M antigens, relative to T_N, in inducing a TSTA response across viral strains.

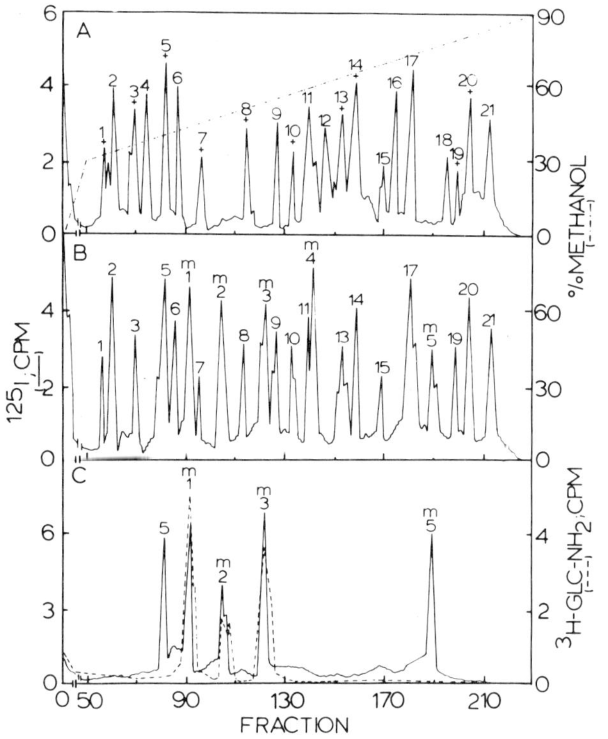

Fig. 3. High pressure liquid chromatography tryptic peptide maps of BKV nuclear (T_N) and plasma membrane (T_M) T antigens isolated from LSH-BR-BK cells. For details see legend of Figure 2. Panel A: [^{125}I]-peptides of BKV T_N. Panel B: [^{125}I]-peptides of BKV T_M. Panel C: [^{125}I]-peptides from surface-radioiodinated cells (——) and [^{3}H]-glucosamine-labeled glycopeptides of T_M (---). Peptides labeled (+) (panel A) are shared with SV40 T_N.

For the immunization of ten Balb/c mice in each group, SV40 T_N and T_M were isolated from plasma membranes and nuclei, respectively, of murine SV3T3 and human SV80 cells and BKV T_M from plasma membranes of LSH-BR-BK hamster cells (Table 1). The cells were metabolically labeled with ^{35}S-methionine for 16 hr. at which time label saturation has occurred for both antigens. The cells were fractionated into nuclei and plasma membranes and, after detergent solubilization (40), T_N and T_M were quantitatively immune precipitated using a rabbit immune sera against SDS-PAGE purified SV40 T antigen (Figure 1). Mice were immunized with SV40 T_M and BKV T_M isolated from membranes of $5 \cdot 10^7$ each SV3T3/SV80 and LSH-BR-BK cells. Based on ^{35}S-methionine radioactivity the

TABLE 1

TUMOR REJECTION OF mKSA SV40-TRANSFORMED MOUSE FIBROBLASTS IN BALB/c MICE

Antigen[a] (Cell type)	Challenge[b]	
	Tumor bearing mice/mice inoculated	
	20 days	40 days
SV40 T_N[c] (SV3T3)	4/10	6/10
SV40 T_M[c] (SV3T3)	0/10	0/10
SV40 T_M (SV80)	0/10	1/10
BKV T_M (LSH-BR-BK)	2/10	4/10
Control	5/10	10/10

[a]Balb/c mice were immunized with purified M_r 94 000 T antigens from SV40- and BKV-transformed cells. The SV40 and BKV T antigens were quantitatively immune precipitated, using rabbit anti-SV40 T serum, from purified plasma membranes and nuclei of $5 \cdot 10^9$ each mouse SV3T3, human SV80 and hamster LSH-BR-BK cells. The T antigens were extracted from polyacrylamide gels with 0.005M Na-phosphate, 0.05% sodium dodecyl sulfate pH 8.0, freed from detergent by dialysis and injected with adjuvant. Mice in the control group were injected with extraction buffer and adjuvant.

[b]TD50 of mKSA cells is $2 \cdot 10^3$ in Balb/c mice after subcutaneous inoculation (sc). For challenge $2 \cdot 10^5$ cells were inoculated sc in 0.1 ml of sterile phosphate-buffered saline.

[c]M_r 94 000 T antigens isolated from nuclei, T_N and plasma membranes, T_M.

identical amount of SV40 T_N antigen was injected. The first subcutaneous immunization was with Freund's complete adjuvant and the booster immunization, 14 days later, with identical quantities of antigen but using incomplete adjuvant. The mice in the control group were injected with saline and adjuvant (Table 1). Ten days after the second immunization the mice were challenged by subcutaneous injection of $2 \cdot 10^5$ mKSA SV40-transformed fibroblasts. (TD50 after 20 days was $2 \cdot 10^3$ cells using a subcutaneous inoculation). Using a 100-fold LD50 dose of mKSA cells all mice in the control were found to have tumors larger than 1.0 cm in diameter after 40 days (Table 1). In contrast, all mice but one were free of tumors after immunization with SV40 T_M antigen. The TSTA response induced lead to a tumor rejection in immunized mice in comparison to animals of the control group (this response was significant for SV40 T_M antigens and BKV T_M with p-values of 0.005 and 0.01, respectively). These data indicate that T_M isolated from membranes of SV40 and BKV transformed cells exhibits a very potent TSTA activity within and across viral strains. Most interesting is that BKV T_M appears to have a higher TSTA activity against SV40-transformed cells than the homologous T_N. Experiments with a larger number of animals are required to determine whether this difference is significant.

CONCLUSIONS

Current evidence indicates that cells neoplastically transformed by SV40 and BKV express highly homologous, virally-encoded T_M antigens in their plasma membranes. T_M represents a M_r 94 000 polypeptide that is exposed at the cell surface and is glycosylated as a result of host-cell processing (40,41,51). T_M exhibits a high degree of molecular homology with the M_r 94 000 T_N that is induced by both SV40 and BKV and shares polypeptide sequences with TSTA (42-44). But in addition, T_M (13,34,35,41) and TSTA (7,50) express antigenic determinants that are unique relative to T_N. The demonstration of the dominant TSTA activity of T_M suggests that this molecule acts as TSTA across viral strains in the surface membrane of cells transformed by SV40 or BKV. The host cell-specific component of TSTA may lie exclusively in the microheterogeneity of the carbohydrate moieties. But minor differences between T_N and T_M in non-glycosylated tryptic peptides may indicate that the two molecules differ also in their polypeptide chains.

Previously published results and our own data suggest that, in the search for BKV transformation-specific markers in human tumors, the most appropriate immunological probes should be derived from human cells transformed by this virus. Since the DNA-DNA hybridization and IIF for T antigen appears to be non-discriminating for lytic infection with or transformation by BKV the more detailed characterization of the unique antigenic determinants on T_M (TSTA) could lead to the development of monospecific immunological reagents to test for transformation _in vivo_.

ACKNOWLEDGEMENTS

Supported by Grant CA-12178 and CA-23642 from the National Institutes of Health. The excellent technical assistance of W. S. Thompson, S. J. Kahn and M. T. M. Monroe and the assistance of R. E. Wilson is gratefully acknowledged.

REFERENCES

1. Tevethia, S.S., Katz, M. and Rapp, F. (1975) Proc. Soc. Exp. Biol. Med. 119, 896.
2. Anderson, J.I., Martin, R.G., Chang, C. and Mora P.J. (1977) Virology 76, 254.

3. Law, L.W., Takemoto, K.K., Rogers, M.J. and Ting, R.C. (1977) J. Natl. Cancer Inst. 59, 1523.

4. Pancake, S.J. and Mora, P.T. (1974) Virology 59, 323.

5. Pancake, S.J. and Mora, P.T. (1976) Cancer Res. 36, 88.

6. Padgett, B.L., Walker, D.L., zu Rhein, G.M., Eckroade, R.J. and Dessel, B.H. (1971) Lancet 1, 1257.

7. Law, L.W., Takemoto, K.K., Rogers, M.J., Hendriksen, O. and Ting, R.C. (1978) Int. J. Cancer 22, 315.

8. Seehafer, J., Downer, D. M., Gibney, D.J. and Colter, J.S. (1979) Virology 95, 241.

9. Pretell, J., Greenfield, R. S. and Tevethia, S.S. (1979) Virology 97, 32.

10. Luborsky, S.W., Chang, C., Pancake, S.J. and Mora, P.T. (1978) Cancer Res. 38, 2367.

11. Dean, J.H., Lewis, D.D., Paderathsingh, M.L., McCoy, J.L., Northing, J.W., Natori, T. and Law, L.W. (1977) Int. J. Cancer 20, 951.

12. Law, L.W., Takemoto, K.K., Rogers, M.J. and Ting, R.C. (1977) J. Natl. Cancer Inst. 59, 1523.

13. Beth, E., Giraldo, G., Schmidt-Ullrich, R., Pater, M.M., Pater, A. and DiMayorca, G. (1981) J. Virol. 40, 276.

14. Martin, R.G. and Chou, J.Y. (1975) J. Virol. 15, 599.

15. Bouck, N., Beales, N., Shenk, T., Berg, P. and DiMayorca, G. (1978) Proc. Natl. Acad. Sci. USA 75, 2473.

16. Simmons, D.T., Takemoto, K.K. and Martin, M.A. (1978) Virology 85, 146.

17. Rundell, K., Tegtmeyer, P., Wright, P.J. and DiMayorca, G. (1977) Virology 82, 206.

18. Simmons, D.T. and Martin, M.A. (1978) Proc. Natl. Acad. Sci. USA 75, 1131.

19. Prives, C., Gilboa, E., Revel, M. and Winocour, E. (1977) Proc. Natl. Acad. Sci. USA 74, 457.

20. Smith, A.E., Smith, R. and Paucha, E. (1978) J. Virol. 28, 140.

21. Shah, K.V., Daniel, R.W. and Strandberg, J. (1975) J. Natl. Cancer Inst. 54, 945.

22. Schmidt-Ullrich, R. and Wallach, D.F.H. (1980) in: Giraldo, G. and Beth, E. (Eds.), The Role of Viruses in Human Cancer, Vol. 1, Elsevier/North Holland Biomedical Press, Amsterdam, pp. 125-139.

23. Padgett, B.L. (1980) in: Giraldo, G. and Beth, E. (Eds.), The Role of Viruses in Human Cancer, Vol. 1, Elsevier/North Holland Biomedical Press, Amsterdam, pp. 117-123.

24. Pater, M.M., Pater, A., Fiori, M., Slota, J. and DiMayorca, G. (1980) in: Essex, M., Todaro, G. and zur Hausen, H. (Eds.), Viruses in Naturally Occurring Cancers. Cold Spring Harbar Conferences on Cell Proliferation, Vol. 7, Cold Spring Harbor Laboratory, pp. 329-341.

25. Paucha, E., Mellor, A., Harvey, R., Smith, A., Hewbrick, R. and Waterfield, M. (1978) Proc. Natl. Acad. Sci. USA 75, 2165.

26. Yang, R.C.A., Young, A. and Wu, R. (1980) J. Virol. 34, 416.

27. Takemoto, K. and Mullarkey, M.F. (1973) J. Virol. 12, 625.

28. Berk, A.J. and Sharp, P.A. (1978) Proc. Natl. Acad. Sci. USA 75, 1274.

29. Manaker, R.A., Khoury, G. and Lai, C.-J. (1979) Virology 97, 112.

30. Law, M.F., Martin, J.D., Takemoto, K.K. and Howley, P.M. (1979) Virology 96, 576.

31. Khoury, F., Howley, P.M., Garon, C., Mullarkey, M.F., Takemoto, K.K. and Martin, M.A. (1975) Proc. Natl. Acad. Sci. USA 72, 2563.

32. Mason, D.H. and Takemoto, K.K. (1976) J. Virol. 17, 1060.

33. Molinaro, G.A., Major, E. O., Bernhardt, G., Dray, S. and DiMayorca, G. (1977) J. Immunol. 118, 2295.

34. Beth, E., Cikes, M., Schleon, L., DiMayorca, G. and Giraldo, G. (1977) Int. J. Cancer 20, 551.

35. Soule, H. R., Lanford, R.E. and Butel, J.S. (1980) J. Virol. 33, 887.

36. Deppert, W., Hanke, K. and Henning, R. (1980) J. Virol. 35, 505.

37. Deppert, W. and Walter, G. (1982) Virology 122, 56.

38. Schmidt-Ullrich, R., Wallach, D.F.H. and Davis, F.D.G. (1976) J. Natl. Cancer Inst. 57, 1107.

39. Schmidt-Ullrich, R., Lin, P.S., Thompson, W.S. and Wallach, D.F.H. (1977) Proc. Natl. Acad. Sci. USA 74, 5069.

40. Schmidt-Ullrich, R., Kahn, S.J., Thompson, W.S. and Wallach, D.F.H. (1980) J. Natl. Cancer Inst. 65, 585.

41. Schmidt-Ullrich, R., Thompson, W.S., Kahn, S.J., Monroe, M.T.M. and Wallach, D.F.H. (1982) J. Natl. Cancer Inst. 69, 839.

42. Tenen, D., Garewal, H., Haines, L., Hudson, V., Woodward, V., Light, S. and Livingston, D.M. (1977) Proc. Natl. Acad. Sci. USA 74, 3745.

43. Chang, C., Martin, R. G., Livingston, D.M., Luborsky, S.W., Hu, C.-P. and Mora, P.T. (1979) J. Virol. 29, 69.

44. Anderson, J.L., Martin, R.G., Chang, C., Mora P.T. and Livingston, D.M. (1977) Virology 76, 420.

45. Lewis, A.M. and Rowe, W.P. (1973) J. Virol. 12, 836.

46. Lebowitz, P., Kelly, T.J., Nathans, D., Lee, T.N. and Lewis, A.M. (1974) Proc. Natl. Acad. Sci. USA 71, 441.

47. Deppert, W. and Walter, G. (1976) Proc. Natl. Acad. Sci. USA 73, 2505.

48. Diamandopoulos, G.T. and Dalton-Tucker, M.F. (1969) Am. J. Pathol. 56, 59.

49. Merk. D.F. and Berg, P. (1980) in: Viral Oncogenes, Cold Spring Harbor Symposia on Quantitative Biology, Vol. 44, Cold Spring Harbor Laboratories, pp. 55-62.

50. Drapkin, M.S., Apella, E. and Law, L.W. (1974) J. Natl. Cancer Inst. 52, 259.

51. Schmidt-Ullrich, R., Verma, S.P. and Wallach, D.F.H. (1975) Biochem. Biophys. Res. Commun. 67, 1062.

The Role of Viruses in Human Cancer. Volume II
G. Giraldo and E. Beth editors

AVIAN SARCOMA VIRUS (ASV) AND KOCH's POSTULATES IN HUMAN VIRAL ONCOLOGY .

G.F. RABOTTI, B. TEUTSCH, J. AUGER, F. MONGIAT and M. MARILLER
Laboratoire de Médecine Expérimentale du Collège de France,
U 112 de l'Institut National de la Santé et de la Recherche
Médicale, 11 Place Marcelin Berthelot, 75231 PARIS Cedex 05 .

The study of the mechanism by which certain retroviruses can cross the species barrier should provide an insight into the etiologic factors of human malignancy .

Several retrovirus of murine (1, 2,3,4) and feline (5) origin have been shown to induce morphologic alterations and changes in growth patterns of human fibroblasts in vitro . With one possible exception (6), not confirmed (7) in the case of fibroblasts derived from some patients with hereditary adenomatosis of the colon transformed by the murine Kirsten sarcoma virus, the link between such transformation in vitro and criteria of malignancy such as tumor production in athymic nude mice need still to be established .

Recently we have reported transformation of cultured human diploid fibroblasts and "supertransformation" of human sarcomatous cells by an avian sarcoma virus . These transformed cell lines were permissive for virus production (8) . It is in this human cell system that we wish to discuss the relation between the criteria of malignancy and the viral transformation induced in vitro .

a) Transformation :

Cultures of human fibroblasts established from skin biopsies of 130 normal donors and cultures of sarcomatous cells originating from 6 different patients have been infected with various subgroups of avian sarcomatogenic virus (ASV) i.e. Schmidt-Ruppin sub-group A (Sr-RSV-A), B (Sr-RSV-2) and Bryan high titer (RSV-RSV-1) strains .

The cultures were infected with 8 - 8,5 FFU of virus filtered through a Millipore membrane (0.22 µm) . The conditions and media used have been described (8) . The diploid fibroblasts cultures from 2 donors only, infected with Sr-RSV-2, showed foci of

TABLE 1 -

TITER OF TRANSFORMING VIRUS RESCUED FROM CELL-FUSION WITH CEF AND OF FILTERED SUPERNATANTS FROM HUMAN RSV-TRANSFORMED CELLS

NORMAL AND RSV-TRANSFORMED FIBROBLASTS	VIRUS OBTAINED BY FUSION OF CELLS WITH CEF (C/E or C/AE)[a] FFU/ML	VIRUS IN FILTERED SUPERNATANTS OF HUMAN RSV TRANSFORMED CELLS[b] FFU/ML
M.F. (CONTROL)	-	-
TR.[+] M.F.(SR-RSV-B)	8×10^3	4×10^5
B.X. (CONTROL)	-	-
TR.B.X.(SR-RSV-B)	2×10^4	3×10^4
SARCOMATOUS CELLS		
PA(CONTROL)	-	-
I.[++] PA(SR-RSV-A)	7.2×10^2	4×10^5
I. PA(SR-RSV-B)	4.0×10^4	-
I. PA(RSV-RAV-1)	5.0×10^3	-

a) RSV-transformed human fibroblasts were cocultivated with CEF at the 4th, 6th, 8th, 12th, 15th, and 19th passages after transformation for the fibroblasts line TR.M.F. and at the 4th, 5th, 12th, 18th and 20th passages for the line TR.B.X. The focus forming activity in CEF, C/E was determined on the supernatants of the 12th passage for both cell lines . The infectivity of the virus rescued from cocultivation of I.PA sarcoma cells was determined 8 times during the first 20 passages of the infected cells and repeated at the 112th passage . The data in the table refer to virus titer obtained at the 12th passage .

b) filtered (Millipore 0.22 µm/pore size) supernatants of human RSV-transformed fibroblasts and of infected sarcoma cells were tested in CEF, C/E after one passage in chicken fibroblasts . These tests were repeated 8 times during the first passages of the infected human cells . The titers in the table correspond to the virus recovered at the 20th passage .

[+]TR. : transformed fibroblasts [++] I. : Infected sarcoma cells .

morphological transformation . These foci appeared similar to those induced by this virus in chicken embryo fibroblasts (CEF) cultures . Transformation was evident either at the 4th or at the 5th passage after infection with Sr-RSV-2 (Fig. 1) but no such lesions were observed in control cultures or in cultures infected with the strains Sr-RSV-A or RSV-RAV-1 .

When 2 X 10^6 transformed fibroblasts of these two lines (B.X and M.F) were inoculated into 5 athymic Swiss nu/nu mice, progressive sarcoma were observed, yet no tumors appeared after inoculation in nude mice of an equal number of uninfected cells from the same donors .

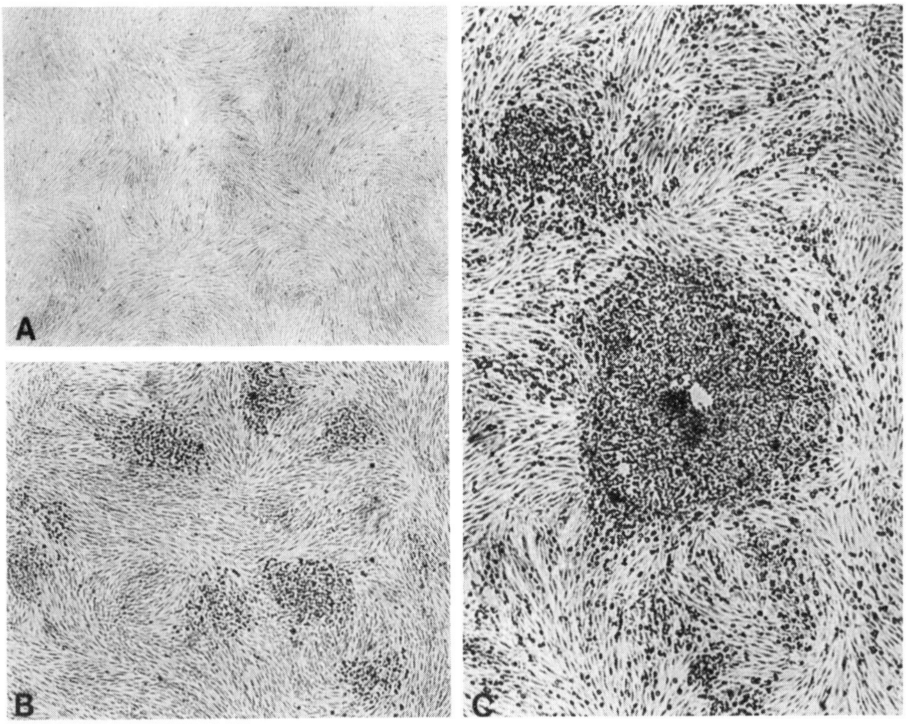

Fig. 1 :

A) Fibroblasts from donor B.X. : Normal aspect .

B) Foci of transformation induced in a culture of this donor by Sr-RSV .
Both cultures at the 7th passage .
X : 30

C) Foci of transformed cells at higher magnification .

b) Viral production :

Cocultivation of RSV transformed human fibroblasts with susceptible CEF induced the recovery of the infectious virus . The recovered virus was preferentially neutralized by chicken anti Sr-RSV-2 antibodies, as expected .

Cell-free filtrates (Millipore 0,22 μm) of the supernatants of these 2 RSV transformed cultures (B.X and M.F) induced foci of transformation in CEF (table 1) .Viral production could still be demonstrated after the 20th passage . Fig. 2 shows the mature type C virus and budding formation from the plasma membrane in one (B.X.) of these two cases . Comparable pictures were obtained in the other case .

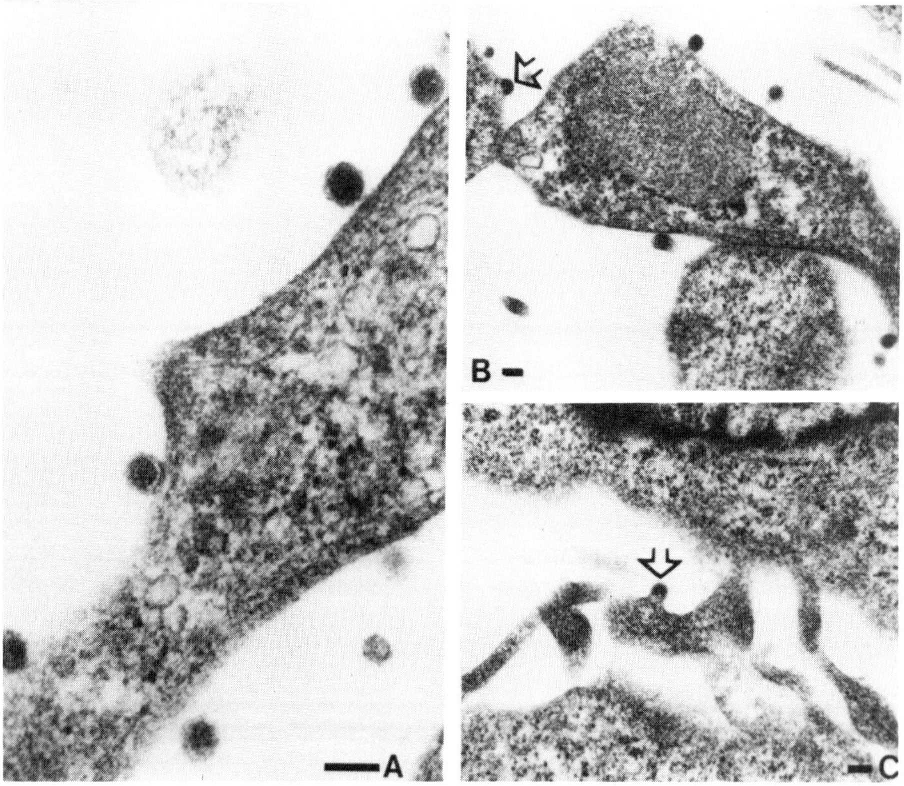

Fig. 2 Viral particles (A, B, Arrows) budding from the plasma membrane in cells from the same donor B.X. infected with RSV . The morphology of the viral particle resembles that of mammalian retroviruses (type C) .

Bar : 100 nm

The morphology of the virus produced by transformed human cells (H-RSV) is different from that of the same virus produced by CEF and resembles that of mammalian retroviruses : the diameter of the core is greater,the intermediate membrane hardly visible (Fig. 3) . The cell cultures of undifferenciated sarcoma cells which originated from 6 different donors were also infected with the same viruses . After 4/5 passages these were co-cultivated with CEF . The cultures from one patient (P.A.) infected with Sr-RSV-A, B or RSV-RAV-1 were virogenic . Each one of the virus utilized to infect these cells could be rescued and neutralized preferentially, using the corresponding chicken antisera . Only did the cells infected with Sr-RSV-2 produce infectious virus found in cell-free filtrated of the supernatants . This infectivity has been tested 8 times at different passages and was still positive after 100 passages (table 1) . In later experiments this line proved to be susceptible also to productive infection with PRAGUE-RSV-C. The morphology of this virus is identical to that shown in Fig. 3.

PARTIAL BIOCHEMICAL CHARACTERIZATION OF THE HUMAN (H)RSV :

Batches of virus from 3/4 liters of supernatant of infected human cells were purified according to published procedures (9).

The specific gravity of the virus (H-RSV) grown in human cells is 1.20 g/cm^3 as compared to 1.15 - 1.16 g/cm^3 for the same virus grown in CEF (Fig. 4) .

Using synthetic homopolymers RT from avian viruses is catalytically more active in the presence of Mg^{2+} than Mn^{2+} . Ions and template-requirements of reverse-transcriptase (RT) of RSV produced in human infected fibroblasts were then studied and compared to those of the RT from the same virus grown in CEF . Cation requirements of RT of retroviruses type C are useful markers for the biochemical characterization of the retroviruses and allow to distinguish between RT and cellular DNA polymerase . In contrast RT from mammalian type C viruses are more active in the presence of Mn^{2+} than Mg^{2+} (10) .

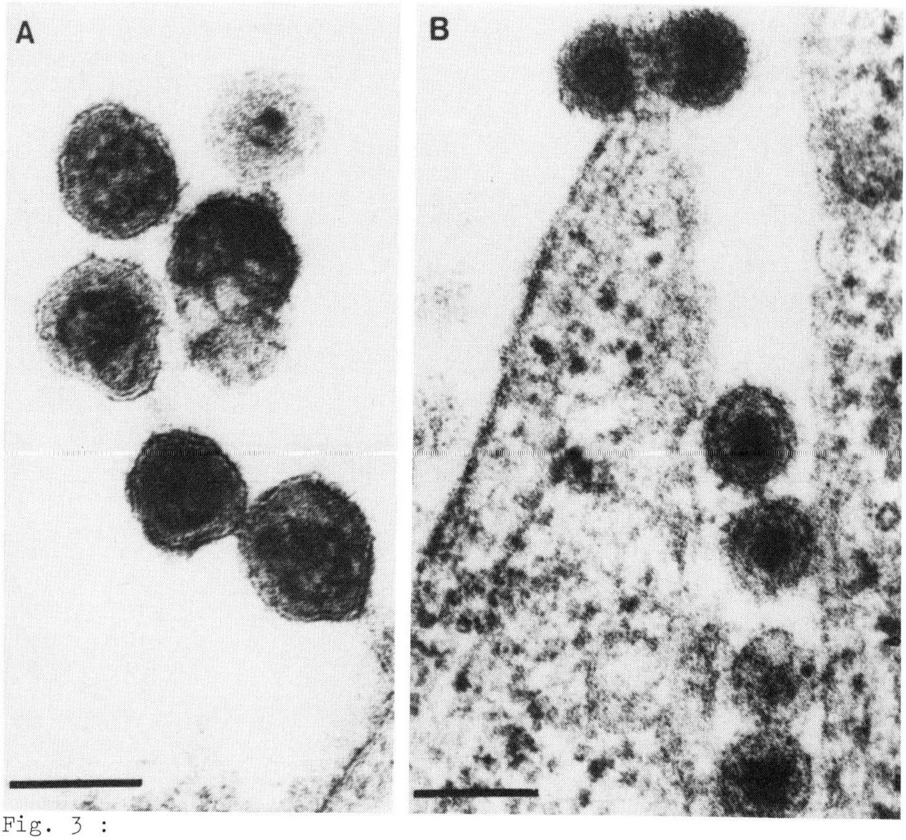

Fig. 3 :

A)Higher magnification of type C viral particles observed in human fibroblasts of the infected B.X. line . Note the coiled aspect of the nucleoid and the lack of visible intermediate membrane . Note also that the dimensions of the viruses are greater than that of the avian type C particles .

B)Type C viruses at the same magnification, in CEF infected with the virus produced in human cells shown in A .

Note that in the chicken the virus particles are smaller, the nucleoïd denser and the intermediate membrane visible (⟶)
This aspect is characteristic of RSV produced in CEF .

Bar : 100 nm

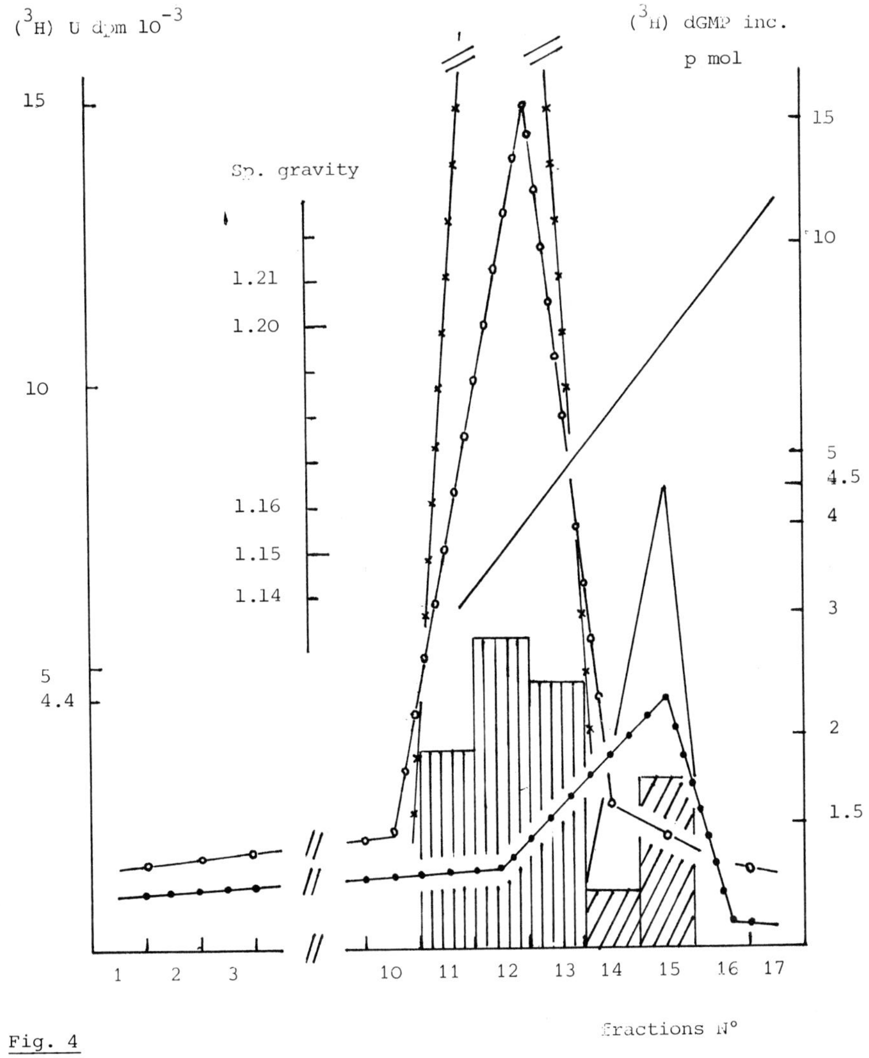

Fig. 4

—o— incorporation of (^{3}H) Uridine by Sr-RSV-2
—●— " " " by PA_s(Sr-RSV)
—✕— polymerase assay Sr-RSV-2 100 p mol
—— " " PA_s(Sr-RSV)
▮▮▮ infectivity of Sr-RSV-2
/// " " PA_s(Sr-RSV)

TABLE 2 Comparison of template-primer requirements between reverse transcriptases from ASV grown in CEF and human cells .

Virus	Template-primer	^3H Substrate	Incorporation (pmol) Mg^{2+} (10mM)	Mn^{2+} (1mM)	Fe^{2+} (0.25mM)
		virus grown in chicken cells .			
AMV	$rA_n \cdot dT_{12-18}$	dTTP	776.2	326	45.6
	$rC_n \cdot dG_{12-18}$	dGTP	1015	271	33
	$rCm_n \cdot dG_{12-18}$	"	1339	1747	441
Sr-RSV-2	$rA_n \cdot dT_{12-18}$	dTTP	199,6	47.3	10.3
	$rC_n \cdot dG_{12-18}$	dGTP	111	33	8
	$rCm_n \cdot dG_{12-18}$	"	128	270	59
Pr-RSV-C	$rC_n \cdot dG_{12-18}$	dGTP	903	42	nd
CEF	$rA_n \cdot dT_{12-18}$	dTTP	0.5	0.1	0.5
	$rC_n \cdot dG_{12-18}$	dGTP	0.2	0.1	0.2

virus grown in human cells .

PA_s(Sr-RSV)	$rA_n \cdot dT_{12-18}$	dTTP	5.8	6.2	6.4
	$rC_n \cdot dG_{12-18}$	dGTP	2.8	4.5	19
	$rCm_n \cdot dG_{12-18}$	"	1	4	2.7
PA_s(Pr-RSV)	$rA_n \cdot dT_{12-18}$	dTTP	7.6	10.7	nd
	$rC_n \cdot dG_{12-18}$	dGTP	4.4	6.4	nd
PA_c(PA_s(Sr-RSV))	$rA_n \cdot dT_{12-18}$	dTTP	1.8	2	17
	$rC_n \cdot dG_{12-18}$	dGTP	0.6	2.7	7
	$rCm_n \cdot dG_{12-18}$	"	1.1	11	10
PM(DF.B(Sr-RSV))	$rA_n \cdot dT_{12-18}$	dTTP	3	6	9
	$rC_n \cdot dG_{12-18}$	dGTP	0.4	2	2.3
	$rCm_n \cdot dG_{12-18}$	"	0.2	11.5	2
PA_s	$rA_n \cdot dT_{12-18}$	dTTP	0.1	0.2	1
	$rC_n \cdot dG_{12-18}$	dGTP	0.1	0.2	0.6

Polymerase assays (37°;1hr) with 120 µl samples containing 20mM Tris, pH 8.3,7.6; 20mM KCl; 20 mM DTT; 0.2% Triton X100; 15 µg of the template-primers; 1.0 µCi of (^{3}H) labeled dGTP (0.1 mM) or dTTP (0.09 mM);and 50 µl of virus .

Specific activity of (^{3}H)dGTP : 185 dpm/p mole;(^{3}H) dTTP : 208 dpm/p mole .

Table II and III show the results of experiments in which viruses were tested with various template-primers in the presence of different cation concentrations . The enzyme activity of Sr-RSV-2 (and of avian myeloblastosis virus , included as a control) was found to be highest in all combination of template primers using Mg^{2+} in agreement with the data of the literature (10) .

In contrast the same virus grown in human cells prefers Mn^{2+} to Mg^{2+} and it utilizes efficiently Fe^{2+} ions .

These results showing a shift in cationic preference from that typical for avian virus to that of mammalian RT suggest a host induced biochemical modification of the virus when it is produced by human cells .

TABLE 3

Activation by divalent cations of poly(rC)oligo(dG) directed reverse transcriptase of ASV grown in avian or human cells .

Virus	(^{3}H) dGMP incorporation (dpm) Mg^{2+} (10mM)	Mn^{2+} (1mM)	Fe^{2+} (0.25mM)	Relative ratio Mn^{2+}/Mg^{2+}	Fe^{2+}/Mg^{2+}
	virus grown in chicken cells				
AMV	187,793	50,185	8,640	0.266	0.045
Sr-RSV-2	20,555	6,111	1,481	0.297	0.072
Pr-RSV-C	167,300	7,700	-	0.045	-
	virus grown in human cells				
PA_s(Sr-RSV)	1,018	1,296	3,515	1.27	3.45
PA_s(Pr-RSV)	815	1,185	-	1.45	-
PA_c(PA_s(Sr-RSV))	120	499	1,295	4.15	10.8
PM(DF.B(Sr-RSV))	70	380	430	5.40	6.14

IMMUNOLOGICAL TESTS :

Preliminary immunological tests show the presence in purified H-RSV of the group-specific antigen of avian sarcoma viruses . The immunodiffusion patterns are shown in Fig. 5 upper left . The center well contains purified virus produced by the human cell line PA_s(Sr-RSV) . The outer well contains : 3)antiserum from Sr-RSV tumor bearing hamster; 4)serum from rabbit immunized with Sr-RSV-2 ; 5)serum from chicken bearing sarcoma induced by Sr-RSV-2 or 6)RSV-RAV-1 .

These tests indicate the presence in the human grown virus of the avian retrovirus group-specific antigen .

The lower left of Fig 5 shows the core purification of the different strains of ASV grown in CEF .

On the right of Fig 5 are shown the results of the bi-dimensional immuno-electrophoresis against rabbit anti Sr-RSV-2 antiserum of the following virus : A) Sr-RSV-2 ; B) PA_s(Sr-RSV); C) tandem comparison between A and B ; D) core proteins of Sr-RSV-2 ; E) as in B ; F) tandem comparison between D and E.

Only one common antigen to the 2 viruses is detected as shown by the peak at the right of the electropherogram and it is clear that this antigen is located in the core of Sr-RSV-2 .

Comparison of CEF grown viral core antigen separated by chromatography with the antigen of human grown RSV indicates that the protein common to the two viruses is the p 27 . In PA_s(Sr-RSV) the presence of envelope antigen could not be demonstrated with rabbit antisera against Sr-RSV-2 .

Immunization of rabbits with purified H-RSV is under way so as to study the envelope antigen (s) of this virus .

TRANSFORMATION OF HUMAN DIPLOID FIBROBLASTS BY H-RSV :

The virus produced by the 2 lines of RSV transformed human diploid fibroblasts and that produced by the sarcoma lines were used to infect the 130 diploid normal fibroblast cultures .

The results are presented in table lV .

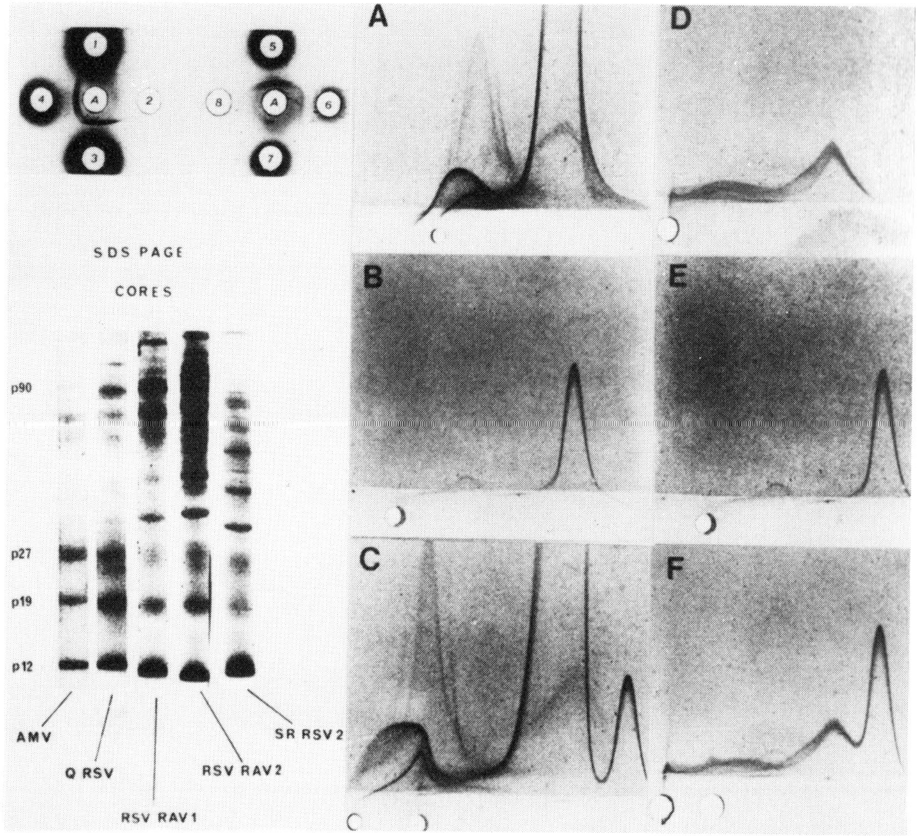

Fig. 5 :

Upper left : double immuno-diffusion .

A : purified RSV produced by human "supertransformed" sarcoma cell line PA_s(Sr-RSV)

3 : antiserum from RSV tumor bearing hamster .

4 : " from rabbit immunized with Sr-RSV-2

5 : " from chicken bearing a tumor induced by Sr-RSV-2

6 : " " " " " RSV-RAV-1

Lower left : SDS-PAGE of purified RSV core proteins .

Right : Bidimentional immuno-electrophoresis of Sr-RSV-2, PA_s(Sr-RSV) and tandem comparison .

a) Sr-RSV-2, b) PA_s(Sr-RSV), c) tandem comparison between a and b,d) core protein of Sr-RSV-2, e) PA_s(Sr-RSV), f)tandem comparison between d and e .

In all cases rabbit anti Sr-RSV-2 antiserum was used .

It may be seen that only one line (≠ 272) out of 130 was transformed by the virus produced by culture ≠ 411 . Again only one line ≠ 27 was found susceptible to the virus produced by the culture ≠ 272, but the virus produced by culture ≠ 27 was found to transform 4 different lines (≠ 32, ≠ 111, ≠ 154, ≠ 428). The virus produced by the infected sarcoma cells PA_s transformed the largest number of cultures . These sarcomatous cells could be also infected with Prague virus (Pr-RSV-C) and produced particles that transformed very efficiently human diploid fibroblasts . It is interesting to note that in the case of the patient PA, donor of sarcomatous cells, we have also tested cultures of normal fibroblasts prepared from his normal skin (PA_c) and that these cultures of normal fibroblasts could not be transformed by virus produced in chicken cells, yet they could be transformed by the virus after growth on its sarcomatous cells (PA_s-(Sr-RSV)) .

If one considers that cultures from only 2 donors were transformed by Sr-RSV-2 grown in CEF, but 26 cultures (including the 2 transformed by Sr-RSV-2) were transformed using the virus after passage in human cells, it is evident that a new tropism has been acquired by the virus after passage in such human cells . Whether this tropism is specific for human cells can be established only after a series of mammalian host cells has been tested for susceptibility to this virus .

The human diploid fibroblasts transformed by Sr-RSV-2, the human sarcoma cell line "supertransformed" by the same virus and some of the lines transformed by the virus after passage in human fibroblasts expressed high level of $pp60^{V\text{-}src}$ protein activity (13) .

It is evident from the data presented that the susceptibility of human cells to infection with avian retrovirus is a rare event , perhaps genetically determined . Having found cases of human permissiveness to virus production it will be possible to test if any of the "spontaneous" human sarcoma might be related to the RSV produced in human cells . The criteria formulated by Koch (1890) in the study of bacterial diseases (14) should provide in this context a rational guideline .

TABLE 4 Transformation of human diploid fibroblasts and super-transformation of human sarcoma cells by ASV produced in chicken (CEF) or human cells (HC)

CEF	HC viruses					CEF
Sr-RSV-2	411(Sr-RSV)	272(Sr-RSV)	27(Sr-RSV)	PA_s(Sr-RSV)	PA_s(Pr-RSV)	Pr-RSV-C
			cells			
*PA_s pp60				*PA_s pp60	*PA_s pp60	*PA_s pp60
				*PA_c pp60		
		*27^-		*27^-	27	
					28	
			32			
				37		
			111	111	111	
				115		
				148		
*149^{pp60}				149^{pp60}	149^{pp60}	
				152		
			154^{pp60}	*154^{pp60}	154	
				155		
				207		nd
				221		
				252	*252^{pp60}	

			266	266
			270	
	*272			
			302	
			309	
			322	
			407	
			*408 pp60	*408 pp60
*411 pp60			*411 pp60	*411 pp60
			422	
		428 pp60	428 OR	428

Summary

virus	diploid fibroblasts	sarcomatous cells	Xeroderma pigmentosum	Bloom syndrome
(CEF) RSV	2/130	1/6	0/2	0/1
(HC) RSV	26/130	1/1	0/2	0/1

The numbers correspond to cultures from individual donors found susceptible to transformation by viruses grown in CEF or human fibroblasts . The first and last column are cultures transformed by viruses produced in CEF; in the 5 central columns are cultures transformed by viruses produced by human cells.

* cultures producing infectious virus .

pp^{60} expression or - absence of phosphorilating protein kinase activity. The cultures without marks were not tested.

The results presented here show that cell-free filtrates from human fibroblasts transformed in vitro by an avian sarcomatogenic virus have increased the transforming capacity for human diploid cells as compared to the same virus grown in CEF . The virus produced by human cells is at least in two respects modified as compared to the original virus produced in CEF .

It remains to be seen if some human sarcoma will show any relationship with RSV produced by human cells . This cannot be excluded since :

1) sarcomatogenic retroviruses of different species may share some of the sequences related to the transforming gene ;

2) recombination between RSV and the human c^{Src}(15) proto-oncogene homologue might occur .

CONCLUSIONS

1 - Sr-RSV-2 can induce malignant transformation in some phenotypes of human diploid fibroblasts and "supertransformation" of human sarcomatous cell lines from selected patients .

11 - The transformed (diploid) or "supertransformed" (sarcoma) cells can be permissive for viral production .

111 - The virus produced by human cells is morphologically, immunologically and biochemically modified as compared to the virus produced by avian cells and has increased tropism for human diploid fibroblasts .

1V - Human cultured cells transformed by the virus grown either in chicken or human cells show the specific activity to the $pp60^{V-src}$ protein .

REFERENCES

1. Bernard, C., Lasneret, J., Boucher, M. and Boiron, M. (1969) C.R. Acad. des Sci. (Paris), 268, 624-627 .

2. Aaronson, S. and Todaro, G. (1970) Nature 225, 458-459 .

3. Klement, V., Freedman, M.V., Mc Allister, R., Nelson-Rees, W. and Huebner, R.J. (1971) J. Nat. Cancer Inst. 47, 65-73 .

4. Rhim, J., Vernon, M., DUH, F. and Huebner, R.J. (1973) Int. J. Cancer 12, 734-741 .

5. Sarma, P., Huebner, R., Basker, J., Vernon, M., Gilden, R. (1970) Science 168, 1098 .

6. Pfeffer, L. and Kopelovich, L. (1977) Cell 10, 313-320 .

7. Rasheed, S. and Gardner, M. (1981) J. Nat. Cancer Inst. 66, 43-49 .

8. Rabotti, G.C., Teutsch, B., Mongiat, F. and Mariller, M. (1983) C.R. Acad. Sci. (Paris) 297, 17-24 .

9. Duesberg,P., Robinson, M., Robinson W., Huebner, R. and Turner,H. (1968) Virology 36, 73-86 .

10. Verma, J. (1977) Biochem. Biophys. Acta 473, 1 - 38 .

11. Rabotti G.C. and Blackham, E. (1970) J. Nat. Cancer Inst. 44, 885-991 .

12. Stromberg, K., Hurley, N., Davis, N., Rueckert, R., and Fleissner, E., (1974) J. of Virol. 13, 513-528 .

13. Rabotti , G.C., Teutsch, B., Auger, J., Mariller, M., and Semmel, M. (in preparation).

14. Koch, R., Verhandlungen des X Internationalen Medizinichen Congress (1890) 1 : 35 - 47 .

15. Spector, D.H., Varmus, H.E. and Bishop, M. (1978) Proc. Nat. Acad. Sci. USA 75, 4102 - 4106 .

The Role of Viruses in Human Cancer. Volume II
G. Giraldo and E. Beth editors

RETROVIRUS-INDUCED LEUKEMIAS OF ANIMALS AND HUMANS

William D. Hardy, Jr.
Memorial Sloan-Kettering Cancer Center
Sloan Kettering Division, Graduate School of Medical Sciences,
Cornell University, 1275 York Avenue, New York, New York 10021 (USA)

INTRODUCTION

The first RNA tumor virus was discovered in 1908 in chickens (1). These viruses are now known to occur in many vertebrate species, for example, mice, cats, cattle, gibbon apes and humans, and to cause a wide variety of naturally occurring neoplastic and non-neoplastic diseases (2). The RNA tumor viruses belong to the subfamily oncovirinae of the family of reverse transcriptase containing viruses known as Retroviridae (3). The Retroviridae are enveloped, spherical viruses, containing a single stranded RNA genome and a RNA dependent DNA polymerase (reverse transcriptase) which synthesizes a complementary DNA from the viral RNA, thereby enabling the virus to integrate its genome into the genome of the infected cell during the viral replication cycle (4,5).

Transmission of Retroviruses

Since retrovirus genes are incorporated into the host cell's genome during the normal replication cycle these viruses can be transmitted vertically (genetically) from parent to progeny if the germ cells become infected (6). Assembled virus particles can also be transmitted infectiously between individuals of a species in the natural environment (2). Genetically transmitted retroviruses are known as endogenous viruses and rarely cause diseases. Exogenous retroviruses are transmitted infectiously and cause naturally occurring tumors and other diseases in many species.

In addition to intraspecies transmission, natural interspecies (transpecies) infection is also known to have occurred (7). For example, the exogenous feline leukemia virus (FeLV) is known to have originated from an endogenous rat retrovirus and another (endogenous) cat retrovirus (8), RD114, is known to have been derived from a primate endogenous retrovirus (9). The bovine leukemia virus (BLV) also originated from another (unknown) species and the gibbon ape leukemia virus (GaLV) originated from an endogenous murine retrovirus (10,11). Although these viruses originated from other species millions of years ago and have been transmitted horizontally or vertically within their second species ever since, there is no reason to suppose that transpecies transmission is no longer possible. In fact there is evidence that BLV has been naturally transmitted to sheep relatively recently (12).

Types of Leukemia Viruses

Leukemia viruses can be subdivided into two classes depending on the rapidity by which they induce tumors (13,14). The chronic leukemia viruses, for example, FeLV, induce disease with a fairly long, but variable, latent period whereas the acute leukemia (transforming) viruses usually induce disease rapidly.

The chronic leukemia viruses are common naturally occurring viruses that are replication competent viruses, that is, they are independently able to replicate in their host after infection because their genome contains all the genes necessary for viral replication, i.e.- the *gag*, *pol*, and *env* genes. Interestingly, the chronic leukemia viruses do not contain transforming genes.

Acute leukemia viruses occur only rarely in nature (Table 1) and are generated by genetic recombination between some of the genes of a chronic leukemia virus and a host cell gene, and may be considered to be vectors of cellular genes (15). The acute leukemia viruses are replication defective viruses because they lack the *pol* gene and, often, parts of the *gag* and *env* genes as well and cannot replicate without the help of a replication competent "helper" leukemia virus which supplies the necessary viral proteins required for virus assembly. In place of some of the viral genes, acute transforming retrovirus genomes contain a transforming *onc* or *leuk* gene (oncogene) derived from host cell sequences (16).

Mechanisms of Viral Leukemogenesis

Oncogenes Acute transforming retroviruses contain transforming genes (oncogenes) that are derived from cellular genes, designated proto-oncogenes, which are highly conserved in animals from Drosophila to humans (Table 1) (16,17). Transformation by acute transforming retroviruses may be a consequence of increased or abnormal expression of normal cell genes which are integrated into the DNA of infected cells, or alternatively,

TABLE 1

Known Oncogenes of Acute Transforming Retroviruses

Oncogene	Viruses	Animal Origin
rel	Avian reticuloendotheliosis Virus	Turkey
	Avian Sarcoma Viruses (ASVs):	
src	Rous, B77, rASV & PR	Chicken
fps(fes)	FSV, PRCII, PRCIV	Chicken
yes	Y73, ESV	Chicken
ros	UR-2	Chicken
	Avian Leukemia Viruses (ALVS):	
myb	AMV, E26	Chicken
myc	MC29, CMII, MH2, OK10	Chicken
erb-A	AEV-A	Chicken
erb-B	AEV-B	Chicken
	Murine Sarcoma Viruses (MuSVs):	
mos	Moloney & Gazdar	Mouse
ras (bas)	Kirsten, Harvey, Rasheed	Rat
bas (ras)	BALB	Mouse
fos	FBJ osteosarcoma virus	Mouse
raf	3611	Mouse
	Murine Leukemia Viruses (MuLVs):	
abl	Abelson	Mouse
	Feline Sarcoma Viruses (FeSVs):	
fes (fps)	Snyder-Theilen (ST), Gardner-Arnstein (GA), Hardy-Zuckerman 1 (HZ1)	Cat
fms	Susan McDonough (SM)	Cat
abl	Hardy-Zuckerman 2 (HZ2)	Cat
sis	Parodi-Irgens (PI)	Cat
frg	Gardner-Rasheed (GR)	Cat
	Simian Sarcoma Viruses (SiSVs):	
sis	Simian Sarcoma Virus	Monkey

( ) denotes homology even though oncogene has different name. See reference 18 for oncogene nomenclature.

transformation might result from mutation of the cellular proto-oncogenes that are transduced into the chronic leukemia virus genome causing structural differences between the viral and cellular protein products. In addition to transduction of cellular proto-oncogenes, transformation may occur by activation of cellular proto-oncogenes: 1) due to adjacent integration of chronic leukemia virues (promoter insertion) (19), 2) due to chemical carcinogenesis (mutation) or 3) due to chromosomal translocations containing proto-oncogenes.

Activated oncogenes have been detected in tumors by DNA mediated transfection using NIH 3T3 mouse fibroblasts as indicator cells (20,21). Transfection introduces multiple copies of the oncogene into the recipient fibroblasts (although a single copy is sufficient for transformation) which then show evidence of transformation, i.e. grow as foci, grow in soft agar and grow as tumors in nude mice. Table 2 lists the oncogenes that have been found to be associated with hematopoietic tumors in animals and humans.

TABLE 2

Cellular Oncogenes Associated with Hematopoietic Tumors*

Cell Type	Species	Mode of Induction	Oncogene**
Promyelocytic leukemia	Human	Spontaneous	c-myc
Chronic myelocytic leukemia	Human	Spontaneous	c-abl
B-cell lymphoma	Chicken	Virus (ALV)	c-myc B-lym
	Human	Spontaneous	c-myc
Plasmacytoma/ myeloma	Human Mouse	Spontaneous Chemical	not determined c-myc
T-cell lymphoma	Human Mouse	Spontaneous Spontaneous, Chemical Radiation or Virus	not determined not determined
	Cat	Virus (FeLV)	c-myc

*Not associated with infection by acute transforming viruses

**Oncogene associated with some but not all of the tumors listed for each cell type. See references 22, 23, 24, 25, 26.

Chromosome translocations and deletions: Great excitement has recently occurred with the findings that link cellular oncogenes to chromosomal translocations and deletions in animal and human cancers (27). The oncogenes c-mos, c-myc and c-abl have been located at the breakpoints in the 8;21, 8;14 and 9;22 translocations associated with acute myeloblastic leukemia, Burkitt's lymphoma and chronic myelogenous leukemia in humans respectively (28,29,30,31). Specific translocations are a prominant characteristic of several types of human and animal leukemias and lymphomas. C-myc is located on human chromosome 8 and mouse chromosome 15 (32). In Burkitt's lymphoma, c-myc is translocated from its normal positon on chromosome 8 to a new location adjacent to the immunoglobulin heavy-chain locus on chromosome 14 (29). The gene rearrangement found in human Burkitt's lymphoma (Table 3) also occurs in plamacytomas of mice whereby the translocation is t(12; 15) and the mouse immunoglobulin heavy-chain gene is on chromosome 12 and c-myc is on chromosome 15 (33).

TABLE 3.

Chromosome Translocations and Cellular Oncogenes in Human Leukemias and Other Cancers.

Chromosome Number	Oncogene	Chromosome Aberration	Disease
6	myb	6q- +6 t(6;14)	Acute lymphoblastic leukemia Lymphoma Ovarian Carcinoma
8	mos myc	t(8;21) t(8;14) +8	Acute myeloblastic leukemia Burkitt's lymphoma Acute nonlymphocytic leukemia
9	abl	t(9;22)	Chronic myelogenous leukemia
11	ras^H	11p-	Wilms tumor
12	ras^K	+12	Chronic lymphocytic leukemia
15	fes	t(15;17)	Acute promyelocytic leukemia
20	src	20q-	Myeloproliferative disease
22	sis	t(9;22) t(8;22)	Chronic myelogenous leukemia Burkitt's lymphoma Meningioma

Adapted from reference 32.

Multi-stage process of oncogenesis: Recent studies suggest that transformation by oncogenes is a multi-step process (34). Pathological analysis shows that tumors pass through distinct stages such as anaplasia, metaplasia and neoplasia. Induction of bursal lymphoma in chickens by ALV requires the activation of two separate oncogenes (19,35). The c-*myc* gene becomes activated by adjacent insertion of an ALV provirus while the B-*lym* gene becomes activated by a second, distinct mechanism (25). The participation of two oncogenes has also been reported in the B-cell lymphoma induced by Abelson MuLV in mice (36,37). Similar findings have been reported in a human promyelocytic leukemia and an American Burkitt's lymphoma (38). In the human tumors, the cells carry altered *myc* genes, as well as an activated *N-ras* oncogene. The NIH 3T3 cells, that are used to demonstrate oncogenes by transfection, are an apparent exception to the multi-step oncogenesis process that occurs in nature probably because the cells have been altered by establishment in long term culture and are thus partially transformed and susceptible to single oncogene transformation. Studies by Weinberg and his coworkers have recently shown that, unlike the NIH 3T3 cells, primary rat embryo fibroblasts are not susceptible to single oncogene transformation by v-*ras* or v-*myc* but can be transformed by transfection of both oncogenes (39). These findings, and those of others, are evidence, at the cellular level, that carcinogenesis is a multistage event requiring either: 1) two oncogenes (*ras* and *myc*) (39), 2) adenovirus early region 1A and polyoma virus middle-T or *Ha-ras-1* (40) and 3) a carcinogen and *Ha-ras* (41). Thus it appears that separate establishment and transforming functions are required for oncogenic transformation of primary cells in culture and probably for leukemogenesis *in vivo*.

Chronic leukemia viruses: It is still not known how chronic leukemia viruses induce leukemia. However, two possible mechanisms have been proposed and are currently being investigated. In one hypothesis, recombinant viruses (42), which are called mink cell focus-inducing (MCF) viruses because of their ability to induce foci on mink cells, are generated by recombination between exogenous and endogenous murine leukemia viruses (MuLVs). MCF viruses were first isolated from preleukemic thymuses of AKR mice and appear to be proximal carcinogens in the induction of a variety of hematopoietic neoplasms. T-cell lymphomas are the major type of neoplasm associated with MCF viruses in mice and Weissman and McGrath have proposed a receptor-mediated model of leukemogenesis in which the MCF virus binds to a subset of T-lymphocytes that have receptors specific for the MCF viral envelope gp70 (42). They postulate that this binding is mitogenic and causes proliferation of T-cells that leads to the development of lymphomas. Central to this hypothesis is the prediction that each thymic lymphoma is made up of the clonal progeny of normal T-cells bearing highly specific receptors for a particular viral glycoprotein.

In the second hypothesis, the avian leukosis virus (ALV), which does not possess a transforming gene, induces B-cell lymphomas by "promoter insertion" whereby the ALV proviral DNA is integrated into the host cell DNA and activates the cellular gene c-myc which in turn activates another cellular sequence (B-lym) that is apparently responsible for inducing lymphoma (19,34,35). There is also evidence for site specific integration of Mo-MuLV induced rat thymic lymphomas (43).

The generation of recombinant MCF viruses and the ALV site specific integration "promoter insertion" mechanisms of leukemogenesis explain the long latent periods required for development of leukemia after infection with chronic leukemia viruses. Either mechanism would require multiple cycles of infection to generate recombinant viruses or site specific integration. In addition, this explains why many infected animals escape leukemia development because of the random nature of the required events leading to leukemogenesis.

Acute leukemia viruses: The acute leukemia virus genome carries the onc or leuk transforming gene which encodes a protein responsible for transformation. The best characterized onc gene product of an acute leukemia virus is that of the Abelson MuLV v-abl gene which causes B-cell leukemias in mice (44,45). This molecule has a protein kinase function which may be related to its ability to cause transformation (45). The functional relationship between the cellular 'onc' genes (proto-oncogenes) and their viral derivatives is unknown at present. It is not known if the cellular homologues of the viral onc genes are able to induce cellular transformation (17). However, it is known that the v-onc genes do differ from their c-onc homologues by both deletions and point mutations. For example the v-abl gene appears to be only an internal fragment of the cellular abl gene and there is at least one base mutation that results in an amino acid substitution (46). The oncogenicity of v-abl may be due to these mutations. Another mechanism by which the v-onc genes may induce transformation is by producing a product that stimulates cell growth. Evidence in support of this hypothesis was recently obtained when it was found that the v-sis oncogene encodes a protein very similar to the platelet derived growth factor (PDGF)(47,48). However, whether or not the v-sis gene product has the physiological functions of PDGF remains to be determined.

ANIMAL LEUKEMOGENIC RETROVIRUSES

Avian Retroviruses: Both chronic and acute avian leukemia viruses exist. The chronic avian leukemia viruses (ALV) are divided into 6 subgroups (49). Subgroup A is ubiquitous in commercial chicken flocks, subgroup B is less common, subgroups C and D are rare and subgroups F and G have only been found in pheasants. Subgroup E is an endogenous genetically transmitted virus of chickens which does not cause disease. The most important ALV is subgroup A, a replication competent virus that is present in most chicken flocks and which causes a B-cell lymphoma (50). The virus is transmitted congenitally via the egg or sperm and infectiously via the feces, saliva and dander. The mortality from ALV varies from 0.1% to 23% and the virus has a major economic impact on the poultry industry (51).

Three subgroups of acute transforming avian leukemia viruses exist: 1) avian erythroblastosis viruses (AEV), 2) avian myelocytomatosis-type viruses such as MC29 and 3) avian myeloblastosis-type viruses such as AMV and E26 (see Table 1) (52). These viruses are replication defective and cause the diseases after which they are named after short latent periods. The transforming sequences of these viruses are different for the different viruses.

Murine Retroviruses: The development of inbred strains of mice revealed that some mice were more susceptible to tumor development than others (53). This observation led to the discovery of the murine leukemia virus (MuLV) in 1951 by Ludwik Gross (54). There are three broad classes of chronic MuLVs: 1) ecotropic viruses which replicate only in mouse cells, 2) xenotropic viruses which replicate only in the cells of non-murine species and 3) amphotropic viruses which can replicate in both murine and non-murine cells (55,56). The amphotropic viruses can be further subdivided into wild mouse viruses and laboratory mouse recombinant (MCF) viruses.

Only the xenotropic viruses do not cause disease in mice. The other chronic MuLVs cause leukemia in susceptible strains of laboratory mice only after a long (9-12 months) latent period. Wild mice possess MuLVs which are now known to be transmitted contagiously and induce leukemia development late in life.

Only one acute MuLV has been found in mice. This virus, the Abelson MuLV, is replication defective and causes B cell lymphoma after a short latent period (44,45). However, no acute MuLVs have as yet been isolated from wild mice.

Feline Retroviruses: Pet cats have 3 known retroviruses: 1) the endogenous, non-disease producing RD114 virus (9), 2) the exogenous leukemogenic FeLV (57) and 3) the feline sarcoma virus (58). In addition, multiple copies of FeLV-related sequences are present in the uninfected cells of all pet cats. FeLV is a chronic leukemia virus that occurs in 1-2% of pet cats and is the major disease inducing killer of cats (59). FeLV is a contagious virus but the spread of the virus can be prevented by detecting FeLV infected cats with a simple blood test and isolating infected cats from uninfected cats (60,61). FeLV induces a tumor-specific antigen known as the feline oncornavirus associated cell membrane antigen (FOCMA) on LSA cells (62,63). FOCMA appears to be a recombinant molecule generated by recombination of exogenous FeLV with endogenous FeLV-C related gp70 sequences (64). This evidence suggests that FeLV may be leukemogenic by virtue of formation of recombinant MCF-like FeLV through recombination of exogenous FeLV with endogenous FeLV sequences. No acute feline leukemia viruses have as yet, been discovered in cats.

FeLV causes a wide variety of neoplastic and non-neoplastic diseases in infected cats (65). The virus causes marked immunosuppression in many cats (65, 66, 67,68). In fact, more cats die of secondary diseases due to this FeLV-induced feline acquired immune deficiency syndrome (FAIDS) than die from leukemia. The secondary diseases that occur commonly in FeLV-infected cats include respiratory diseases, chronic stomatitis, feline infectious peritonitis and bacterial septicemias (65). The means by which FeLV induces immunosuppression is unknown. However, it is known that FeLV infection can result in lymphopenia, neutropenia, the formation of immune complexes which are immunosuppresive (67). In addition, one of the FeLV envelope components, p15E, decreases lymphocyte blastogenesis *in vitro*

(69).

About one-third of cats that develop lymphosarcoma (LSA) do not have FeLV in their tumors while two-thirds are FeLV positive (70). The FeLV-negative cats with LSA have a history of exposure to FeLV-infected cats and express the FeLV-induced FOCMA on their LSA cells. Since FeLV-negative leukemic cats do not show evidence of integration of FeLV sequences into the LSA cell DNA, malignant transformation may occur by a "hit and run" mechanism, whereby the provirus is not maintained in the cell, or by FeLV deranging host-cell production of an essential regulatory protein (71). Similar virus-negative human leukemias may also occur for which no virus has yet been isolated.

Bovine Retroviruses: The only known retrovirus of cattle, the bovine leukemia virus (BLV), is widespread in cattle herds in many parts of the world and causes the most common form of bovine leukemia-adult or enzootic leukemia (72,73). However, unlike FeLV in cats, BLV does not appear to cause non-neoplastic diseases. BLV is an exogenous contagiously transmitted chronic leukemia virus that differs in several respects from the other leukemogenic retroviruses such as FeLV (74). For example, it does not replicate *in vivo* in infected cattle because BLV infected cattle produce antibodies to several BLV structural proteins that prevent virus replicaton (72,75). BLV can only be isolated after the infected cells are grown in cell culture for several days (72). The structural proteins of BLV are also different from those of other retroviruses since BLV proteins do not cross react with other mammalian retrovirus proteins (76). Although BLV is transmitted contagiously between cattle it appears to be transmitted in a different manner from that of the avian, murine and feline leukemia viruses. Since it is a cell associated virus, transfer of infected lymphocytes rather than free virus seems to be required and, as a result, the virus is transmitted slowly (77). In addition, genetic susceptibility to infection appears to be important. The newly isolated human leukemia virus (HTLV) is reportedly more similar to BLV than to any of the other retroviruses, although HTLV did not arise from BLV infection of humans.

Primate Retroviruses: Both endogenous and exogenous primate retroviruses exist, but of these viruses, the gibbon ape leukemia virus (GaLV) is the only exogenous, disease inducing leukemia virus (78). GaLV is a chronic leukemia virus with a wide host range that induces myelogenous leukemia in infected apes within 2-4 months (79). Large amounts of the virus have been found in the blood and saliva of infected apes and epidemiologic studies indicate that GaLV is spread contagiously (80). Most investigations into GaLV have been done using ape colonies and the occurrence, if any, of GaLV in wild gibbons is unknown.

HUMAN RETROVIRUSES

Since retroviruses were known to induce leukemia in many animal species it seemed that it would only be a matter of time before a human leukemogenic retrovirus was discovered. The discovery was finally made mainly as a result of the persistent efforts of Robert Gallo and his co-workers at the National Cancer Institute. In order to better understand human leukemias, Gallo studied the FeLV and BLV animal leukemia virus models, where retroviruses induce leukemias but are not always found in the leukemic cells of the animals. As was mentioned earlier, FeLV can induce FeLV-negative lymphosarcomas in 30% of cats that develop these tumors. No FeLV sequences exist in these virus-negative lymphosarcomas but latent FeLV can be reactivated and isolated from bone marrow cells but not from tumor cells (81). Bovine leukemia is an even more relevant model system for HTLV than feline leukemia. BLV does not share antigenic cross-reactivity (76) or genomic homology (73) with any known animal retrovirus. In addition, unlike all other animal retroviruses, BLV does not replicate in vivo in leukemic cells but can be isolated from these cells after several days in cell culture (72). With the BLV model in mind Gallo realized that if he could grow human leukemic cells in culture he might be able to isolate a human retrovirus.

The first major accomplishment by Gallo's group in their search for a human retrovirus was the discovery of T-cell growth factor (TCGF) which enabled them to grow neoplastic T-cells (82,83). The first human retrovirus, human T-cell leukemia virus (HTLV) was isolated from a T-cell line derived from a patient with an aggressive variant of cutaneous T-cell lymphoma (84,85). HTLV was also isolated by Miyoshi and his colleagues in Japan from people with adult T-cell leukemia (ATL) (86), a disease that occurs in clusters in the southern islands of Japan (87). Hinuma and his colleagues named their isolate ATLV but it is known that all ATLV isolates are identical to the known isolates of HTLV (88). To date, the virus has been discovered in Japanese with adult T-cell leukemia and in patients with T-cell lymphosarcoma cell leukemia (TLCL) but not in childhood cancers, nor in most cutaneous T-cell and all non-T-cell leukemias and lymphomas, myeloid leukemias, Hodgkin's disease and solid tumors (89,90).

There are 3 subtypes of HTLV: I, II and III. (91,92). HTLV genomic sequences are not found in the DNA of healthy people which indicates that HTLV is an exogenous contagiously transmitted virus (93). Based on the size, immunological reactivity (92), and the amino acid sequences of the major core protein of the virus (94), as well as the preference of its reverse transcriptase for magnesium over manganese as a divalent cation (95), HTLV most closely resembles BLV suggesting that the two viruses might have had a common ancestral origin. However, this does not suggest current interspecies transmission of BLV from cattle to human.

Serologic reagents and assays were developed to detect HTLV antigens and it was discovered that, like BLV infected cattle, most people infected with HTLV produce antibodies to the viral core proteins (96, 97, 98, 99). Seroepidemiological studies of the occurrence of HTLV antibody by Gallo's group and Japanese investigators revealed that HTLV was spread contagiously and that there are endemic areas in Japan and the Caribbean (96,97,98,99). Most patients with HTLV-induced tumors have antibody to HTLV proteins. In addition about 35% of healthy relatives of HTLV-disease patients have antibodies against HTLV whereas less than 5% of normal healthy non-contacts in non-endemic areas have HTLV antibody (100). Epidemiologic data suggest that close, prolonged or repeated contact, as occurs in the ingestion of mother's milk or in sexual relations, is required for natural HTLV transmission. Other possible modes of transmission might be via insect vectors, as with BLV, or via blood transfusions as with FeLV.

HTLV can transform cord blood T cells in vitro (86) which is unusual since no other chronic leukemia virus can transform cells in vitro. HTLV lacks an onc gene and, as with other chronic leukemia viruses, induces monoclonal tumors (101,102). How then does HTLV cause leukemia? Gallo's group has found that the site of HTLV integration appears to be restricted (103) and that a cellular gene is highly expressed in all HTLV-positive neoplastic T cells and in normal cord blood T cells transformed by HTLV (104). These preliminary findings suggest that promoter insertion, as occurs in ALV-induced chicken lymphoma, may be the operative mechanism in

HTLV-leukemogenesis. Thus, Gallo postulates that HTLV infection may activate, by promoter insertion, a cellular gene encoding for production of TCGF or TCGF receptors (or both) that gives rise to a cell that produces TCGF and TCGF receptors leading to autostimulation and T-cell proliferation (neoplasia) (99).

Recent observations suggest that HTLV may also be etiologically associated with the human acquired immune deficiency syndrome (AIDS). HTLV antibodies have been found in about one-third of AIDS patient and has been isolated from several AIDS patients (105,106,107,108). An almost identical syndrome, FAIDS, occurs in many FeLV-infected pet cats and the human leukemogenic retrovirus may cause a similar immunosuppressive syndrome in people.

CONCLUSION

Many animal species have contagiously transmitted retroviruses that cause leukmias, sarcomas and even carcinomas. Studies of animal retroviruses, especially FeLV and BLV, showed that even though retroviruses cause leukemia they may not be evident in the tumor cells in vivo. For years no retrovirus was found in human leukemic cells. However, through the pioneering efforts of Robert Gallo and his colleagues a contagiously transmitted human retrovirus has been discovered in a subset of T-cell neoplasms of humans. It should be noted, however, that, unlike most animals with leukemia, most human leukemia patients still do not show evidence of a retroviral etiology. As in some cats who develop FeLV-negative leukemias after infection with FeLV, these human leukemias may be caused by a retrovirus that is no longer detectable in any form in the leukemic cells. Further studies of activation of retroviruses from non-leukemic tissues of these patients may result in finding additional human retroviruses.

ACKNOWLEDGEMENTS

The author thanks Dr. A.J. McClelland for assistance in the preparation of this manuscript and Dr. Robert Gallo for generously making data available before publication. Supported by National Institutes of Health grants CA-16599 and CA-08748.

REFERENCES

1. Ellerman, V. and Bang, O: Zentralbl. F. Bakt, Abt.I (Orig) 46: 595, 1908.

2. Hardy, W.D., Jr.: Cancer Invest. 1: 67, 1983.

3. Fenner, F.: Intervirology 7: 61, 1976.

4. Baltimore, D.: Nature 226: 1209, 1970.

5. Temin, H.M. and Mizutani, S.: Nature 226: 1211, 1970.

6. Benveniste, R.E. and Todaro, G.J.: Nature 252: 170,1974.

7. Todaro, G.J.: In: Klein, G. (Ed.), Viral Oncology, Raven Press, New York, 1980, p. 291.

8. Benveniste, R.E., Sherr, C.J. and Todaro, G.J.: Science 190: 886, 1975.

9. Livingston, D.M. and Todaro, G.J.: Virology 53: 142, 1973.

10. Callahan, R., Lieber, M.M., Todaro, G.J., Graves, D.C. and Ferrer, J.F.: Science 192: 1005, 1976.

11. Lieber, M.M., Sherr, C.J., Todaro, G.J., Benveniste, R.E., Callahan, R. and Coon, H.G.: Proc. Natl. Acad. Sci. 72: 2315, 1975.

12. Paulsen, J., Rohde, W., Pauli, G., Harms, E. and Bauer, H.: In: Clemmesen, J. and Yohn, D.S. (Eds.), Comparative Leukemia Research, Karger, Basel, 1976, p. 190.

13. Weiss, R., Teich, N., Varmus, H. and Coffin, J.: RNA Tumor Viruses, Cold Spring Harbor Laboratory, Cold Spring Harbor, N.Y. 1982.

14. Shih, T.Y. and Scolnick, E.M.: In: Klein, G. (Ed.) Viral Oncology, Raven Press, New York, 1980, p. 135.

15. Roussel, M., Saule, S., Lagrou, C., Rommens, C., Beug, H., Graf, T. and Stehelin, D.: Nature 281: 452, 1979.

16. Bishop, J.M.: Cell 23: 5, 1981.

17. Duesberg, P.H.: Nature 304: 219, 1983.

18. Coffin, J.M., Varmus, H.E., Bishop, J.M., Essex, M., Hardy, W.D., Jr., Martin, G.S., Rosenberg, N.E., Scolnick, E.M., Weinberg, R.A. and Vogt, P.K.: J. Virol. 40: 953, 1981.

19. Hayward, W.S., Neel, B.G. and Astrin, S.M.: Nature 290: 475, 1981.

20. Aaronson, S.A. and Rowe, W.R.: Virology 42: 9, 1970.

21. Hill, M. and Hillova, J.: Nature 237: 35, 1972.

22. Cooper, G.M.: Science 218: 801, 1982.

23. Dalla-Favera, R., Wong-Staal, F. and Gallo, R.C.: Nature 299: 61, 1982.

24. Lane, M.A., Sainten, A. and Cooper, G.M.: Cell 28: 873, 1983.

25. Cooper, G.M. and Neiman, P.E.: Nature 287: 656, 1980.

26. Levy, L.S., Gardner, M. and Casey, J.: Personal communication.

27. Klein, G.: Nature 294: 313, 1981.

28. Rowley, J.D.: Science 216: 749, 1982.

29. Hamlyn, P.H. and Rabbitts, T.H.: Nature 304: 135, 1983.

30. deKlein, A., vanKessel, A.G., Grosveld, G., Bartram, C.R., Hagemeijer, A., Bootsma, D., Spurr, N.K., Heisterkamp, N., Groffen, J. and Stephenson, J.R.: Nature 300: 765, 1982.

31. Dalla-Favera, R., Martinotti, S., Gallo, R.C., Erikson, J. and Croce, C.M.: Science 219: 963, 1983.

32. Rowley, J.D.: Nature 301: 290, 1983.

33. Stanton, L.W., Watt, R. and Marcu, K.B.: Nature 303: 401, 1983.

34. Cairn, J.: Nature 289: 353, 1981.

35. Payne, G.S., Courtneidge, S.A., Crittenden, L.B., Fadly, A.M., Bishop, J.M. and Varmus, H.E.: Cell 23: 311, 1981.

36. Baltimore, D., Shields, A., Otto, G., Goff, S., Besmer, P., Witte, O. and Rosenberg, N.: Cold Spring Harbor Symp. Quant. Biol. 44: 849, 1980.

37. Lane, M.A., Neary, D. and Cooper, G.M.: Nature 300: 659, 1982.

38. Murray, M.J. et al. (in press).

39. Land, H., Parada, L.F. and Weinberg, R.A.: Nature 304: 596, 1983.

40. Ruley, H.E.: Nature 304: 602, 1983.

41. Newbold, R.F. and Overell, R.W.: Nature 304: 648, 1983.

42. McGrath, M.S. and Weissman, I.L.: Cell 17: 65, 1979.

43. Tsichlis, P.N., Strauss, P.G. and Hu, L.F.: Nature 302: 445, 1983.

44. Witte, O.N., Rosenberg, N., Paskind, M., Shields, A. and Baltimore, D.: Proc. Natl. Acad. Sci. 75: 2488, 1978.

45. Witte, O.N., Dasgupta, A. and Baltimore, D.: Nature 283: 826, 1980.

46. Goff, S.P. and Baltimore, D.: In: Klein, G. (Ed). Advances in Viral Oncology, Raven Press, New York, 1982, p. 127.

47. Doolittle, R.F., Hunkapillar, M.W., Hood, L.E., Devare, S.G., Robbins, K.C., Aaronson, S.A. and Antoniades, H.N.: Science 221: 275, 1983.

48. Waterfield, M.D., Scrace, G.T., Whittle, N., Stroobant, P., Johnsson, A., Wasteson, A., Westermark, B., Heldin, C-H., Huang, J.S. and Deuel, T.F.: Nature 304: 35, 1983.

49. Duff, R.G. and Vogt, P.K.: Virology 39: 18, 1969.

50. Crittenden, L.B.: In: Essex, M., Todaro, G. and zurHausen, H. (Eds), Viruses in Naturally Occurring Cancers, Cold Spring Harbor Laboratory, New York, 1980, p. 529-541.

51. Purchase, H.G., Okazaki, W. and Burmester, B.R.: Avian Dis. 16: 57, 1972.

52. Stehelin, D., Saule, S., Roussel, M. and Pluguet, N.: In: Essex, M., Todaro, G. and zurHausen, H. (Eds), Viruses in Naturally Occurring Cancers, Cold Spring Harbor Laboratory, New York, 1980, p. 577.

53. Gross, L.: Oncogenic Viruses, 2nd Ed., Pergamon Press, New York. 1970, p. 229.

54. Gross, L.: Proc. Soc. Exp. Biol. Med. 76: 27, 1951.

55. Levy, J.A.: Science 182: 1151, 1973.

56. Hartley, J.W. and Rowe, W.P.: J. Virol. 19: 19, 1976.

57. Jarrett, W.F.H., Crawford, E.M., Martin, W.B. and Davie, F.: Nature 202: 567, 1964.

58. Hardy, W.D., Jr.: In: Hardy, W.D., Jr., Essex, M. and McClelland, A.J. (Eds.) Feline Leukemia Virus, Elsevier, New York, 1980, p. 79.

59. Hardy, W.D., Jr.: In: Hardy, W.D., Jr., Essex, M. and McClelland, A.J. (Eds.) Feline Leukemia Virus, Elsevier, New York, 1980, p. 33.

60. Hardy, W.D., Jr., Old., L.J., Hess, P.W., Essex, M. and Cotter, S.M.: Nature 244: 266, 1973.

61. Hardy, W.D., Jr., McClelland, A.J., Zuckerman, E.E., Hess, P.W., Essex, M., Cotter, S.M., MacEwen, E.G. and Hayes, A.A.: Nature 263: 326, 1976.

62. Hardy, W.D., Jr., Zuckerman, E.E., MacEwen, E.G., Hayes, A.A. and Essex, M.: Nature 270: 249, 1977.

63. Essex, M., Cotter, S.M., Stephenson, J.R., Aaronson, S.A. and Hardy, W.D, Jr.: In: Hiatt, H.A., Watson, J.D., and Winston, J.A. (Eds),

Origins of Human Cancer, Cold Spring Harbor Laboratory, New York, 1977, p. 1197.

64. Snyder, H.W., Jr., Singhal, M.C., Zuckerman, E.E., Jones, F.R. and Hardy, W.D., Jr.: Virology in press.

65. Hardy, W.D., Jr.: In: Hardy, W.D., Jr., Essex, M. and McClelland, A.J. (Eds). Feline Leukemia Virus, Elsevier, New York, 1980, p. 3.

66. Perryman, L.E., Hoover, E.A. and Yohn, D.S.: J. Natl. Cancer Inst. 49: 1357, 1972.

67. Hardy, W.D., Jr.: In: Klein, G., (Ed.),Springer Seminar Immunopathology, Springer-Verlag, New York, 1982, p. 75.

68. Essex, M., Hardy, W.D., Jr., Cotter, S.M. and Jakowski, R.M.: In: Ito, Y. and Dutcher, R.M. (Eds)., Comparative Leukemia Research 1973, Karger Basel, 1975, p. 483.

69. Mathes, L.E., Olsen, R.G., Heberband, L.C., Hoover, E.A. and Schaller, J.P.: Nature 274: 687, 1978.

70. Hardy, W.D., Jr., McClelland, A.J., Zuckerman, E.E., Snyder, H.W., Jr., MacEwen, E.G., Francis, D. and Essex, M.: Nature 288: 90, 1980.

71. Koshy, R., Wong-Staal, F., Gallo, R.C., Hardy, W.D., Jr. and Essex, M.: Virology 99: 135, 1979.

72. Miller, J.M., Miller, L.D., Olson, C. and Gillette, K.G.: J. Natl. Cancer Inst. 43: 1297, 1969.

73. Burny, A., Bruck, C., Chantrenne, H., Cleuter, Y., DeKegel, D., Ghysdael, J., Kettman, R., Leclercq, M., Leunen, J., Mammerickx, M. and Portelle, D.: In: Klein, G. (Ed.) Viral Oncology, Raven Press, New York, 1980, p. 231.

74. Kettmann, R., Portetelle, D., Mammerickx, M., Cleuter, Y., Dekegel, D., Galoux, M., Ghysdael, J., Burny, A. and Cantrenne, H.: Proc. Natl. Acad. Sci. 73: 1014, 1976.

75. Miller, J.M. and Olson, C.: J. Natl. Cancer Inst. 49: 1459, 1972.

76. Ferrer, J.F.: Cancer Res. 32: 1871, 1972.

77. Bech-Nielsen, S., Piper, C.E. and Ferrer, J.F.: J. Am. Vet. Med. Assoc. 39: 1089, 1978.

78. Kawakami, T.G., Huff, S.D., Buckley, P.M., Dungworth, D.L., Snyder, S.P. and Gilden, R.V.: Nature 235: 170, 1972.

79. Snyder, S.P., Dungworth, D.L., Kawakami, T.G., Callaway, E. and Lau, D. T-L.: J. Natl. Cancer Inst. 51: 89, 1973.

80. Gallo, R.C., Gallagher, R.E., Wong-Staal, F., Aoki, T., Markham, P.D., Schetters, H., Ruscetti, F., Valerio, M., Walling, M.J., O'Keefe,

R.T., Saxinger, W.C., Guy Smith, R., Gillespie, D.H. and Reitz, M.S., Jr.: Virology 84: 359, 1978.

81. Rojko, J.L., Hoover, E.A., Quackenbush, S.L. and Olsen, R.G.: Nature 298: 385, 1982.

82. Morgan, D.A., Ruscetti, F.W. and Gallo, R.C.: Science 193: 1007, 1976.

83. Poiesz, B.J., Ruscetti, F.W., Mier, J.W., Woods, A.M. and Gallo, R.C.: Proc. Natl. Acad. Sci. 77: 6815, 1980.

84. Poiesz, B.J., Ruscetti, F.W., Gazdar, A.F., Bunn, P.A., Minna, J.D. and Gallo, R.C.: Proc. Natl. Acad. Sci. 80: 7415, 1980 .

85. Poiesz, B.J., Ruscetti, F.W., Reitz, M.S., Kalyanaraman, V.S. and Gallo, R.C.: Nature 294: 268, 1981.

86. Miyoshi, I., Kubonishi, I., Yoshimoto, S., Akagi, T., Ohtsuki, Y., Shiraishi, Y., Nagata, K and Hinuma, Y.: Nature 294: 770, 1981.

87. Takatsuki, K., Uchujama, J., Sagawa, K. and Yodoi, J.: In: Seno, S., Takaku, F., Irino, S. (Eds.), Topics in Hematology, Oxford, Amsterdam, Excerpta Medica, 1977, p. 73.

88. Yoshida, M., Miyoshi, I and Hinuma, Y.: Proc. Natl. Acad. Sci. 79: 2031, 1982.

89. Gallo, R.C., Kalyanaraman, V.S., Garngadharan, M.G., Sliski, A., Vonderheid, E.C., Maeda, M., Nakao, Y., et al.: Cancer Research 43: 3892, 1983.

90. Popovic, M., Sarin, P.S., Robert-Guroff, M., Kalyanaraman, V.S., Mann, D., Minowada, J., Gallo, R.C. : Science 219: 856, 1983.

91. Kalyanaraman, V.S., Sarngadharan, M.G., Robert-Guroff, M., Blayney, D., Golde, D. and Gallo, R.C.: Science 218: 571, 1982.

92. Kalyanaraman, V.S., Sarngadharan, M.G., Poiesz, B., Ruscetti, F.W. and Gallo, R.C.: J. Virol. 38: 906, 1981.

93. Reitz, M.S., Poiesz, B.J., Ruscetti, F.W. and Gallo, R.C.: Proc. Natl. Acad. Sci. 78: 1887, 1981.

94. Oroszlan, S., Sarngadharan, M.G., Copeland, T.D., Kalyanaraman, V.S., Gilden, R.V. and Gallo, R.C.: Proc. Natl. Acad. Sci. 79: 1291, 1982.

95. Rho, H.M., Poiesz, B.J., Ruscetti, F.W. and Gallo, R.C.: Virology 112: 355, 1981.

96. Kalyanaraman, V.S., Sarngadharan, M.G., Bunn, P.A., Minna, J.D. and Gallo, R.C.: Nature 294: 271, 1981.

97. Posner, L.E., Robert-Guroff, M., Kalyanaraman, V.S., Poiesz, B.J., Ruscetti, F.W., Fossieck, B., Bunn, P.A., Minna, J.D. and Gallo, R.C.: J. Exp. Med. 154: 333, 1981.

98. Robert-Guroff, M., Nakao, Y., Notake, K., Ito, Y., Sliski, A. and Gallo, R.C.: Science 215: 975, 1982.

99. Kalyanaraman, V.S., Sarngadharan, M.G., Nakao, Y., Ito, Y., Aoki, T. and Gallo, R.C.: Proc. Natl. Acad. Sci. 79: 1653, 1982.

100. Gallo, R.C.: Hosp. Pract.: 79, 1983.

101. Wong-Staal, F., Hahn B., Manzari, V., Colombini, S., Franchini, G., Gelmann, E.P. and Gallo, R.C.: Nature 302: 626, 1983.

102. Seiki, M., Hattori, S., Hirayama, Y. and Yoshida, M.: Proc. Natl. Acad. Sci 80: 3618, 1983.

103. Hahn, B., Manzari, V., Colombini, S., Franchini, G., Gallo, R.C. and Wong-Staal, F.: Nature 303: 253, 1983.

104. Manzari, V., Gallo, R.C., Franchini, G., Westin, E., Ceccherini-Nelli, L., Popovic, M. and Wong-Staal, F.: Proc. Natl. Acad. Sci. 80: 11, 1983.

105. Gallo, R.C., Sarin, P.S., Gelmann, E.P., Robert-Guroff, M., Richardson, E., Kalyanaraman, V.S., Mann, D., Sidhu, G.D., Stahl, R.E., Zolla-Pazner, S., Liebowitch, J. and Popovic, M.: Science 220: 865, 1983.

106. Barre-Sinoussi, F., Chermann, J.C., Rey, F., Nugeyre, M.T., Chamaret, S., Gruest, J., Dauguet, C., Axler-Blin, C., Vezinet-Brun, F., Rouzioux, C., Rozenbaum, W. and Montagnier, L.: Science 220: 868, 1983.

107. Gelmann, E.P., Popovic, M., Blayney, D., Masur, H., Sidhu, G., Stahl, R.E. and Galo, R.C.: Science 220: 862, 1983.

108. Essex, M., McLane, M.F., Lee, T.H., Falk, L., Howe, C.W.S., Mullins, J.I., Cabradilla, C. and Francis, D.P.: Science 220: 859, 1983.

The Role of Viruses in Human Cancer. Volume II
G. Giraldo and E. Beth editors

ADULT T-CELL LEUKEMIA VIRUS (ATLV): AN OUTLOOK

YORIO HINUMA
Institute for Virus Research, Kyoto University, Kyoto 606 (Japan)

INTRODUCTION

In 1977, Takatsuki et al. (1) described a new disease entity, adult T-cell leukemia (ATL) which is endemic in southwest of Japan and suggested a viral etiology for this disease. In 1979, Van der Loo et al. (2) detected type C retrovirus from patients with cutaneous T-cell lymphoma (CTCL) including mycosis fungoides and Sézary syndrome in Holland. In 1980, Gallo and his colleagues (3) isolated and characterized the type C retrovirus (HTLV) in a T-cell line (HUT-102) derived from a U.S.A. patient with mycosis fungoides, and later (1981) they (4) isolated the same virus from T-cells cultured from a patient with Sézary syndrome. In 1981, Hinuma et al. (5) found an antigens in a T-cell line (MT-1) derived from an ATL patient, that was reactive with sera of all ATL patients and about 25% of healthy adults in ATL-endemic areas, but not those of healthy adults in ATL-nonendemic areas, as shown by indirect immunofluorescence. The antigen was called ATLA (ATL-associated antigen) and was found to be related to a C type retrovirus, which was detected in the same cell line. This strongly suggested a causal correlation between the retrovirus and ATL. These studies eventually led to the biochemical characterization of a retrovirus in ATL-related cell lines, which was named ATL virus (ATLV)(6). The data strongly suggested ATLA was ATLV-specific. These serological data were confirmed by Gallo's group, demonstrating their virus (HTLV) is reactive with sera of Japanese ATL patients (7).

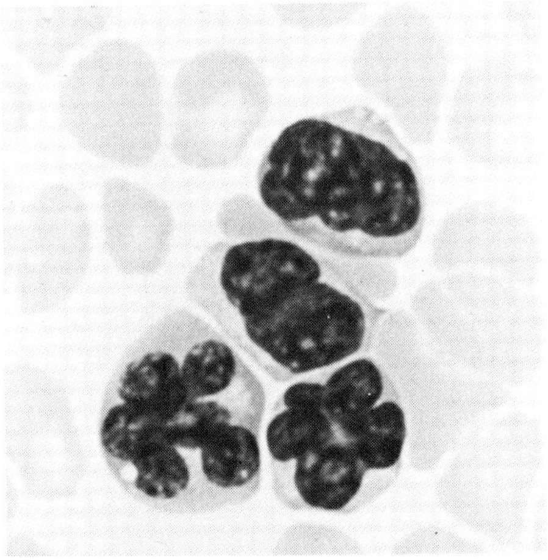

Fig. 1. Leukemic cells in peripheral blood of a patient with ATL. May-Giemsa.

Characteristic clinical and hematological features of typical ATL (1) are: (i) onset in adulthood, but not childhood; (ii) acute leukemia with rapid progression and most patients die within 2 years after onset;

(iii) resistance to treatment with current anti-leukemic agents; (iv) appearence in peripheral blood of pleomorphic leukemic cells that have makedly deformed nuclei (Fig. 1) and OKT-4 surface marker; (v) frequent association with lymphoadenopathy, hepatosplenomegaly, and hypercalcemia; (iv) frequent skin lesions such as erthrodermia and nodule formation. ATL aslo has been called adult T-cell leukemia/lymphoma (ATLL) because of its lymphomatous leukemia nature (8). Recent observations (9) after the discovery of ATLV have shown various atypical forms of ATL other than the typical acute form. They are: i) smoldering type; ii) lymphoma type; iii) chronic type; and iv) crisis type.

During the last 2 years, more than 500 ATL patients have been diagnosed in Japan (unpublished data).

DETECTION AND ISOLATION OF ATLV

ATLV has been detected in fresh ATL leukemic cells by biochemical, immunological and morphological procedures.

Proviral DNA is easily detectable by dot-blot hybridization with a probe of ATLV cDNA in leukemic cells of peripheral blood of ATL patients. When DNA of the leukemic cells was analysed by the Southern blotting procedure, all ATL patients tested showed discrete positive bands indicating the presence of ATLV proviral DNA in their cellular DNA (6). The presence of singly (rarely double) band of each DNA with different sizes in different patients suggests that different cellular sites of ATLV provirus integration. This implies that ATLV is not a widely distributed endogenous virus but is aquired exogenously.

ATLA could also be detected in short-term mass cultures of peripheral leukocytes by indirect immunofluorescence with the use of a monoclonal antibody (to p19 and p28 ATLV polypeptides) or anti-ATLA positive human serum (Fig. 2) (10). However, no ATLA-bearing cells were detected in cell preparations examined before culture, probably due to an unknown mechanism which suppresses ATLV antigen expression *in vivo*.

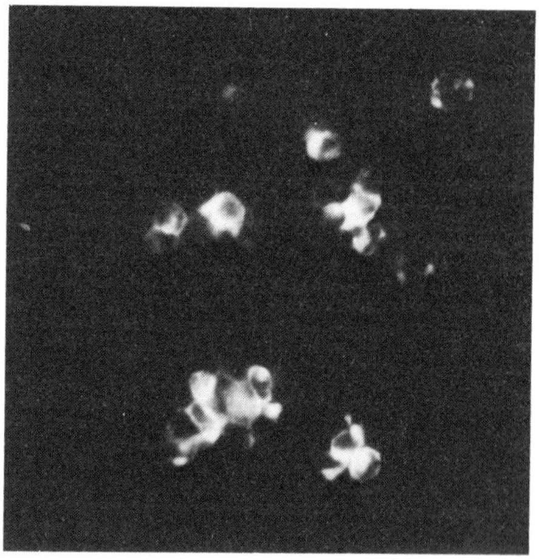

Fig. 2. ATLA-bearing cells in a 3 days-culture of peripheral leukocytes of an ATL patient. Indirect immunofluorescence with use of an anti-ATLA positive human serum.

One to 7 day of *in vitro* culturing is needed in order to detect ATLA-positive cells, in peripheral leukocytes of ATL patients.

Electron microscopic examination of ATL-derived cultures with a high percentage (>20%) of ATLA-positive cells showed the presence of C type virus particles (10). The virus particles observed in the short-term cultures resembled those observed in established ATLV-carrying cell lines, such as MT-1, in that they varied greatly in size and that very few typical budding forms were detected (Fig. 3).

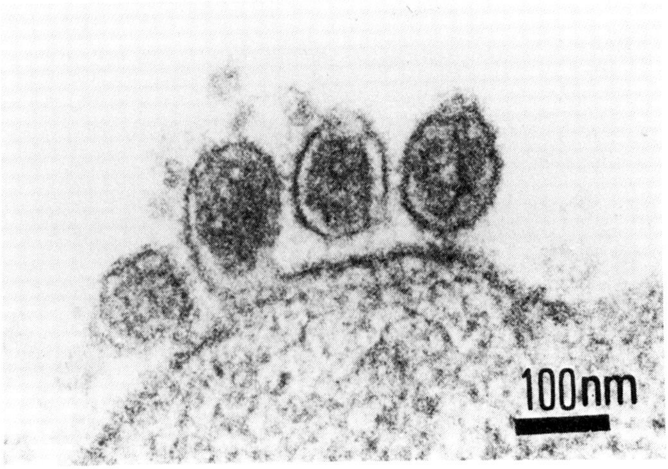

Fig. 3. Budding of ATLV from cell surface of MT-2 cells. Electron micrograph (by N. Nakai).

ATLA-bearing cells were also detected in the cultures of peripheral leukocytes from healthy anti-ATLA-positive adults (10, 11). The frequency of ATLA-positive cultures, are not high as that from ATL patients, which is about 50% of subjects tested. Fewer ATLA positive cells are found in cultures from healthy virus carriers than are found in cultures from ATL patients. However, it has been clearly shown that clonally derived T-cell lines established from such carriers often show abnormal karyotypes in chromosomes (12), similar to those observed in ATL cells.

ATLV has not been isolated from cell-free materials. However, ATLV-carrying cell lines have been successfully isolated. Two types of cultures of peripheral ATL cells have been used. One is a culture of leukocytes containing leukemic cells supplemented with IL-2. From this type of culture, ATLV harboring cell line could be established. This may be originated from leukemic cells, although it could not be completely ruled out that certain normal T-cells were transformed by transmission of ATLV or its genome in leukemic cells in a given leukocyte preparation from ATL patients. The ATLV-carrier cell line, MT-1 (13) and the HTLV-carrier cell line, HUT-102 (3) might be in this category.

Another type of isolation technique for ATLV-carrying cells is by a co-culturing procedure. Co-cultures of ATL cells with normal umbilical cord leukocytes established ATLV-positive T-cell lines, after probable transformation of cord cells by transmission of ATLV from ATL cells such as MT-2 (14), which is a high producer of ATLV, as represented that 100% of cells were ATLA-positive.

INFECTIVITY AND TRANSFORMING ACTIVITY OF ATLV

ATLV produced from the ATLV-positive cell line, MT-2 infects normal human lymphocytes. However, it is an abortive infection, as evidenced by formation of ATLA in the target cells (15). However, continuous growth of ATLA-positive cells were not observed, suggesting cell-free ATLV has no capacity to transform normal cells. On the other hand, efficient transformation of normal human lymphocytes is consistently obtained when the cells are co-cultured with X-ray irradiated MT-2 cells (16, 17). At least two possiblities are considered why cell-associated but not cell-free MT-2 ATLV is capable of transforming normal cells: i) MT-2 cells may produce both transforming and non-transforming viruses, but the amount of the former may be too small to induce efficient infection of cells. However, with co-cultivation of MT-2 cells, a minor population of a transforming virus pool may exert its function in the cells or ii) Some modification (e.g. DNA rearrangement) of virus genetic materials, which are transmetted to the recipient cells by cell to cell infection or fusion may cause transformation. Recently Seiki et al. (18) reported no sequences correspond to "onc" of known animal retroviruses was observed in ATLV cDNA obtained from leukemic cells of an ATL patient. However, this does not exclude a possible presence of either a transforming strain of ATLV or of a gene responsible for transformation, that is different from known "onc".

By co-cultivation procedures, it was found that not only T-cell but also B-cells (unpublished data) and non-T non-B cells were susceptible to transformation (17). Furthermore, lymphocytes from Japanese monkey (Macaca fuscata)(19), and the Japanese white rabbit (20) were found to be transformed by this procedure. More recently, it was found that various type of non-lymphoid cells including simian embryo, Vero, FL and HeLa cells were abortively infected by ATLV (unpublished data). In addition, the formation of syncytia was observed not only in these cells but also XC, Kc, CC-81 and RC cells co-cultured with ATLV carrying cells (unpublished data).

It is interesting to note that unique B-cell lines harbouring both Epstein-Barr virus (EBV) and ATLV have been established (21). We have established several B cell lines from peripheral blood of ATL patients, and these cells as well as their cloned cells consistently express EBV specific EBNA and ATLV specific ATLA. These facts indicate that ATLV infection apparently is not restricted to T-cells. Other types of hematopoietic cells and non-lymphoid cells may be susceptible to ATLV infection and these cells may act as the ATLV reservoir in an infected individual.

TABLE 1
ATLV-SPECIFIC POLYPEPTIDES

Polypeptide	Suggested function
p76	core ?
gp68	envelope precursor
p53	core precursor
gp46	envelope
p28	core
p24	core
p21	core
p19	core
p15	envelope
p10	core

ATLA in MT-2 and YAM (derived from normal adult T-cells transformed by co-culture with irradiated MT-2 cells) cells lines has been analysed by a immunoprecipitation procedure, using anti-ATLA-positive human sera (22, 23, 24, 25). At least, 10 polypeptides suggested to be specific for ATLV have been detected (TABLE 1). Among them, both gp46 and p15 may be associated with envelope and others (p28, p24, p21 and p19) with ATLV core. By analysis with a mouse monoclonal antibody to ATLV (27), it was observed that MT-2 cell line and T-cell lines after transformation by co-culture with MT-2 cells produce both p28 and p19, whereas MT-2 unrelated cell lines, such as MT-1 produced p19 but not p28. A function of p28, that is possibly related to cell transformation, is undermined yet. Recent studies (unpublished data) indicated that an envelope glycoprotein, gp46 may play an essential role for virus binding to cells. This glycoprotein may be an important candidate for a sub-unit ATLV vaccine for prevention of ATLV.

Inorder to develop an animal model for ATLV infection, the experimental ATLV infection of crab-eating monkeys was attempted (unpublished data). In all animals inoculated with living MT-2 cells or cell-free virus, ATLV-specific antigens appeared and persisted in peripheral lymphocytes. All animals developed antibody response to ATLV polypeptides. However, up to 50 weeks after inoculation no animal developed any symptoms of leukemia.

It has been found that most troops of Japanese monkeys distributed from northern to southern parts of Japan showed a high incidence of anti-ATLA positive individuals (26). This geographic distribution is quite different from that in man, since the latter was mainly restricted in the southwest region of Japan. Later it was found that anti-ATLA is detectable not only in Japanese monkeys, but also in many other macaque monkeys in southeast of Asia, and green monkeys and chimpanzees in Africa (unpublished data). However, none of positive sera were found in prosimians and New World monkeys. These obser-varions strongly suggest that ATLV or ATLV like agents is naturally spreading in many species of monkeys.

IMMUNE RESPONSE TO ATLV

ATLV is naturally infecting retrovirus in man. Various immune responses to ATLV in the host have been demonstrated, by either humoral antibody or cellular immune responses.

Antibodies against ATLV in human sera were first demonstrated by indirect immunofluorescence by use of ATLV-carrier cell lines (5, 14). The antigen, called ATLA, which was later shown to be an ATLV specific polypeptide complex. ATLA was found mostly in cytoplasma of cells fixed with acetone. Antibody to surface of ATLV-positive cells has been demonstrated by membrane immunofluores-cence, and called anti-ATLMA (unpublished data). However, anti-ATLA and anti-ATLMA were not distinguishable from each other.

PAGE analysis after immunoprecipitation demonstrated that all serum from ATL patients and healthy virus-carriers reacted with ATLV-specific gp68 and gp46 (possible *env*-products). But polypeptides, p28, p24, p19 and p15, which may be "gag" products, core polypeptides, only reacted with sera having high titers (over 1:80) of anti-ATLA as determined by immunofluorescence (25). Thus, anti-ATLA antibodies in human sera are predominantly directed against glycopolypeptides of ATLV. The geometric mean titer of anti-ATLA of ATL patients was higher than that of healthy donors (28), but ATLV-polypeptides reactive with sera does not differentiate between ATL patients and healthy carriers (24).

It is puzzling that leukemic T-cells of ATL patients have helper/inducer markers (OKT-4 or Leu 3a), but the cells actually exert activity to suppress immunoglobulin synthesis in B-cells in a *in vitro* test (29, 30). Although this apparent discrepancy between cell surface marker and function of the leu-kemic cells has not been resolved, it may be interpreted as follows: these leukemic cells may have activity to induce supprecer T-cells, because ATL leu-lemic cells are known to react with juvenite arthritis patients sera that define a marker of suppressor inducer T-cells.

It has been reported (31, 32) that most ATL patients give a negative reaction in the purified protein derivative (PPD) skin test, suggesting an impairment of cell-mediated immunity. We were interested in the immune status of ATL patients and apparently healthy ATLV-carriers. Therefore, we (33) examined the pattern of antibodies to EBV in ATL patients in comparison with these in anti-ATLA-positive and -negative healthy controls in ATL-endemic areas. All sera including those from ATL patients and healthy controls examined were anti-EBV-positive. The results obtained were as follows; The anti-viral capsid antigen (VCA) titers in ATL patients were were within the range observed in healthy controls. Anti-early antigen (EA) was found in 27% of sera from ATL patients but in only 8% of the sera from anti-ATLA negative controls. Antibodies to EBV-associated nuclear antigen (EBNA), present in almost all healthy adults, were not found in 30% of the sera from ATL patients. Of the sera of anti-ATLA positive healthy adult donors, 11% were negative for EBNA antibody. These results suggest that functional impairment of the T-cell system in most ATL patients, and also to a lesser extent in anti-ATLA-positive controls, who might be healthy ATLV-carriers, may cause an unusual immune response of antibodies to EBV-associated antigens. Indeed, our recent studies has shown that ATL-specific killer T-cells could be induced in *in vitro* system (34) from ATL patient in regression or healthy ATLV-carriers, but not from ATL patients during the progressive stage of the disease (unpublished data).

SPREADING AND TRANSMISSION OF ATLV

As mentioned earlier, it was documented that most, if not all, of anti-ATLA-positive individuals including ATL patients and healthy donors, are ATLV-carriers, since ATLV-antigen-positive cells were detectable by short-term culture or clonal cell culture of their peripheral lymphocytes (10, 11).

Examination of sera from healthy donors aged to 6 to 80 years old in ATL-endemic areas showed (28) that the frequency of seropositive individuals increased gradually with age, starting from a minimum (about 2%) at 6 to 10 years old and reaching a maximum (about 30%) at 40 to 60 years old. However, it is not yet clear why the incidence of anti-ATLA gradually increases with age, especially in a relatively late period of life. Based on this finding, the prevalence of ATLA antibody of residents in a given place, was determined from adults of 40 years old or over (28).

A nation-wide seroepidemiologic survey of ATLV, based on anti-ATLA detection, has been made in Japan. Up to date, sera from adults in 48 different locations were screened for anti-ATLA. The total number of sera tested was 7,675, and 648 (9%) were anti-ATLA positive (unpublished data). As shown in Fig. 4, high incidences of antibody-positive donors were found in almost all locations in southwestern Japan. These areas are ATLV (virus)-endemic areas corresponding to ATL (disease)-endemic areas. However, it is notable that there are some locations in northern Japan where a fairly high incidence of anti-ATLA-positive donors were detected. However, ATLV is not widely distributed in the general population. The serological survey strongly suggested that ATLV may be found in a minority of residents in certain quite restricted regions of each of the Japanese island. It is remarkable that the ATLV-endemic regions were mostly rural small islands or coastal regions (28).

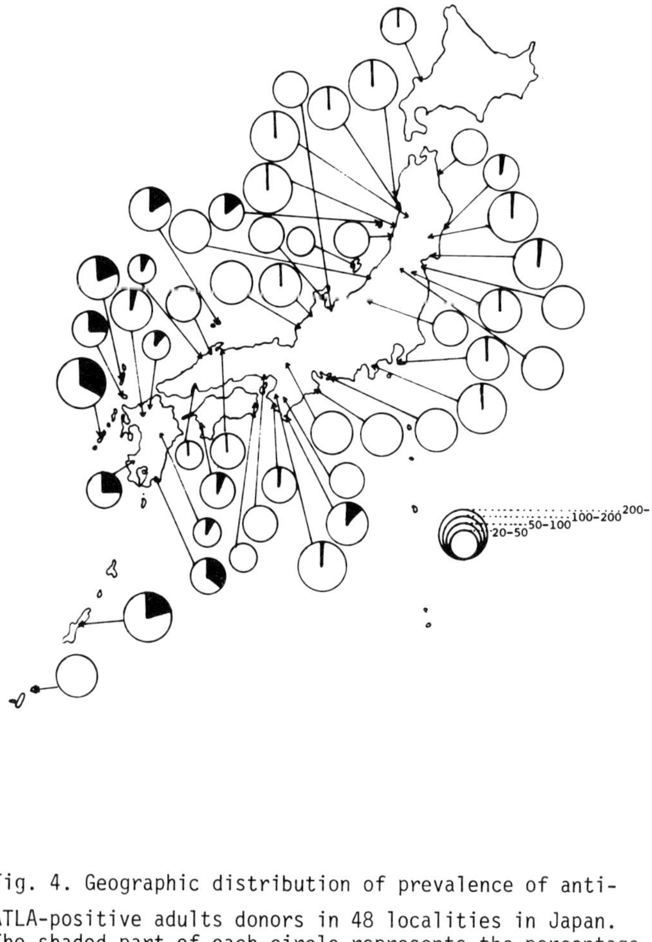

Fig. 4. Geographic distribution of prevalence of anti-ATLA-positive adults donors in 48 localities in Japan. The shaded part of each circle represents the percentage of antibody-positive donors.

We have observed (35) that two patients with ATL had anti-ATLA for at least 5 years and 10 years respectively before onset of ATL. This fact strongly suggests that ATL may develop in seropositive healthy adults. According to the estimation by Tajima (personal communication), one ATL patient may develop from about 2,000 healthy virus-carriers. This estimate is similar to that found in acute poliomyelitis following poliovirus infection or Japanese B encephalitis following infection, in which one patient developed disease from one or two thousands infected individuals.

Studies (36) of the family trees of anti-ATLA-positive and -negative persons showed a high incidence of anti-ATLA-positive individuals families in an ATLV-endemic area. Many women from families, where most members were antibody-positive, were married to men from antibody-negative families and their children were antibody-positive. Some children of anti-ATLA-positive parents were anti-ATLA-positive but other children were anti-ATLA-negative. There were few husband-positive/wife-negative couples, but husband-positive/wife-positive and husband-negative/wife-positive couples were fairly common. From these serological observation two main routes of natural transmission of ATLV were suggested. One route is a vertical transmission.

As already mentioned, ATLV is transmitted by exogenous infection, and not vertically at the genetic level (6). Intrauterine, perinatal or milk infection from virus carrier-mother to child has been considered but not yet proven. The other route to be considered is horizontal transmission, especially from husband to wife, possibly by sexual contact. On the other hand, ATLV transmission caused by blood transfusion has been documented (37). We observed in a retrospective study, that 26 (63%) of 41 recipients of whole blood or blood components containing cells from donors having antibodies to ATLA produced the anti-ATLA. However, no anti-ATLA was detected in all 14 recipients of fresh frozen plasma prepared from anti-ATLA-positive donors. None of 252 recipients of blood with cell components from anti-ATLA-negative donors produced anti-ATLA. The development of anti-ATLA in these recipients may correspond to antibody production after establishment of primary infection with ATLV that is associated with cells in blood from ATLV carriers.

A retrovirus, HTLV isolated from sporadic CTCL cases in U.S.A. is reported to be similar to ATLV in Japan (18). Patients with ATL-like disease and some healthy anti-HTLV-positive individuals were found among Black populations in the West Indies (39), suggesting HTLV- or ATLV-endemic areas outside Japan. We have examined sera collected from adult donors in several places (or races) outside Japan for anti-ATLA (40). As far as studies, no regions were found in which the incidence of anti-ATLA-positive donors was high as that in Japan, and few (less than 2%), if any, positive donors were found in some regions, such as Taiwan, China, The Solomons, North America, The West Indies and South America. It is interesting to note that three positive-donors in Taiwan were confirmed to be Chinese, but one positive donor in Nanjing, China was found to be a Japanese born in an ATL-endemic area of Japan.

RELATION OF ATLV TO DISEASES

Of 250 patients with ATL, 240 were anti-ATLA positive (unpublished data). No clinical and hematological difference have been found between anti-ATLA-positive and -negative ATL. It is possible that certain ATL patients infected with ATLV do not develop detectable antibody to ATLV, because of either tolerance or in-sufficiency, or some other failure, of their immune response. Indeed, it was reported (41) that ATLA-bearing cells were detected in the culture of peripheral lymphocytes from several anti-ATLA-negative ATL patients. However, it can not be excluded that there are certain cases of ATL caused by agents or mechanisms other than ATLV.

Concerning this point, CTCL including mycosis fungoides and Sézary syndrome should be considered carefully, because Gallo's group first isolated HTLV from the patients with CTCL (2, 3). CTCL are known to be distributed throughout the world and not endemic in specific areas. In Japan, none of 12 CTCL patients (10 with mycosis fungoides and 2 with Sézary syndrome) in ATL-nonendemic areas had anti-ATLA. Other investigators in Japan have shown data similar to ours. Moreover, it was demonstrated (unpublished data) that no cDNA of ATLV was detected in peripheral leukocytes from two patients with Sézary syndrome by molecular hybridization procedures. These data strongly suggest no causal relation of ATLV to CTCL.

We have tested 76 sera from homosexual patients with aquired immunodefficiency syndrome (AIDS) which were supplied from 3 different medical institutions in U.S.A. (unpublished data) for anti-ATLA against fixed MT-1 cells smears. All of them were negative. None of 12 AIDS sera analysed by PAGE after immunoprecipitation using MT-2 cells demonstrated bands corresponding to ATLV-associated polypeptides. These results suggest no relationship of ATLV to AIDS.

CONCLUDING REMARKS

ATLV has been characterized extensively within the last few years. The further analysis of genes and proteins of ATLV and also immunologic studies may disclose the mechanism of oncogenesis by ATLV which leads to onset of ATL in man. World-wide epidemiological studies by detection of antibody to ATLV in sera and the detection of viral genome in cells by cloned probes of ATLV most likely will elicit the ecology of ATLV itself and its relation to that of man. In this context, monkey ATLV or ATLV like agents might shed more lights on the evolutionary interaction of ATLV with man.

By the elucidation of exact routes of ATLV transmission, we may establish plans to prevent infection of man with ATLV. Furthermore we may expect to have a vaccine effective for prevention of the viral infection. Concerning the therapy of ATL patients, I don't think that it is too optimistic to predict that certain immunologic treatments linked to ATLV-specificity, such as monoclonal antibodies or killer T-cells may prove to be effective.

ACKNOWLEDGEMENTS

The author is supported, in part, by Grants-in-Aid for Cancer Research from the Ministry of Education, Science and Culture, and from the Ministry of Health and Welfare of Japan, and Grants from Mitsubishi Foundation and from Japan Sciety of Promotion of Science. Thanks are due to Dr. James R. Blakeslee Jr. for review of manuscript and to Miss Kazz Ohta for her secretarial assistance in preparation of manuscript.

REFERENCES

1. Takatsuki, K., Uchiyama, T., Sagawa, K., and Yodoi, J.: In: *Topics in Hematology,* S. Seno, F. Takaku, and S. Irino (eds.), Excerpta Medica, Amsterdam, 1977, p.73.
2. Van der Loo, E. E., Van Muijen, G. N. P., Van Vloten, W. A., Beens, W., Cheffer, E., and Meijer, C. J. L. M.: *Virchow Arch. B Cell Path.* 31: 193, 1979.
3. Poiesz, B. J., Ruscetti, F. W., Gazdar, A. F., Bunn, P. A., Minna, J. D., and Gallo, R. C.: *Proc. Natl. Acad. Sci. U.S.A.* 77: 7415, 1980.
4. Poiesz, B. J., Ruscetti, F. W., Reitz, M. S/, Kalyanaraman, V. S., and Gallo, R. C.: *Nature* 294: 268, 1981.
5. Hinuma, Y., Nagata, K., Hanaoka, M., Nakai, M., Matsumoto, T., Kinoshita, K., Shirakawa, S., and Miyoshi, I.: *Proc. Natl. Acad. Sci. U.S.A.* 78: 6476, 1981.
6. Yoshida, M., Miyoshi, I., and Hinuma, Y.: *Proc. Natl. Acad. Sci. U.S.A.* 79: 2031, 1982.
7. Robert-Guroff, M., Nakao, Y., Natake, K., Ito, Y., Sliski, A., and Gallo, R. C.: *Science* 215: 975, 1982.
8. The T- and B-cell Malignancy Study Group. *Jpn. J. Clin. Oncol.* 11: 15, 1981.
9. Yamaguchi, K., Nishimura, H., Kohrogi, M., Jono, M., Miyamoto, Y., and Takatsuki, K.: *Blood* in press 1983.
10. Hinuma, Y., Gotoh, Y., Sugamura, K., Nagata, K., Goto, T., Nakai, M., Kamada, N., Matsumoto, T., and Kinoshita, K.: *Gann* 73: 341, 1982.
11. Gotoh, Y., Sugamura, K., and Hinuma, Y.: *Proc. Natl. Acad. Sci. U.S.A.* 79: 4780, 1982.
12. Fukuhara, S., Hinuma, Y., Gotoh, Y., and Uchino, H.: *Blood* 61: 205, 1983.

13. Miyoshi, I., Kubonishi, I., Sumida, M., Hiraki, S., Tsubota, T., Kimura, I., Miyamoto, K., and Sato, I.: *Gann* 71: 155, 1980.

14. Miyoshi, I., Kubonishi, I., Yoshimoto, S., Akagi, T., Ohtsuki, Y., Shiraishi, Y., Nagata, K., and Hinuma, Y.: *Nature* 296: 770, 1981.

15. Chosa, T., Yamamoto, N., Tanaka, Y., Koyanagi, Y., and Hinuma, Y.: *Gann* 73: 844, 1982.

16. Miyoshi, I., Yoshimoto, S., Kubonishi, I., Taguchi, H., Shiraishi, Y., Ohtsuki, Y., and Akagi, T.: *Gann* 72: 997, 1981.

17. Yamamoto, N., Okada, M., Koyanagi, Y., Kannagi, M., and Hinuma, Y.: *Science* 217: 737, 1982.

18. Seiki, M., Hattori, S., Hrayama, Y., and Yoshida, M.: *Proc. Natl. Acad. Sci. U.S.A.* 80: 3618, 1983.

19. Miyoshi, I., Taguchi, H., Fujishita, M., Yoshimoto, S., Kubonishi, I., Ohtsuki, Y., Shiraishi, Y., and Akagi, T.: *Lancet* i: 1016, 1982.

20. Miyoshi, I., Yoshimoto, S., Taguchi, H., Kubonishi, I., Fujita, M., Ohtsuki, Y., Siraishi, Y., and Akagi, T.: *Gann* 74: 1, 1983.

21. Yamamoto, N., Matsumoto, T., Koyanagi, Y., Tanaka, Y., and Hinuma, Y.: *Nature* 299: 367, 1982.

22. Yamamoto, N., and Hinuma, Y.: *Int. J. Cancer* 30: 289, 1982.

23. Yamamoto, N., Schneider, J., Hinuma, Y., and Hunsmann, G.: *Z. Naturforsch.* 37: 731, 1982.

24. Yamamoto, N., Schneider, J., Koyanagi, Y., Hinuma, Y., and Hunsmann, G.: *Int. J. Cancer* in press 1983.

25. Schneider, J., Yamamoto, N., Hinuma, Y., and Hunsmann, G.: *Virology* in press, 1983.

26. Ishida, T., Yamamoto, K., Kaneko, R., Tokita, E., and Hinuma, Y.: *Microbiol. Immunol.* 27: 297, 1983.

27. Tanaka, Y., Koyanagi, Y., Chosa, T., Yamamoto, N., and Hinuma, Y.: *Gann* 74: 327, 1983.

28. Hinuma, Y., Komoda, H., Chosa, T., Kondo, T., Kohakura, M., Takenaka, T., Kikuchi, M., Ichimaru, M., Yunoki, K., Sato, I., Matsuo, R., Takiuchi, Y., Uchino, H., and Hanaoka, M.: *Int. J. Cancer* 29: 631, 1982.

29. Uchiyama, T., Sagawa, K., Takatsuki, K., and Uchino, H.: *Clin. Immunol. Immunopathol.* 10: 24, 1978.

30. Yamada, Y.: *Blood* 61: 192, 1983.

31. Matsumoto, M., Nomura, K., Matsumoto, T., Nishioka, K., Hanada, S., Furusho, H., Kikuchi, H., Kato, Y., Utsunomiya, A., Uematsu, T., Iwahashi, M., Hashimoto, S., and Hunoki, K.: *Jpn. J. Clin. Oncol.* 9: 325, 1979.

32. Shimoyama, M., Minato, K., Saito, H., Kitahara, T., Konda, C., Nakazawa, M., Ishihara, K., Watanabe, S., Inada, N., Nagatani, T., Deura, K., and Mikata, A.: *Jpn. J. Clin. Oncol.* 9: 357, 1979.

33. Imai, J., and Hinuma, Y.: *Int. J. Cancer* 31: 197, 1983.

34. Kannagi, M., Sugamura, K., Sato, H., Okochi, K., and Hinuma, Y.: *J. Immunol.* 130: 2942, 1983.

35. Kinoshita, K., Hino, S., Amagasaki, T., Morita, S., Yamada, Y., Ikeda, S., Kamihira, S., Ichimaru, M., Munehisa, T., and Hinuma, Y.: *Gann* 73:684,1982.

36. Tajima, K., Tominaga, S., Suchi, T., Kawagoe, T., Komoda, H., Hinuma, Y., Oda, T., and Fujita, K.: *Gann* 73: 893, 1982.

37. Okochi, K., Sato, H., and Hinuma, Y.: *Vox Sang.* in press 1983.

38. Popovic, M., Reitz, M. S., Sarngadharan, M. G., Robert-Guroff, M., Karyanalaman, V. S., Nakao, Y., Miyoshi, I., Minowada, J., Yoshida, M., Ito, Y., and Gallo, R. C.: *Nature* 300: 63, 1982.

39. Blattner, W. A., Karyanalaman, V. S., Robert-Guroff, M., Lister, T. A., Galton, D. A. G., Srin, P. S., Crawford, M. H., Catovsky, D., Greaves, M., and Gallo, R. C.: *Int. J. Cancer* 31: 275, 1982.

40. Hinuma, Y., Chosa, T., Komoda, H., Mori, I., Suzuki, M., Tajima, K., Pan, I-H. and Lee, M.: *Lancet* i: 825, 1983.

41. Shimoyama, M., Minato, K., Tobinai, M., Nagai, T., Setoya, S., Watanabe, S., Hoshino, H., Miwa, M., Nagoshi, H., Ichiki, N., Fukushima, N., Sugiura, K., and Funaki, N.: *Jpn. J. Clin. Oncol.* 13: 245, 1983.

The Role of Viruses in Human Cancer. Volume II
G. Giraldo and E. Beth editors

VIRUSES IN HUMAN LEUKAEMIA

Abraham Karpas
Dept. of Haematological Medicine, Cambridge University Medical School, Hills Road, Cambridge, UK

Of the large number of micro-organisms directly associated with the development of disease only viruses have been shown to be involved in malignant transformation of cells. Representatives of several groups of viruses, including both DNA and RNA viruses, have been found to cause cell transformation in vivo and in vitro. Malignancy of blood forming cells, namely leukaemia/lymphoma, is mostly caused by retroviruses. Only visceral neurolymphomatosis of chicken (Marek's disease) is caused by a herpes (DNA) virus and Burkitt's Lymphoma in man is associated with another herpes virus, Epstein-Barr virus (EBV).

In the past, several authors reported the isolation of human retroviruses from cultured leukaemia cells, but these turned out to be laboratory contaminants by one or several animal retroviruses.

Until recently there has been no conclusive evidence for retroviral involvement in any type of human leukaemia. Although this remains true for most types of human haematological malignancies, considerable interest has been shown recently in the possible involvement of a retrovirus in an uncommon type of human leukaemia. This followed reports from the USA and Japan of studies concerning the isolation of new C-type retroviruses from cultured leukaemia cells. In both cases the viruses were isolated from in vitro growing haemic cells (1-3). In both cases the viruses were obtained from thymus (T-cell) derived malignancies. However, the sources of origin of the cells and virus, and the evidence so far available for a pathogenic role differ sharply.

In this chapter I shall try to provide a critical review of those viruses which have been proven to be closely associated with human leukaemia/lymphoma, i.e. HTLV, ATLV and Epstein Barr virus (EBV). However, since the two chapters by Dr D. Burkitt and Dr A. Epstein will be dealing with Burkitt's lymphoma and EBV I shall confine myself briefly to EBV research as far as it constitutes a fundamental contribution to viral oncology. In addition I shall discuss our own work on the possible role of virus-like particles (VLP) in the more common types of human leukaemia, namely acute lymphoid and myeloid leukaemia. Before reviewing the possible role of these retroviruses in human malignancy, it might be useful to outline the conditions and criteria necessary to establish a virus as a causative agent of a malignancy.

SUGGESTED CRITERIA TO ESTABLISH VIRAL INVOLVEMENT IN HUMAN MALIGNANCY

The Koch postulates which were originally formulated in order to establish the causative role of a micro-organism for a specific infectious disease stipulate that the isolated agent will induce the same or similar disease following inoculation into a susceptible host and that it would be recoverable from the diseased tissue. These criteria have been used to establish the role of newly isolated animal retroviruses in the aetiology of leukaemia. However, the situation is more complex when dealing with viruses that could be oncogenic in man. A great volume of data have been obtained from experimental studies with the various animal leukaemias. A range of conditions are encountered where a leukaemogenic retrovirus could not always be recovered or even readily demonstrated in the malignant cells for the following reasons: either (i) there is a failure of the virally transformed cells to release a complete infectious virus as a result of a block in virus assembly, or (ii) incomplete integration of the proviral genome into the nuclear DNA of the cells has occurred. When the entire viral genome is integrated it is often, but not always, possible to induce the production of a complete virus and its relsease by treating leukaemic cells with a range of chemical or physical agents such as halogenated pyrimidine nucleosides (IUDR or BUDR), Me_2SO or radiation. However, if the cells fail to produce a complete virus following such treatment, the only way to establish the involvement and presence of the viral genome is by demonstrating the presence of proviral nucleotide sequences in the host DNA. This can only be done by molecular hybridization experiments which will detect even single proviral genes within the malignant or phenotypically normal cell DNA. Any of these methods will help us to evaluate the possible viral involvement in human leukaemia, since a direct experimental approach is not feasible for obvious reasons. Therefore knowledge gained from the study of naturally occurring and experimentally induced retroviral leukaemias of animals allows us to put forward the following postulates as an absolute requirement to establish a causal relationship between a newly isolated human retrovirus and a malignancy. Postulates 1 and 2 apply to all situations.

1. Proviral nucleotide sequences should always be present and detectable in the DNA of the fresh malignant cell.
2. A high percentage, if not all of the affected patients, should have antiviral antibodies.
3. In the case of an exogenous (horizontally) transmitted retrovirus, a significant percentage of the normal population in the endemic areas should also have antibodies.

4. When a complete retrovirus is involved, as in the case of exogenous infection, in vitro infection of cells at least from the same lineage as the malignant cells should be demonstrable.
5. The retrovirus or proviral nucleotide sequences and viral antigen should be demonstrable in the cells of at least some of the serum positive normal individuals.

Transformation and "immortalization" of human haemic cells in vitro by a retrovirus is not an essential requirement to prove the etiological role of the virus because of the inherent difficulty of establishing prolonged growth of malignant cells from leukaemic patients.

By applying the above criteria it should be possible to ascertain whether or not a newly isolated retrovirus such as the human T-cell leukaemia virus (HTLV) from the USA or the adult T-cell leukaemia virus (ATLV) from Japan is the causative virus of the respective diseases.

A. JAPANESE ADULT T-CELL LEUKAEMIA VIRUS (ATLV) AND DISEASE

Japanese adult T-cell leukaemia (ATL) is the only known form of endemic human leukaemia. It is an epidemiologically and clinically unique type of disease found in south-west Japan and reported first in 1977 as a distinct type of leukaemia (4). The clinical disease is characterised by onset in adulthood and short survival, approximately 90% of patients mortality within a year (5). T-cell lines established from ATL patients were found to produce C-type retrovirus particles (Table 1).

Serological and epidemiological studies

Initial serological studies showed already a close immunological association between ATLV and the disease. The study revealed that 100% (44/44) of the patients' sera had antibodies that reacted with the virus producing cultured leukaemic cells. In addition, antibodies with similar activity were detected in 26/100 healthy adults from the endemic areas, whereas such antibodies occurred in only 2/105 normal subjects from non-endemic areas (3). A study of a further 138 ATL patients' sera revealed that only 2 patients did not have antibodies. With 26% serum-positive normal individuals in the endemic areas it was not surprising that many patients with other forms of haematological malignancies in south-west Japan also had antibodies to ATLV (6). However, in leukaemic patients with forms of leukaemia other than ATL outside the endemic area such antibodies could not be found. It was also interesting to note that of the 13 persons with circulating antibodies to ATLV who lived in the non-endemic areas, 10 had been born in the endemic areas. Other individuals with antibodies to ATLV in non-endemic areas had a history of blood transfusion, which might be a source of infection and spread ATLV (7-9). In vivo transmission of the virus has been reported (10).

Examination of sera from healthy individuals in the ATL endemic area showed an age-dependent increase of sero-positive persons with a maximum of about 37% in females and 21% in males at 40 years of age and over (11). Study of family trees of families with serum-positive individuals showed a high incidence of anti-ATLV-positive persons in such families (12), suggesting that a transmission from mother to child and husband to wife might be common (11). The sera of 12 Japanese patients with mycosis fungoides and Sezary Syndrome (MF/SS) from non-endemic areas, did not contain antibodies to ATLV (7). Likewise, I found that the sera of 11 British patients with MF/SS were free from antibodies to ATLV (Table 2).

Virological studies

The initial virological studies were performed with the malignant cell line which had been established by Miyoshi et al. (13) from a patient with adult T-cell leukaemia. The cells which were found to produce C-type particles were used for the serological surveys (3). Immunofluorescence (IF) studies with positive sera from ATL patients revealed that only 1-5% of the cells were positive. This increased by approximately a factor of 5 on growing the cells in the presence of IUDR. Once the MT-2 cell line (see below) was developed most virological studies were performed with these cells since 100% of the cells were positive for the viral antigen (14). C-type virus particles released by budding into the culture fluid were purified on a sucrose gradient. They were found to band at a density of 1.52-1.55 g/ml. The particles contained RNA and specific proteins with molecular weights of 11,000, 14,000, 17,000, 24,000 and 45,000 (15), the last being a glycoprotein (16). The virus also contained RNA-directed DNA-polymerase with a higher activity in the presence of Mg^{2+} than Mn^{2+} (15).

Cell free preparations of ATLV can infect normal lymphocytes both from adults and from cord blood, but transformation and immortalization of the infected culture did not occur. A mouse monoclonal antibody GIN-14 (17) against ATL antigen (ATLA), a core protein of ATLV was used in the IF test to monitor the infection of the target cells by the virus. ATLA positive cells were first detected 3 days after virus innoculation. When PHA was used to stimulate the proliferation of the T-cells, the percentage of ATLA-positive cells reached a maximum of 10-20% within one week after infection; thereafter it decreased to 1% and all cells were dead by the fourth week. When cultures not stimulated by PHA were exposed to cell-free particles only a very small percentage (0.1%) of the cells became ATLA-positive (17). Although the majority of the virus-infected cells were T-cells, other lymphoid cells also became infected by ATLV (Table 3). In vivo infection of human B-cells was reported. All seven B-lymphoblastoid cell lines derived from in vivo infection with EBV, from patients with ATL, contained ATLV-positive cells (0.1-7%). This interesting observation indicates that B-lymphoblasts can be infected by two unrelated viruses and that the B-cells may act as a reservoir for ATLV infection in vivo (18).

Infection and transformation of normal human T-cells was first achieved following the co-cultivation of fresh leukaemic cells from a patient with ATL with white blood cells (WBC) from umbilical cord blood. This gave rise to the MT-2 cell line with 100% of the cells positive for the viral antigen (14). When several irradiated ATLV-positive T-cell lines were used in co-cultivation with normal adult peripheral blood or cord WBC, transformation occurred only with the MT-2 line. No transformation occurred with MT-1 or MT-4 cells (19).

ATLV/ATLA positive cell clones were isolated from peripheral blood lymphocytes of all 5 anti-ATLA-sera positive healthy adults (20). The continuous in vitro proliferation of the ATLV positive cells was dependent on the addition of T-cell growth factor (TCGF) to the growth medium, but after four months' culture they acquired the ability to grow without TCGF (20). This might indicate that ATLV infected T-cells from normal individuals can become malignant. Similar preparation of cells from 6 serum-negative individuals which were also grown in the presence of TCGF did not express ATLV/ATLA, and the cell growth remained dependent on TCGF.

Infection and transformation of Japanese monkey lymphoid cells by ATLV was achieved following the co-cultivation of irradiated MT-2 cells with WBC obtained from serum negative monkeys (21). Over 90% of the monkey cells (Si-1 line) were found to be positive for ATLV.

Molecular hybridization

cDNA was prepared from sucrose gradient purified virus using the endogenous reaction of reverse transcriptase. The cDNA hybridized with the viral RNA as well as with cytoplasmic RNA of the virus producing lines (MT-1 and 2). However, it failed to hybridize with RNA from virus negative human cells. To test for the presence of ATLV-proviral sequences in fresh ATL cells, high molecular weight (chromosomal) DNA was digested with restriction enzymes. It was then separated in agarose gel and analyzed by the Southern blotting technique. The DNA from the leukaemic cells of all 5 patients with ATL showed discrete positive bands, indicating the presence of ATLV proviral DNA in their cellular DNA. The presence of only one band in each DNA with different sizes in different patients indicates differences in the cellular sites of integration of ATLV proviral DNA. This supports the conclusion that ATLV is not a widely distributed endogenous virus, but an exogenous one. In the case of endogenous retroviruses, there are multiple proviral copies (22). DNA from lymphocytes of normal adults and from human cell lines from other malignancies did not contain proviral sequences of ATLV (15).

Molecular cloning

Both the integrated proviral DNA as well as viral cDNA have been cloned. The clone (ATM-1) of an integrated provirus DNA in the MT-2 cell line contained DNA of about 13,000 base pairs (bp) long, which included flanking host DNA. The proviral sequence was about 8,000 bp long with long terminal repeats (LTR) at both ends. These two LTR sequences were linked to cellular sequences with direct repeats of 7 bp. Each LTR consisted of 754 bp. Analysis of the sequence of the proviral DNA showed that the LTR had the same structural organisation as that of other animal retroviruses - that is, U3-R-U5. Hence the mechanism of ATLV replication appeared to be the same as that of other known animal retroviruses. However, the lengths of the small terminal repeats at the end of the RNA genome, 228± bases, is longer than those of several other animal retroviruses, but similar in length to the LTR of BLV (23). Recently the complete nucleotide sequence of ATLV provirus was found to have 9,032 bases (24).

In vivo transmission

A study of a serum positive woman during the ninth month of pregnancy showed that 0.5% of her lymphocytes were positive for ATLA. Immediately after delivery cord blood was found to contain the same level of antibodies (IgG) and lymphocytes cultured from cord blood revealed 0.5% positive reaction for ATLA. This observation points to a possible transplacental transmission of the virus (10), although the possibility of vertical transmission, namely through a proviral integration in the ovum, needs to be studied.

Blood transfusion has been suggested as a possible mode of ATLV transmission (8).

Animal Studies

Serum samples obtained from eighteen of twenty Japanese macaca contained antibodies to ATLV (25). Cultured lymphocytes (1%) from all four serum positive monkeys which were tested for ATLV were found to be positive for the virus. Electron microscope examination of one of those cultures revealed c-type particles (26). The 20 animals were captured in different localities in Japan. In the Japanese monkey, horizontal transmission of ATLV was demonstrated from virus positive male to virus negative female monkey (27).

Serum samples from 27 African Green monkeys (Cercopithecus aethiops), 13 rhesus monkeys (M. mutatta) and 12 chimpanzees housed in Germany were tested for ATLA antibodies. 82% of the African Green monkey sera were positive for ATLA, but none of the sera from the rhesus and chimpanzee. ATLA-positive cells were found in cultured lymphocytes from the serum-positive African green monkeys. These African Green monkeys had been in captivity for 2-6 years. However, none of more than 1000 sera from West German leukaemic patients and healthy individuals (including the animal keepers of the serum-positive African Green monkeys) possessed antibodies to ATLA (28).

The most important question relating to this interesting observation is whether the African Green monkeys in the wilds of Kenya are also serum positive. If they are, it will be important to determine whether there are also endemic areas of human infection in Kenya and if so, how the African strain relates to the Japanese strain.

B. HUMAN T-CELL LEUKAEMIA VIRUS (HTLV) AND MYCOSIS FUNGOIDES/SEZARY SYNDROME, ADULT T-CELL LEUKAEMIC LYMPHOMA OF BLACKS

Mycosis fungoides is a T-cell lymphoma that primarily involves the skin. Sezary syndrome is considered to be the leukaemic variant of the disease. The same type of neoplastic T-cells are involved in both diseases. Although mycosis fungoides (MF) and Sezary syndrome (SS) are the most common haematological malinancies which involve the skin, they are uncommon but spread world wide (Table 1). Survival time is often measured in years. In a series reported by Lennert (29), MF represent 0.5% and SS 0.26% of the malignant lymphomas. One per 2-3 million of the European population is affected by these forms of cutaneous T-cell leukaemia/lymphoma. In 1979 a Dutch group reported the presence of c-type virus-like particles in patients with mycosis fungoides and Sezary syndrome. However, no budding virus could be seen and the particles were found only in the cytoplasm of the Langerhans cells (dendritic histiocytes of the dermis), but never in the T-cells (30).

Adult T-cell leukaemia/lymphoma (ATLL) of blacks is a disease which was recognised as a distinct entity by Catovsky et al. in 1982 (31). It appears to be a malignancy of helper T-cells which also resemble morphologically the leukaemic cells of Japanese ATL; the clinical course resembles that of ATL of Japan, rather than MF/SS. Initially the disease was thought to be endemic in the Caribbean, but more recent studies revealed that it affects also USA blacks (32) and that ATLV serum positive leukaemic and normal individuals were found in Nigeria (33). Since ATLL of blacks was not recognised until recently as a disease, distinct from T-cell acute lymphoid leukaemic (T-ALL) and MF/SS, many of the reported studies about HTLV and the disease might be confusing. Nevertheless, the various studies will be reviewed in temporal sequence as they appeared in the literature and I shall try to put them in the right perspective in the discussion.

In December 1980 it was reported from the USA that a retrovirus had been isolated from a cell line (HUT 102) which had been established the previous year (34) from a 26 years old, black US patient with mycosis fungoides (1). A second isolate of HTLV from uncultured cells of a patient with Sezary T-cell leukaemia was later reported by the same laboratory (2). The latter patient was a 64 year old black female who was born in the West Indies and went to the USA as an adult. Using T-cell growth factor (TCGF), a cell line (CTCL-2 line) was also established from this patient (2). In 1982 a new subtype of HTLV (HTLV-II) (35) was recovered from the Mo cell line which was established in 1978 from a west coast patient with a T-cell variant of hairy cell leukaemia (36).

Serological and epidemiological studies

The initial and most of the subsequent serological studies were performed with a protein of 24,000 MW (p.24) which had been purified from HTLV (37). The p.24 was used in radioimmuno-assays of sera for antibodies in US patients with T-cell malignancy and in normal individuals, including some of the patients' relations. Of the sera from 23 patients with T-cell malignancy, only 2 were positive; of 11 normal family members, one was positive. Antibodies to HTLV p.24 could not be found in any of the other sera samples from 50 normal US individuals (38,39). Further tests of several hundred randomly selected normal donors in the US and Europe did not reveal antibodies to HTLV (40); the sera of US patients with a wide range of other leukaemias were also negative (Table 2). On the other hand, many Japanese sera from patients with ATL cross-reacted with HTLV, although of the 39 sera from healthy donors from the ATL endemic region of Japan, none had antibodies to HTLV (41). In another report the sera of 7 black British patients of Caribbean origin with T-cell malignancy were found to be positive for HTLV (42), whereas another 9 black patients from the Caribbean with a range of haematological malignancies, including one with mycosis fungoides, did not have detectable antibodies to HTLV. Of the 336 sera samples obtained from normal Caribbeans, 11 (3%) were found to be serum positive. Sera of 6 British patients with MF/SS were found to be negative (42). The same group (35) also found that the sera of one patient with a T-cell variant of hairy cell leukaemia had antibodies to HTLV.

Anti-sera to the purified p.12 of bovine leukaemia virus (BLV) reacted not only with BLV p.12 but also cross-reacted with its HTLV homologue. Amino acid sequence of HTLV p.24 revealed statistically significant sequence homology only to BLV (43).

In 1981 the development and study of a monoclonal antibody to the p.19 of HTLV was reported (44). Later studies with this monoclonal antibody revealed that it reacts with the epithelial component of normal (uninfected) human thymus (45). Hence it is not a viral but a cellular p.19.

Virological studies

Initially Poiesz et al. (1,2) reported the isolation of HTLV from cell lines derived from one patient with MF (HUT 102 line) and another with SS (CTCL-2 line). A further report mentioned the isolation of HTLV from 7 black British patients of Caribbean origin, although no details are given in the paper (42). Further isolations of HTLV from the east and west coast of the USA, the Caribbean, Venezuela, Brazil, Guyana, Ecuador, Israel and Alaska were mentioned in a recent review article (46), but no details were given. The HTLV isolated from Israel was from diffuse histiocytic lymphoma patients (47). According to Gallo et al. (48), HTLV is "T-cell tropic" and B-cells are not infected by the virus, since HTLV was present in the HUT-102 cell line, but not in EBV-infected B-lymphoblasts derived from the same patient. Also other human B-cell lines could not be infected with HTLV in vitro (49).

In 1982 a new subtype of HTLV (HTLV II) (35) was recorded from the Mo cell line which was established from a west coast patient with a T-cell variant of hairy cell leukaemia (36). However, in a recent paper Reitz et al. (50) reported that HTLV II rate of hybridization with HTLV I is less than 10%. It therefore appeared that HTLV II is an entirely different virus from HTLV I. The paper does not provide any data which could explain why the isolate from the Mo line is grouped together with HTLV.

Molecular hybridization

HTLV produced by the HUT 102 cell line was purified and the endogenous reverse transcriptase activity was used to generate ^{3}H-cDNA (51). This cDNA was used in hybridization studies with DNAs prepared from 13 species of mammals with known endogenous retroviruses. No significant level of hybridization was recorded with any of the DNA preparations. Likewise no hybridization was recorded between HTLV cDNA and 18 different 70S RNAs from various animal retroviruses grown in a wide range of human and animal cells. Hybridization between HTLV cDNA and DNAs from 6 normal human tissue and one human ovarian adenocarcinoma showed no HTLV sequences in any of the human tissue tested (51). It was therefore concluded that HTLV is not an endogenous human retrovirus. HTLV-cDNA does hybridize to the DNA of the HUT 102 cell line. On the other hand, no studies of hybridization to DNA from fresh malignant T-cells from either mycosis fungoides, Sezary syndrome or Caribbean blacks with T-cell leukaemic or lymphoma has been included in the reported study (51). The only paper reporting hybridization of HTLV cDNA with the DNA of fresh leukaemic cells concerned a black patient with SS who gave rise to the second HTLV isolate; 17% hybridization was recorded with the leukaemic cell DNA, while 3-11% hybridization with normal control human DNA was reported (2).

The reverse transcriptase of HTLV was reported to "slightly prefer" Mg^{2+} over Mn^{2+} and to be serologically distinct from the reverse transcriptase of several animal retroviruses (52). Using molecular hybridization, it was also reported that the Japanese ATLV is identical with HTLV (53).

Transmission studies

Attempts to transmit HTLV to human T-cells obtained from randomly selected normal adult donors were unsuccessful. However, when white blood cells were established from relatives of patients with HTLV-positive T-cell neoplasm, successful transmission of the virus was reported in 5 out of 10 individuals (49), suggesting genetic susceptibility to infection by HTLV. However, isolation of HTLV from leukaemic T-cells from the USA, Israel, West Indies and Japan and the infection by the virus of cord T-cells was recently reported (47). Genetic susceptibility was no longer important. The continuous proliferation of HTLV infected T-cells depends on the addition of TCGF to the culture medium.

C. EPSTEIN-BARR VIRUS (EBV) AND BURKITT'S LYMPHOMA

EBV was first observed in lymphoblastoid cell lines from African patients with Burkitt's lymphoma (54). Henle et al. (55) who were the first to demonstrate the infectivity of the virus to normal human lymphoid cells also established that it is the causative agent of glandular fever (56). These findings opened the door to the understanding of the role of EBV infection in man and to the observation that most individuals become infected by the virus, develop antibodies against it and usually remain latent carriers of the virus for the rest of their lives. Zur Hausen and Schulte-Holthausen (57) were the first to report the presence of EBV nucleic acid in "virus free" cells from Burkitt's lymphoma. This observation explained the latent carrier state of the virus in man. Klein and his associates reported the finding of EBV-genome negative cells in cases of African Burkitt's lymphoma, some of which gave rise to cell lines (58). Karyotype analysis revealed identical cytogenetic abnormalities (translocations) in EBV-positive as well as in EBV-negative Burkitt's lymphoma cells (59). B-cells, whether normal or malignant, contain numerous receptors for EBV.

D. VIRUS-LIKE PARTICLES IN HUMAN LEUKAEMIC CELLS

In our laboratory over 400 samples of fresh leukaemic cells were examined by electron microscopy for budding c-type virus particles. None could be found in any sample of fresh leukaemic cells.

However, over the past 20 years several investigators have reported virus-like particles (VLP) aggregated within so-called inclusion bodies in the cytoplasm of fresh human leukaemic lymphoblasts and myeloblasts (reviewed by Karpas and Fischer (60)). A systematic search for inclusion bodies in the white blood cells of children with acute lympohoid leukaemia revealed that such VLP-containing inclusions could be found in the fresh leukaemic cells of 12 of the 28 patients studied (61). We were first to establish long term human leukaemic cell lines in tissue culture which express such VLP within cytoplasmic inclusion bodies (62-64). Following culture of the cells in the presence of bromodeoxyuridine (BUDR) or Me_2SO, a 40-50 fold increase in the number of VLP-containing cytoplasmic inclusion bodies occurred (60). We have tried but failed to infect a wide range of target cells by a cell-free filtrate containing VLP. However, in one of 102 experiments the co-cultivation of lethally irradiated (10,000 rads) VLP-producing leukaemic cell line with fresh cultures of normal bone marrow cells resulted in active proliferation of the bone marrow cells (65). This single case of success could have resulted from infection of the bone marrow cells by the VLP or as a result of transfection by DNA. The difficulty in reproducing the above study could be explained by the equally infrequent success in obtaining continuous proliferation of human leukaemic cells. In my own experience from samples of fresh leukaemic cells from 900 patients, I have been able to obtain only six continuous cell lines.

DISCUSSION

Epidemiological studies from Europe and the USA into the incidence of leukaemia have not provided any evidence of contact transmission, nor could a distinct geographical area with a higher incidence of leukaemia be located (see recent review (66)). In 1977 the first study of an endemic incidence of leukaemia (ATL) was reported from Japan (Table 1) (4). A C-type retrovirus ATLV was found to be produced by a leukaemic cell line (MT-1) from a patient with ATL (3). The review of the published data concerning ATLV indicates that it fulfils all the suggested postulates for a possible aetiological role in the development of a human malignancy (Tables 2 & 3). Proviral nucleotide sequences were present in each of the five DNA preparations derived from fresh ATL cells. Likewise nearly all (152/154) of the patients' sera had antiviral antibodies. As one would expect of exogenous virus, a significant percentage (one fourth) of normal persons in the endemic area of south-west Japan had antibodies to the viral antigen (ATLA). Viral antigen could be demonstrated in the leukaemic cells, but only after in vitro culture (67). Likewise the presence of virus-positive cells in normal serum-positive individuals could be detected only after culture. Cell-free preparations of ATLV infected normal human lymphocytes. In short term culture it was shown that ATLV also infected a mixed population of lymphocytes (T, B and Null cells) (17).

Differences between Adult T-cell Leukaemia (ATL) of Japan, ATL Lymphoma (ATLL) of blacks, Mycosis Fungoides/Sezary Syndrome (MF/SS) and Acute T-cell Leukaemia (ATL)

	Epidemiology	Family Clustering	Disease	% mortality in 12 months	Skin lesions	Retrovirus isolated
ATL	Endemic south-west Japan	+	acute	90%	+-	ATLV
ATLL	Blacks US, UK, Caribbean	?	acute	50%	+-	HTLV
MF/SS	Uncommon, non-endemic world-wide	-	chronic	0	+-	?
ALL	Non-endemic world-wide	-	acute	50%	-	-

TABLE 1

Serological Studies

C-type retrovirus	Japanese sera Normal in endemic area	ATL	MF/SS	UK and US sera Normal	MF/SS	ATLL	Caribbean normal serum	Italian (I) German (G) sera
ATLV	26/100	44/44	0/12	0/60*	0/11 UK	2/2	0-1%	0/26 I 0/1000 G
HTLV	0/39	5/6	?	1/66 US	2/23 US 0/6 UK	7/7	3%	?

* British blacks of West Indies origin.

I The Italian sera were from children with acute lymphcid leukaemia.

G The German sera were from leukaemic and normal individuals ( )

TABLE 2

TABLE 3 - Virological Studies

C-type retrovirus	Infection of normal human cells	Hybridization of viral cDNA to human DNA			Animal infection
		Normal	Leukaemic		
			ATL/ATLL	MF/SS	
ATLV	T, B, Null	-	+	-	M. Fuscata C. Aethiops
HTLV	T only	-	+*	?	?

* The hybridization was done with DNA from cultured leukaemic cells

In vivo infection of B-cells with ATLV was also reported in 7/7 EBV-infected lymphoblastoid lines from patients with ATL (18). The high percentage of Japanese monkeys (Macaca fuscata) infected with the virus points to a possible reservoir of infection for man by ATLV. Whether a vector is involved in the transmission of the virus needs to be explored. The demonstration of contact transmission between monkeys indicates that contact transmission may occur in man. The epidemiological studies from the endemic areas of Japan show clustering of serum-positive individuals in families of ATL patients. The presence of anti-viral antibodies in normal individuals does not necessarily assure immunity from the development of the disease, since it was reported that 5 and 10 year old sera of two new patients with ATL had antibodies to the virus (68).

Epidemiological studies indicate that viral transmission from mother to child might be an important route of infection (10). However, no clue is provided as to the reasons why the disease appears only in adults.

The absence of antibodies to ATLV was established for the following groups:

(1) 12 patients with mycosis fungoides (MF) and Sezary's syndrome (SS), from the non-endemic areas of Japan (7).

(2) 11 British patients with MF/SS (Karpas, unpublished).

(3) 1,000 sera of German leukaemic patients and healthy controls (28).

These findings suggest that ATLV is associated only with ATL. Furthermore, the absence of proviral sequences in the DNA from two Caucasian British patients with SS and from four Japanese patients with MF makes it unlikely that ATLV is associated with those forms of human leukaemia (69).

What are the connections between HTLV and MF/SS, other forms of T-cell leukaemia/lymphoma, T-cell hairy cell leukaemia and diffuse histiocytic lymphoma (47) in the Western hemisphere? From careful reading of the published data about HTLV and T-cell malignancies and cross-checking the evidence referred to in these papers, it appears to the reviewer that HTLV has so far been shown to be associated only with one form of human malignancy, namely with T-cell leukaemia/lymphoma (ATLL) of blacks. ATLL of blacks was first described as a distinct entity by Catovsky et al. in 1982 (31). The clinical manifestation of the disease resembles ATL of Japan. We were also able to confirm that the sera of two British black patients with ATLL contained antibodies which reacted with ATLV infected cells (MT-1 line). It is therefore likely that the earlier reported diagnosis of MF and SS of the black (US) patients (1,2) whose cells gave rise to the original HTLV isolate was incorrect. The failure to find ATLV proviral sequences and the absence of antibodies in Japanese and British patients with MF/SS supports our view that ATLV or a similar virus (HTLV) is not involved in the development of those malignancies. We have also tested for the presence of antibodies to ATLV in a case of T-cell chronic lymphoid leukaemia (T-CLL) of an adult patient from Israel, but failed to detect antibodies or ATLV antigen in the malignant cells. This case is of particular significance since the malignant cells were helper T-cells, namely the same population of cells as in ATL cases. A recent report from Holland about the expression of HTLV in neoplastic T-cells froma patient with T-CLL (70) is based on the reaction of the monoclonal antibodies to p.19 (44) with the malignant cells. However, since it has been shown that the monoclonal antibodies react also with non-infected (normal) cells (45), such a reaction cannot be taken as evidence for the presence of HTLV in T-CLL.

As to the presence of HTLV proviral sequences in fresh leukaemic cells, only DNA from fresh ATL cells from Japan were used in hybridization studies (50,71). All other reported hybridizations with DNA from ATLL of blacks were performed with cell lines rather than with fresh cells (47,72). Therefore the most important evidence necessary to link HTLV with any human malignancy other than the Japanese ATL is still missing from the literature.

The results of the infectivity studies with HTLV differ from the studies with ATLV in several important respects:

(a) HTLV failed to infect B-cells (48,49), while ATLV infected B, T and Null lymphocytes from unrelated donors (18,19).

(b) The continuous in vitro proliferation of the HTLV infected cultures was dependent on the addition of TCGF (49). However, the addition of TCGF to the growth medium will allow also non-infected T-cells from normal individuals to proliferate in vitro.

The serological data from Japan support the epidemiological studies on the endemic nature of ATL in south-west Japan. In contrast, the epidemiological studies of ATLL in the Western hemisphere have not provided as yet convincing evidence to point to any endemic area. Screening of anibodies to HTLV or ATLV in the West Indies does not support the notion that the Caribbean is an endemic area. The highest rate recorded is 3% (11/336) of HTLV serum positive individuals in St. Vincent (42). When ATLV was used, 1% (2/191) serum positive were found in Dominica while no serum positive case could be found in Puerto Rico (0/104) (73). None of the sera from 60 normal British blacks of Caribbean origin I tested had antibodies to ATLV. It appears that certain blacks are likely to be carriers of the virus since ATLV serum positive individuals have been found in Africa (33). Therefore, it is likely that an endemic area(s) will be found on the African continent. Those individuals whose ancestors may have come from the endemic areas to the American continent could harbour the virus. Intercontinental transfer of ATL and virus was documented recently in the case of a 49 year old Japanese woman who was born in the endemic area of south-west Japan and emigrated to the USA 26 years prior to the development of malignancy (ATL) (74). The failure, so far, to detect an endemic area in the USA and UK with a significant percentage of HTLV/ATLV serum positives would suggest that the virus - though exogenous - would only rarely be transmitted by contact. Probably, as in the few documented cases in Japan, the transmission is mainly from mother to child. The small percentage of serum positive blacks among the UK blacks of Caribbean origin suggests that even a mother to child transmission is not a frequent event.

The exact similarities and differences between the virus of ATL of Japan and the virus of ATLL of blacks (HTLV) as well as origin and spread of virus carriers should emerge in the coming years. I venture to predict that the nucleic acid sequence of HTLV will reveal some differences from ATLV.

These retroviruses (ATLV/HTLV) are not involved in the common T-cell lymphoid leukaemia/lymphoma nor are they associated with the other forms of lymphoid and myeloid malignancies. So far only virus-like particles (VLP) have been observed in fresh leukaemic cells from both acute lymphoid and myeloid leukaemia. The nature of the VLP in human leukaemic cells and their possible association with malignant transformation has not yet been established. However, the 40-50 fold increase in VLP following growth of the cells with BUDR and IUDR supports the hypothesis that the inclusion bodies are the site of synthesis of viral structural proteins (60). The reasons for the expression of what can be interpreted as incomplete viral particles in the human leukaemic cells is not yet known. Particle formation may be incomplete because of incomplete proviral information in the cells, or there may be a block in particle assembly. Cytoplasmic inclusion bodies which contain similar VLP were reported in mouse cells chronically infected with Moloney leukaemia virus (MLV) following culture of the cells with interferon (75). Interferon was found to arrest MLV assembly and release, resulting in the accumulation and breakdown of RNA-deficient intracellular virus particles (76). The similarity of the MLV VLP with those seen in our human leukaemic cell lines further supports the notion that the VLP may be a morphological expression of human viral oncogenes. Purification of the structural proteins might help to identify the oncogenes involved in the malignant transformation and will also enable the preparation of specific antibodies. With such antibodies it will be possible to establish whether the proteins of the VLP are present in fresh human leukaemic cells and whether they are leukaemia specific.

CONCLUSION

The geographical distribution of the various types of human leukaemia excludes endemic forms of the disease. There is one exception: Japanese adult T-cell leukaemia (ATL). The retrovirus (ATLV) isolated from ATL cells fulfils all the criteria required for a causal relationship with this haematological malignancy:

(1) Proviral sequences were found in the DNA of the leukaemic cells of each of the patients studied. No such sequences were found in the DNA of normal human haemic cells.

(2) Antibodies to the virus were present in nearly all ATL patients and in 26% of the normal adults in the endemic areas.

These observations indicate that ATLV is an exogenous retrovirus. On the other hand, proviral sequences could not be found in the DNA from fresh leukaemic cells of patients with mycosis fungoides/Sezary syndrome. Although the various studies of ATLV and disease suggest a causal relationship between the virus and ATL in Japan, it is not yet clear why only adults develop the disease.

The reported isolation of HTLV from cultured cells obtained from patients with mycosis fungoides/Sezary syndrome or hairy cell leukaemia is misleading. HTLV is probably associated only with adult T-cell leukaemia/lymphoma (ATLL) of certain black populations of African origin. Antibodies to the virus were found in most ATLL patients. So far proviral sequences were demonstrated only in cell lines, but not yet in fresh leukaemic cells. In contrast to the high incidence of antibodies to ATLV among the normal population of the endemic areas of southwest Japan, only 0-3% HTLV serum positives have been found in the supposedly endemic area of the Caribbean. Although it was claimed that HTLV is identical with ATLV, it is difficult to explain several basic differences such as the failure of HTLV to infect cells other than thymus derived lymphocytes. The global distribution, similarity or differences of ATLV and HTLV should become apparent in the coming years.

The virus like particles observed in some fresh human leukaemic cells and human leukaemic cell lines may be a morphological expression of endogenous structural viral proteins, since their production can be amplified in the same way as virus-formation.

1. Poiesz, B.J. Ruscetti, F.W., Gazdar, A.F., Bunn, P.A., Minna, J.D. & Gallo, R.C. Proc.Natl.Acad.Sci.USA. 77:No.12,7415-7419.
2. Poiesz, B.J., Ruscetti, F.W., Reitz, M.S., Kalyanaraman, V.S. & Gallo, R.C. Nature 294,268-271,1981.
3. Hinuma, Y., Nagata, K., Hanoka, M., Nakai, M., Matsumoto, T., Kinoshita, K-I. Shirakawa, S. & Miyoshi, I. Proc.Natl.Acad.Sci.USA. 78,6476-6480,1981.
4. Uchiyama, T., Yodoi, J., Sagawa, K., Takatsuki, K. & Uchino, H. Blood 50,481-492,1977.
5. Hanaoka, M., Takatsuki, K., Shimoyama, M. (1982). Adult T cell leukaemia and related diseases. Gann monograph on Cancer Research. No. 28.
6. Hinuma, Y., Komoda, H., Chosa, T., Kondo, T., Kohakura, M., Takenaka, T., Kiruchi, M., Ichimaru, M., Yunoki, K., Sato, I., Matsuo, R., Takiuchi, Y., Uchino, H. & Hanaoka, M. Int.J.Cancer 29,631-635,1982.
7. Hinuma, Y. Gann Monograph on Cancer Research 28,211-218,1982.
8. Miyoshi, I., Fujishita, M., Taguchi, H., Ohtsuki, Y., Akagi, T., Morimoto, Y.M. & Nagasaki, A. Lancet i,683-684,1982.
9. Shimoyama, M., Minato, K., Tobinai, K., Horikoshi, N., Ibuka, T., Deura, K., Nagatani, T., Ozaki, Y., Inada, N., Komoda, H., Hinuma, Y. Jpn.J.Clin.Oncol. 12,109-116,1982.
10. Komuro, A., Hayami, M., Fuji, H., Miyahara, S., Hirayam, M. Lancet i,240, 1983.
11. Tajima, K., Tominaga, S., Suchi, T., Kawagoe, T., Komoda, H., Hinuma, Y., Oda, T. & Fujita, K. Gann 73,893-901,1982.
12. Miyoshi, I., Taguchi, H., Fujishita, M., Niiya, K., Kitagawa, T., Ohtsuki, Y. & Akagi, T. Gann 73,339,340,1982.
13. Miyoshi,I., Kubonishi, I., Suida, M., Yoshimoto, S., Hiraki, S., Tsubota, T., Kobashi, H., Lai, M., Tanaka, T., Kimura, I., Miyamoto, K., and Sata, J. Jpn.J.Clin.Oncol.9(suppl.),485-494,1979.
14. Miyoshi,I., Kubonishi, I., Yoshimoto, S., Akagi, T. Ohtsuki, Y., Shiraishi, Y., Nagata, K. & Hinuma, Y. Nature 294,770-771,1981.
15. Yoshida, M., Miyoshi, I. & Hinuma, Y. Proc.Natl.Acad.Sci.USA. 79,2031-2035, 1982.
16. Yamamoto, N., Schneider, J., Hinuma, Y. & Hunsmann, G. Naturforsch. 37,731-2, 1982.
17. Chosa, T., Yamamoto, N., Tanaka, Y., Koyanagi, Y., Hinuma, Y. Gann, 73,844-7, 1982.
18. Yamamoto, N., Okada, M., Koyanagi, Y., Kannagi, M., Hinuma, Y. Nature 299, 367-369,1982.
19. Yamamoto, N., Okada, M., Koyanagi, Y., Kannagi, M., Hinuma, Y. Science 217, 737-739,1982.
20. Gotoh, Y-I., Sugamura, D. & Hinuma, Y. Proc.Natl.Acad.Sci.USA. 79,4780-4782, 1982.
21. Miyoshi, I., Taguchi, H., Fujishita, M., Yoshimoto, S., Kubonshi, I.,Ohtsuki, Y. Shiraishi, Y. & Akagi, T. Lancet i,1016,1982.
22. Varmus, H.E. Science, 216,812-820,1982.
23. Seiki, M., Hattori, S. & Yoshida, M. Proc.Natl.Acad.Sci.USA, 79,6899-6902, 1982.
24. Seiki, M., Hattori, S., Hirayama, Y. and Yoshida, M. Proc.Natl.Acad.Sci.USA. 80,3618-3622,1983.
25. Miyoshi, I., Yoshimoto, S., Fujishita, M., Taguchi, H., Kubonishi, I. & Minezawa, M. Lancet ii, 658, 1982.
26. Miyoshi, I., Ohtsuki, Y., Fujisita, M., Yoshimoto, S., Kubonishi, I. & Minezawa, M. Gann 73,848-849,1982.
27. Miyoshi, I., Fujishita, M., Taguchi, H., Niiya, K., Kobayashi, M., Matsubayashi, K., Miwa, N. Lancet i, 241,1983.
28. Yamamoto, N., Hinuma, Y., Zur Hausen, H., Schineider, J., Hunsmann, G. Lancet i, 240-241,1983.

29.Lennert, K. Springer-Verlag,Berlin, pg. 186.
30.Van der Loo, E.M., van Muijen, G.N.P., van Vloten, W.A., Beens, W., Sceffer, E. & Meijer, C.J.L.M. Virchows Arch. B Cell Path. 31,193-203,1979
31.Catovsky, D., Greaves, M.F., Rose, M., Galton, D.A.G., Goolden, A.W.G., McCluskey, D.R., White, J.M., Lampert, I., Bourikas, G., Ireland, R., Brownell, A.I., Bridges, J.M., Blattner, W.A., Gallo, R.C. Lancet i,639-643, 1982.
32.Bunn, P.A., Schechter, G.P., Jaffe, E., Blayney, D., Young, R.C., Matthews, M.J., Blattner, W., Broder, S., Robert-Gurolf, M. & Gallo, R.C. New Eng.J. Med. 309,257-264,1983.
33.Fleming, A.F., Yamamoto, N., Bhusnurmath, S.R., Maharajan, R., Schneider, J., Hunsmann, G. Lancet II,334-335,1983.
34.Gazdar, A.F., Carney, D.N., Bunn, P.A., Russell, E.D., Jaffe, E.S., Schechter, G.P. & Guccion, J.G. Blood 55,409-417,1980.
35.Kalyanaraman, V.S., Sarngadharan, M.G., Robert-Guroff, M., Miyoshi, I., Blayney, D., Golde, D. & Gallo, R.C. Science 218,571-573,1982.
36.Saxon, A., Steven, R.H. & Golde, D.W. Ann.Intern.Med. 88, 323,1978.
37.Kalyanaraman, V.S., Sarngadharan, M.G., Poiesz, B., Ruscetti, F.W. & Gallo, R.C. J. of Virology 38,906-915,1981.
38.Kalyanaraman, V.S., Sarngadharan, M.G., Bunn, P.A., Minna, J.D. & Gallo, R.C. Nature 294,271-273,1981.
39.Posner, L.E., Robert-Guroff, M., Kalyanaraman, V.S., Poiesz, F.W., Ruscetti, F.W., Fossieck, B., Bunn, Jr., P.A., Minna, J.D. & Gallo, R.C. J.Expt.Med. 154,333-346,1981.
40.Kalyanaraman, V.S., Sarngadharan, M.G., Nakao, Y., Ito, Y., Aoki, T. & Gallo, R.C. Proc.Natl.Acad.Sci.USA 79,1653-1657,1982.
41.Robert-Guroff, M., Nakao, Y., Notake, K., Ito, Y., Sliski, A., Gallo, R.C. Science 215,975-978,1982.
42.Blattner, W., Kalyanaraman, W., Robert-Guroff, M., Lister, T., Galton, D., Sarin, P., Crawford, M., Catovsky, D., Greaves, M., Gallo R. Int.J.Cancer, 30,257-264,1982.
43.Oroszlan, S., Sarngadharan, M.G., Copeland, T., Kalyanaraman, V.S., Gilden, R.V. & Gallo, R.C. Proc.Natl.Acad.Sci.USA 79,1291-1294,1982.
44.Robert-Guroff, M., Ruscetti, F.W., Posher, L.E., Ppoiesz, B.J. & Gallo, R.C. J.Expt.Med. 154,1957-1964,1981.
45.Haynes,B.F., Robert-Guroff, M., Metzger, R.S., Franchini, G., Kalmanaraman, V.S., Palker, T.J. & Gallo, R.C. J.Expt.Med. 157,907-920,1983.
46.Gallo, R.C., Reitz, S. Jr. J.Nat.Cancer Inst. 69,1209-1214,1982.
47.Popovic, M., Sarin, P.S., Robert-Guroff, M., Kalyanaraman, V.S., Mann, D., Minowada, J. & Gallo, R.C. Science 219,856-859,1983.
48.Gallo,R.C., Mann,D., Broder,S., Ruscetti, F.W., Maeda, M., Kalyanaraman, V.S., Robert-Guroff, M. & Reitz, Jr. Proc.Natl.Acad.Sci.USA 79,5680-5683, 1982.
49.Ruscetti, F.W., Robert-Guroff, M., Ceccherini-Nelli, L., Minowada, J., Popovic, M. & Gallo, R.C. Int.J.Cancer 31,171-180,1983.
50.Reitz, M.S., Popovic, M., Haynes, B.F., Clark, S.C. & Gallo, R.C. Virology, 126,688-692,1983.
51.Reitz, M.S., Bernard, J., Poiesz, F., Ruscetti, W. & Gallo, R.C. Proc.Natl. Acad.Sci.USA, 78, No. 3,1887-1891,1981.
52.Rho, H.M. Poiesz, B.J., Ruscetti, F.W. & Gallo, R.C. Virology 112,355-360, 1981.
53.Popovic, M., Reitz, Jr., M.S., Sarngadharan, M.G., Robert-Guroff, M., Kalyanaraman, V.S., Nakao, Y., Miyoshi, I., Minowadall, J., Yoshida, M., Ito, Y. & Gallo, R.C. Nature 300,63-66,1982.
54.Epstein, M.A., Barr, Y.M. Lancet i, 252-253,1964.
55.Henle, G., Dichl, V., Kohn, G., zur Hausan, Henle, G. Science 157,1064-1065, 1967.

56.Henle, G., Henle, W., Diehl, V. Proc.Natl.Acad.Sci. 59,94-101,1968.
57.Zur Hausen, H., Schulte, Holthausen, H. Nature 227,245-248,1970.
58.Klein,G., Lindahl, T., Jondal, M., Lubond, W., Menezes, J., Nilsson, K., Sundstrom, C. Proc.Natl.Acad.Sci. 71,3283-3286,1974.
59.Lenior, G.M., Preud'homme, J.L., Bernheim, A. & Berger, R. Nature 298,474-476,1982.
60.Karpas, A., Fischer, P. Leukaemia Res. 4,315-329,1980.
61.Smith, H. Leukaemia Res. 2, 133-140,1978.
62.Karpas, A. Lancet ii, 110,1978.
63.Karpas, A., Worman, C.P., Khalid, G., Neuman, H., Hayhoe, F.G.J., Newell, D., Stewart, J.W. Brit.J.Haemat. 44,415-424,1980.
64.Fischer, P., Karpas, A., Nachera, E., Hass, O., Winterleither, H., Krepler, P. Brit.J.Haemat. 46,23-31,1981.
65.Karpas, A., Wrighitt, T.G., Nagington, J. Lancet ii,1026-1019,1978.
66.Karpas, A. American Scientist, 70,277-285,1982.
67.Hinuma, Y., Gotoh, Y-I., Sugamura, K., Nagata, K., Goto, T., Nakai, M., Kamada, N., Matsumoto, T. & Kinoshita, K-I. Gann. 73,341-344,1982.
68.Kinoshita,K., Hino, S., Amagasaki, T., Yamada, Y., Kamihira, S., Ichimaru, M. Munehisa, T. & Hinuma, Y. Gann. 73,684-685,1982.
69.Karpas, A., Hatanaka, M. & Hinuma, Y. Leukaemia Reviews Int., Ed. M.A. Rich, Marcel Dekker, 86-87,1983.
70.Vyth-Dreese, F.A. & De Vries, J.E. Int.J.Cancer 32,53-59,1983.
71.Hahn, B., Manzari, V., Colombini, S., Franchini, G., Gallo, R.C., Wong-Staal, F. Nature 303,253-256,1983.
72.Wong-Staal, F., Hahn, B., Manzari, V., Colombini, S., Franchini, G., Gelmann, E.P. & Gallo, R.C. Nature 302,620-628,1983.
73.Hinuma, Y., Chosa, T., Komoda, H., Mori, I., Suzuki, M., Tajima, K., Hung, P.I., Lee, M., Lancet i, 824-825, 1983.
74.Haynes, B.F., Miller, S.E., Palker, T.J., Moore, J.O., Dunn, P.A., Bolognesi, D.P. & Metzger, R.S. Proc.Nat.Acad.Sci. USA. 80,2054-2058,1983.
75.Aboud, M., Shoor, R., Bari, S., Hassan, Y., Malik, Z. & Salzberg, S. J.Gen. Virology 62,219-225,1982.
76.Aboud, M. & Hassan, Y. J.Virology 45,489-495,1983.

The Role of Viruses in Human Cancer. Volume II
G. Giraldo and E. Beth editors

THE ETIOLOGY OF HUMAN BREAST CANCER: RELATED VIRAL AND NON-VIRAL ANTIGEN EXPRESSION IN MAMMARY TUMORS OF MICE AND MAN.

NURUL H. SARKAR
Laboratory of Molecular Virology, Memorial Sloan-Kettering Cancer Center, 1275 York Avenue, New York, N Y 10021, USA.

During the last 15 years the possibility that human breast cancer has a viral etiology has been investigated using an animal model, murine mammary cancer. This model was chosen because the murine mammary tumor virus (MuMTV), a type B retrovirus, is the only known virus that causes mammary cancer in mice (See reference 1 for review). The results obtained from many studies of human breast cancer can be summarized as follows: 1) A small number of particles morphologically similar to type B viruses have been found in some human milk samples. 2) Retrovirus-like reverse transcriptase and RNA were detected in the milk and breast cancer tissues of some women. 3) A high percentage of breast tumors were found to contain antigen cross-reacting with gp52, the envelop glycoprotein of MuMTV. 4) Serum from breast cancer patients appeared to contain MuMTV related antigens and/or antibodies. 5) Leukocytes of many breast cancer patients were found to be sensitized to MuMTV related antigen(s). These findings suggest that a virus closely related to MuMTV may be etiologically associated with breast cancer in women. This view, however, remains controversial since the validity of some of the studies has been questioned.

A major concern has been the differences in results obtained from different laboratories. One of the possible reasons for such inconsistency could be differences in the preparation of MuMTV antigens and cDNAs representing the viral genome, and in the method of detection of MuMTV proteins, RNA or DNA in human samples using these probes. The potential for contamination in preparing the probes and the use of such probes in some of the studies also may have contributed to the difficulty of interpreting some of the findings. Thus, one may question the significance of such studies. Even though a putative mammary tumor virus has not been isolated from human breast tumors and the controversy about the role of MuMTV in the development of breast cancer in women continues, some recent studies have generated renewed interest in continuing investigations into the association and significance of MuMTV with human breast cancer.

Some of the studies that are summarized below became possible because of the extraordinary progress that has been made during the last three years in the field of retrovirus biology and molecular virology. Several technical advances, such as cloning of MuMTV proviruses (2-6) and DNA transfection assays (7-10) designed to identify transforming genes associated with neoplasms have contributed much to the rapid rate of progress. In this paper I will limit my discussion of the research that has been done during the last 3 years on the putative relationship between murine and human breast cancer.

GEOGRAPHIC DISTRIBUTION OF ANTI MuMTV-gp52 ANTIBODY IN HUMAN SERA.

The occurrence of antibody reactive to MuMTV-gp52 in the sera of patients with breast cancer from 4 different countries have been determined using an enzyme linked immunoabsorbent assay (11). In this assay, highly purified MuMTV membrane antigen gp52 (12) was used. The breast cancer patients used in that study were 145 Caucasian women from the United States, 21 black women from Kenya, East Africa, 24 Chinese women from Peking, China, and 53 women from Bombay, India. Age matched healthy individuals were also included as controls. It was found that an increased level of anti MuMTV-gp52 was present in the sera of about 18.6% (27/145) of the patients from the United States, 61.9% (13/21) from Kenya and 37.7% (20/53) from India. By contrast, less than 5% (1/24) of the sera from Chinese breast cancer patients were positive for MuMTV-gp52 reactivity. The percent of healthy women from the United States and Kenya that had anti MuMTV-gp52 antibody in their serum was significantly lower than the matched groups of women with breast cancer It should be emphasized that the sera of women from Kenya with breast cancer had the highest titers of reactivity to MuMTV-gp52 followed, in turn, by the sera of breast cancer patients from East India, America and China Consistent with these findings, Levine et al. (13) have demonstrated that the incidence of MuMTV-gp52 related antigen in breast cancer tissues of women from Tunisia was much higher (70%) than that in the breast cancer tissues of American women (47%).

These results indicate that both the expression of gp52 in breast cancer tissues and antibody production in women with breast cancer varies significantly in patients from one country to the other. Although extensive work is necessary to determine if the geographic distribution of anti MuMTV-gp52 is correlated with the epidemiolgoy of breast cancer, results based on limited observations suggest that the occurrence of MuMTV-gp52 is positively correlated with anti MuMTV-gp52 production. Thus, the association of MuMTV related product(s) with human breast cancer cannot be ignored, and no matter what the etiologic significance of MuMTV in human breast cancer is, it might still be possible for MuMTV-gp52 to be an acceptable reagent for diagnostic purposes, particularly for identifying normal women with a high risk of breast cancer.

DISTRIBUTION OF ANTI MuMTV-gp52 ANTIBODY IN THE SERA OF FAMILY MEMBERS WITH A HISTORY OF BREAST CANCER

Epidemiological studies have shown that women from families with a history of breast cancer have twice as much risk of breast cancer than women without familial history of breast cancer (13,14). Thus, if MuMTV-gp52 has prognostic value one might expect to find high titer MuMTV-gp52 antibody at a greater frequency in the sera of healthy women whose family have had a history of breast cancer than the members of those families that have not suffered from breast cancer. To test this hypothesis sera from two American families and one parsi family from India with genetic propensity for breast cancer were tested for MuMTV-gp52 antibody. It was found that 33% and 71% of the healthy members of the two American families and 23% of the healthy family members of the parsi family, respectively, had high levels of antibody to MuMTV-gp52. Surprisingly, one of the five male members in one family, and one of the 3 male members in the other had increased titer of antibody. The obvious implication of these findings is that MuMTV components may be associated with male breast cancer as well with cancer of the female breast.

PRESENCE OF MuMTV-gp52-RELATED ANTIGEN IN MALE BREAST CANCER

Because of our findings of antibody reactive to MuMTV-gp52 in the sera of some male members of the families with histories of breast cancer we examined male breast cancer tissues for the presence of MuMTV-gp52. It is known that in America the incidence of breast cancer in males is very low, about 0.8 per 100,000 as compared to 85.4 per 100,000 females (16). It should be pointed out that even in mice, male breast cancer is relatively rare except in the H strain where some spontaneous tumors develop (17). However, it is not known whether or not MuMTV is inovolved in the development of the male mammary tumors in this strain of mice, although estrogenic hormones have been postulated to be involved in tumor inducation. Fortunately, at the Memorial Sloan-Kettering Cancer Center 36 male mammary carcinomas that have been previously embedded in paraffin between 1956 and 1976 were available for studies. These samples were examined using anti MuMTV-gp52 serum in the immunoperoxidase assay as described by Mesa-Tejada et al (18). Breast cancer tissues from women were also included in the study To our surprise almost all (32/26, 89%) male mammary carcinoma were positive (19) for the MuMTV-gp52 related antigen (Figure 1). By contrast about 28% of

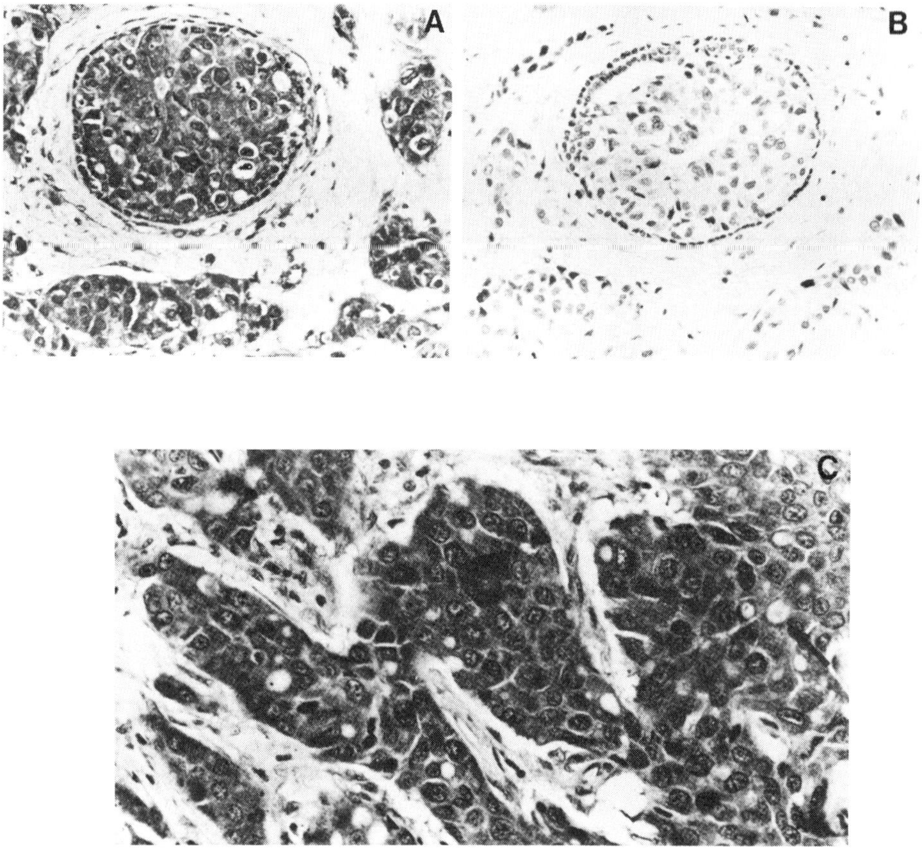

Fig. 1. Immunoperoxidase staining of intraductal and infiltrating mammary carcinoma in a female (A and B) and a male (C) patient. A and C were stained with anti MuMTV-gp52 whereas B was stained with the same antibody preparation but was absorbed with purified MuMTV; A and B were serial sections. Note that the positive diffuse cytoplasmic staining of tumor cells appears gray-black. Hematoxilin was used to counterstain the sections; 130X. From Lloyd et al. (19) with the permission of American Cancer Society, Inc.

the 50 female breast cancer tissues examined showed the presence of MuMTV-gp52 related antigen. The reason for the significantly lower incidence of MuMTV-gp52 reactivity in cancer of the female breast than in the male breast is unknown. However, it appears that among other factors, the histologic type of tumor may be responsible for the differences since, when both intraductal and infiltrating duct carcinoma were present, 75% of male and 40% of female carcinomas stained positively. Breast cancer cells that metastasized to the axillary lymph nodes were positive for MuMTV-gp52 in five of the male breast cancer tissues that were available for testing. It should be pointed out that the detection of MuMTV-gp52 in tumor cells after metastasis to axillary lymph nodes may be of practical diagnostic value in that in cases of unknown primary tumors, a positive gp52-reactivity may indicate that the metastatic tumor is a mammary carcinoma. The etiological significance of MuMTV in human breast cancer remains to be elucidated, but the finding that MuMTV-gp52 is present in both female and male breast cancer tissues points out further the important prognostic value of gp52 in human breast cancer.

DETECTION OF MuMTV-RELATED SEQUENCES IN HUMAN CELLULAR DNA

In three earlier studies, attempts were made to detect MuMTV-related sequences in primary breast tumors and in established tumor cell lines (20-22). In these studies DNAs (cDNA) complementary to MuMTV RNA were prepared from detergent disrupted MuMTV and used as probes in a liquid hybridization assay in which the degree of hybridization was estimated by measuring the sensitivity of the annealed cDNA to a single strand specific S1 nuclease digestion (22). Unfortunately, these studies failed to detect MuMTV-related sequences in the genome of human tumor cells. As mentioned earlier, the quality of the cDNA probe and the relative lack of sensitivity of liquid hybridization techniques may have contributed, among other factors, to the negative results that were obtained. The most compelling evidence of the involvement of a retroviruses in human breast cancer has been the demonstration of MuMTV-gp52 related antigen in cancer tissues and antibodies in the sera of patients. Thus it is reasonable to assume that at least MuMTV envelope gene related sequences are present in the genomic DNA of breast tumor cells. This assumption has to be documented before an MuMTV-like retrovirus can be considered to have a role in human mammary tumorigenesis irrespective of whether or not the virus is transmitted via the germ line, as an endogenous provirus, or as an exogeneous virus infection. In the murine mammary tumor system it is a general rule that most mammary tumors induced by endogeneous or exogeneous MuMTVs produce MuMTV particles (24,25) and that they contain one or more additional MuMTV proviruses as compared to uninfected tissues (26-31). Exceptions to this rule, however, have been found in some rare, mammary tumors of mice in which either no MuMTV provirus was detected or amplification of the existing endogeneous provirus(es) could be demonstrated (R. Callahan and C. Cohen, personal communication).

The current technology for the detection of MuMTV proviruses in mouse tissues involves the digestion of genomic DNA by restriction endonucleases, Southern transfer of DNA onto nitrocellulose paper, and hybridization with specific cloned probes. Restriction enzymes such as EcoRI, that have a single recognition site close to the middle of the MuMTV proviruses (26,31) generate two characteristic fragments representing the 5' and 3' cell-virus junctions, have often used for the detection and estimation of the number of endogenous and newly integrated MuMTV proviruses. The sizes of the proviral fragments thus obtained will depend on the location of the neighbouring restriction sites in the cellular DNA. For example, examination of C3Hf mamary tumor DNA revealed the presence of new EcoRI fragments of different sizes in the different tumors in addition to the existing endogenous proviruses. It is of interest to note that EcoRI generates at least two restriction fragments (13.0 kb and 5.0 kb) from the DNA of high percentage of C3Hf mammary tumors implying that many C3Hf mammary tumors contain at least a new endogenous provirus integrated at a common site in the cell genome (31).

The availability of clones comprising the various segments of the MuMTV provirus and the fact that the blot hybridization technique is more sensitive than liquid hybridization prompted two groups of investigators to examine human DNA to determine if proviral copies of a retrovirus related to MuMTV are present (32,33). Callahan and his colleagues found that under low stringent conditions of hybridization an MuMTV probe representing the entire provirus, a gag probe and an envelope probe detected various size EcoRI restriction fragements from human breast tumor and normal cellular DNAs (32). Recently May et al. (33), using similar technology have found that DNA from a number of human breast cancer cell lines, and an epithelial cell line derived from a normal breast, and placentas, to hybridize with cloned MuMTV rep, gag-pol, env and LTR probes. The MuMTV gag-pol region was the most homologous (>80%)withthe human DNA (Figure 2). These observations suggest that the majority of the human MuMTV-related sequences are genetically transmitted.

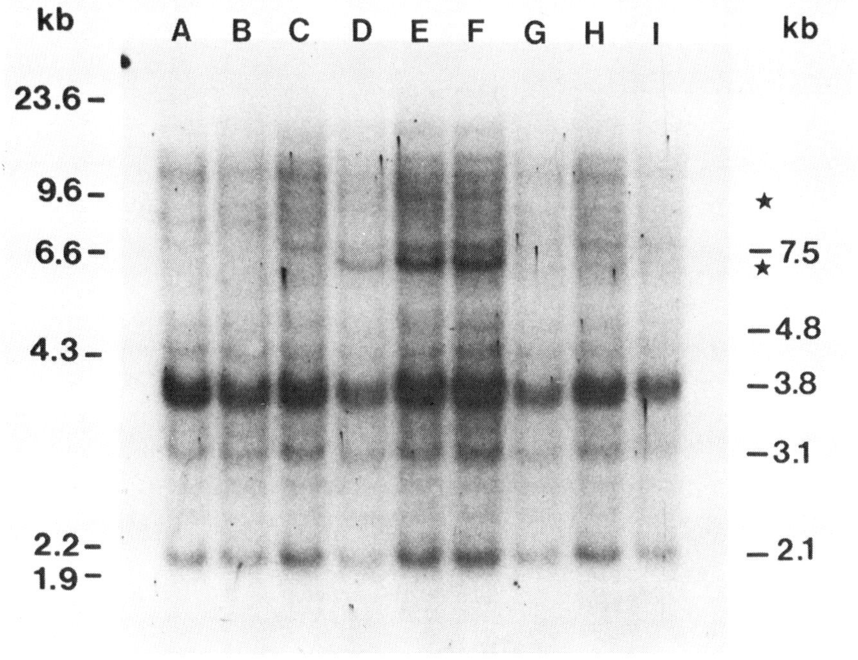

Fig. 2. Detection of MuMTV gag-pol specific sequences in the EcoRI digests of DNA derived from a number of human breast tumor cell Lines (A-G), a normal breast cell line (H) and human placenta (I) under low-stringent hybridization conditions. The asterisks indicate the positions of two additional fragments that were present only in the DNA from sublines of one of the cell lines. The positions of the molecular weight markers are indicated on the left. From May et al. (33) with the courtesy of Dr. Felicity May and permission of IRL Press Limited.

It should be mentioned, however, that May et al. also detected MuMTV gag-pol related sequences in two additonal EcoRI fragments of the DNA from one of the human breast cancer cell lines, MCF-7. The results of these two studies taken together, provide a basis for the following discussion:

1. Human DNA contains MuMTV-related sequences, and the gag-pol sequences share a greater homology with the human DNA than the other sequences. This may appear to be surprising and raises critical questions about the significance of the MuMTV related sequences in humans since, as mentioned earlier, both male and female human breast tumors have been shown to express MuMTV env related antigens. However, this apparent contradiction can possibly be explained by the finding that sera from breast cancer patients do indeed react with MuMTV-A particles, that are composed primarily of the gag gene products of MuMTV (34). In one study about 33% of the 24 human breast tumors examined by radioimmunoassay was found to contain MuMTV -p28 (a major product of the gag gene) - related antigen (35).

In another study MuMTV-p28 related antigen has been detected in 40% of the 40 human breast cyst fluids samples that were examined (36). Furthermore the retrovirus-like particles produced occasionally by the human breast cancer cell line, MCF-7, have been shown to contain antigens cross reactive to MuMTV-p28 (21). Thus the presence of MuMTV gag-pol related sequences in human DNA is consistant with these results. The fact that more studies have been done with MuMTV-gp52 reagents than MuMTV-p28 may have resulted in a false impression of relative expression of MuMTV-gp52 related products in humans compared to that of p28 or other components of the virus. The demonstration of the gag-pol sequences in human tissues may justify an expanded examination of human breast tumors for the expression of MuMTV-p28 related antigens and of serum samples for antibody reactive to MuMTV-p28. Such an analysis may lead to the development of a better assay system for investigations into the relationship between MuMTV expression and human breast cancer than has been possible with MuMTV-gp52.

2. The finding that several EcoRI restrictions fragments of human DNA are partially homologous to the MuMTV env sequences implies that the putative polypeptides encoded by the MuMTV-env-related sequences in human DNA, should not cross react strongly with MuMTV-gp52. Such a consideration may explain why anti MuMTV sera detects cross reactive antigen more often in human breast tissues than antibody prepared against purified MuMTV-gp52. (R. Mesa-Tejada, M. Tomana, personal communications). Similarly, in our experience human sera reacts more strongly with intact MuMTV than with purified MuMTV-gp52. It is conceivable that only a few antigenic sites of MuMTV-gp52 are cross reactive

to a putative human breast cancer associated antigen. Thus, during purification of gp52 from MuMTV some of the antigenic sites of MuMTV-gp52 could be destroyed or the molecule could assume an altered configuration making the antigenic sites no longer available for binding. Similarly, antibody prepared against such an altered MuMTV-gp52 should not be expected to detect antigen in breast cancer tissues. This could explain the variation in the results obtained by different investigators from different laboratories. Attempts must now be made to purify the MuMTV-gp52 related antigen from human tumors using immuno-chemical procedures and to compare the peptide maps of the MuMTV-gp52 related human antigen and MuMTV-gp52, so that the degree of relatedness between the mouse and human MuMTV env gene products can be determined. Since the nucleotide sequence of the MuMTV-gp52 gene is known (37,38) and MuMTV related sequences from human DNA have been cloned and are available (32), sequence analysis of the human envelope related gene will enable the mouse MuMTV env gene and its human counterpart to be compared.

3. The evidence that one of the several breast tumor cell lines tested (MCF-7) contain additional MuMTV related EcoRI restriction fragments is suggestive of genetic polymorphism. Whether the extra sequences seen in this cell line were derived as a consequence of a putative exogenous virus infection, or by rearrangement or amplification of endogenous sequences is not known.

NON-VIRAL ANTIGEN EXPRESSION IN MAMMARY TUMORS OF MICE AND MAN

A major goal of contemporary cancer research is the identification of specific genes and/or gene products that may be responsible for the transformation of normal cells into neoplastic cells. For a long time a major enterprise for tumor immunologists has been the search for tumor specific or tumor associated cell surface antigens (TCSAs) in various neoplasms of both experimental animals and humans. As a result of this work well defined TCSAs in the avian, murine and feline RNA tumor virus systems have been described (see reference 39 for early work). Recent investigations into the biology of transforming genes associated with retroviruses have led to the discovery of a number of transformation specific proteins that, unlike the TCSAs, are located intracellularly but are associated with the cell membrane. A great deal of information concerning the structure and biological functions of some of the transforming proteins, particularly the transforming proteins of Rous sarcoma, Harvey sarcoma and Abelson leukemia virus are available (40). The discovery of these transforming proteins has added new a dimension into the understanding of the biology of cancer.

Although the expectation of discovering immunogenetic determinants characteristics of the transformed phenotype in human cancer has not been realized (41), some evidence for the presence of immunogens associated with some human tumors, for example, malignant melanoma, astrocytoma, renal carcinoma and leukemia have been found in a few patients using a system of autologous typing (42-45). With the advent of the powerful hybridoma technology (46) there is a promise of discovering tumor specific antigens in various human neoplasms. In the MuMTV-induced murine mammary tumor system, transplantation studies and cellular cytotoxic tests have provided evidence for the expression of tumor associated and MuMTV specific antigens in mammary tumor cells (47-53).

For the past several years we have been interested in identifying and characterizing immunochemically a putative mammary tumor associated antigen (MTAA). Our approach has been to immunize mice and rabbits with irradiated mammary tumor cells from 3 established cell lines MuMTV-73 (54), GR-3A (55) and Mm5mt (56) and from cells prepared from primary mammary tumors of C3H and GR mice, and to use the sera of these immunized animals for the immunoprecipitation of MTAA, from ^{35}S-methionine labeled MuMTV-73, GR-3A or Mm5mt cells and primary cultures of C3H and GR tumor cells. It should be mentioned that the cell lines and the tumors used in these studies produced large number of MuMTV particles. In our experiments, sera from non-immunized animals as well as from C3H and GR mice bearing primary or transplanted tumors were also used. It is disappointing to report that that our initial attempts to detect specific antigen(s) that could be implicated as an MTAA have not been successful. However, many of the sera raised in the mice and rabbits against the irradiated tumor cells as well as sera from mice bearing primary or transplanted tumors detected mainly MuMTV-gp52 in the tumor cells. These observations are consistent with the findings of Westernbrink and Koornstra (57). Furthermore, it has been known for some time that an antibody to MuMTV-gp52 exists in MuMTV infected mice (58-60). One of the possible reasons for our inability to raise an antibody against the putative MTAA may be the fact that MTAA is weakly antigenic and/or is expressed in much lower amounts in mammary tumors than is MuMTV-gp52. As a result, immunized animals respond primarily to MuMTV-gp52 rather than to the putative MTAA. For these reasons we investigated the non-MuMTV-producing mammary tumors of BALB/c mice. Another reason for our choice of BALB/c mammary tumors as a model was that the etiology of these tumors like human mammary tumors is unknown (61).

Approximately 3-5% of breeding BALB/c mice spontaneously develop mammary tumors, mostly adenocanthomas, late in life. About 18% of these tumors are adenocarcinomas (62). Because of the low incidence of spontaneous mammary adenocarciomas and lack of transplantable tumor or established cell lines derived from such tumors, we started our studies by establishing an epithelial cell line (BALB/c-ST) from an adenocarcinoma that developed spontaneously in a 20-month-old BALB/c mouse. Electron microscopic and immunological studies showed that the BALB/c-ST cells do not produce MuMTV

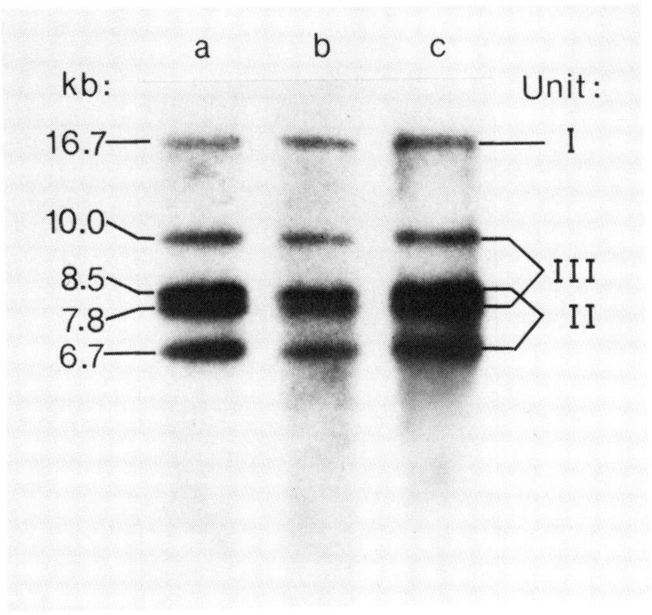

Fig. 3. EcoRI restriction pattern of MuMTV-specific proviral DNA of BALB/c mouse liver (Lane a), BALB/c-ST cells (lane b) and BALB/c-ST cell induced tumor (lane c). Hybridization was done to a ^{32}P-labeled MuMTV cDNA-representative probe. The molecular weights of the EcoRI fragments are given on the left side of the Figure; the Roman numerals I, II and III on the right represent the three endogenous proviruses found in BALB/c mice.

particlesor MuMTV-specific polypeptides (63). Furthermore, restriction endonuclease analysis of the tumor cell DNA revealed that the cells did not contain newly integrated MuMTV proviruses (Figure 3). The cells were, however, found to be tumorigenic in syngeneic mice (Table 1). We, therefore, considered this cell line to be suitable for our search for a putative tumor specific antigen.

TABLE 1. TUMORIGENECITY OF THE BALB/c-ST CELL LINE

Cells	Number of Cells Inoculated[1]	% Tumor Incidence[2]	Latent Period (Weeks)
BALB/c-ST	4×10^6	100 (10/10)	3-4
	6×10^6	100 (20/20)	3-4
BALB/c-ST infected with C3H-MuMTV	4×10^6	80 (16/20)	3-4
	6×10^6	100 (12/12)	3-4
BALB/c-ST infected with RIII-MuMTV	6×10^6	0 (0/15)	-
	8×10^6	0 (0/15)	-

[1] $4\text{-}8 \times 10^6$ cells in 100-150 µl medium were injected subcutaneously into mice.

[2] Numbers in parentheses denote number of animals which developed tumors/ number of animals inoculated.

[3] BALB/c-ST cells were infected in vitro with MuMTV isolated from the milk of C3H or RIII mice.

Using the technique of immunoprecipitation and analysis of the immunoprecipitates by gel electrophoresis, we found that 11 of the 23 sera from mice bearing BALB/c-ST cell-induced tumors precipitated predominantly a protein with a molecular weight of 86,000 dalton (p86) from the extracts of ^{35}S-methionine-labeled BALB/c-ST cells (Figure 4). Small amounts of a low molecular weight protein (p19) was sometimes found to be co-precipitated with p86. Immunoprecipitates obtained by some of the tumor bearing mouse sera from the extracts of BALB/c-ST cells that were labeled with ^{3}H-glucosamine were also found (Figure 5A) to contain p86 and, in addition, a glycoprotein of 73-kilodalton (gp73). It is of interest to note that the glycoprotein p86, unlike gp73, could not be detected in the extracts of cells that were surface labeled with ^{125}I (Figure 5B) suggesting that this protein is

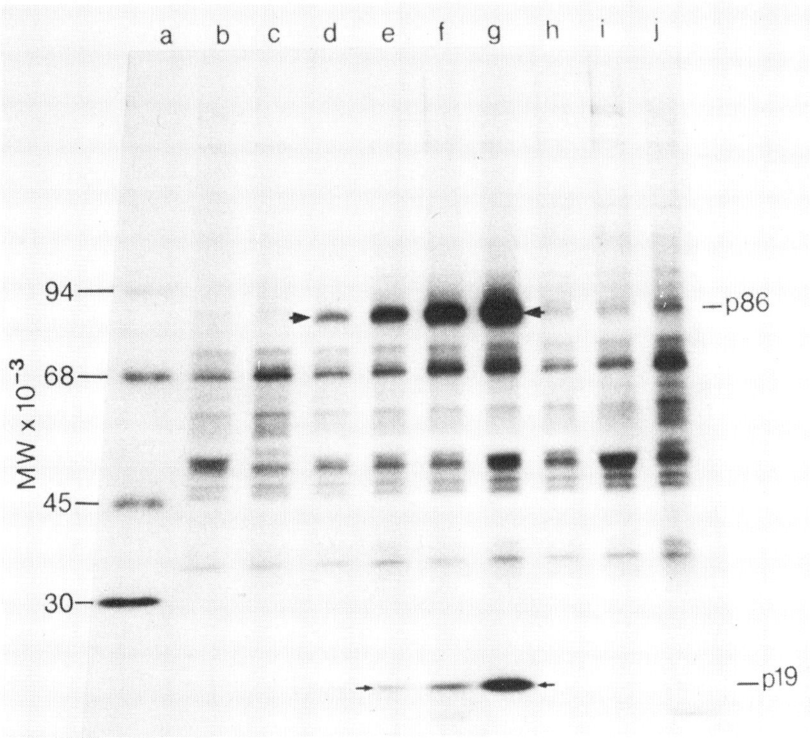

Fig.4. Detection of a major (p86) and a minor (p19) protein in the cells of a BALB/c-ST cell line that was established from a spontaneous mammary adenocarcinoma of a 20-month-old BALB/c mice. BALB/c-ST cells were labeled with 35S-methionine for 60 min., cell extracts were prepared and immunoprecipitated with serum from a normal BALB/c mouse (lane b), anti MuMTV serum (lane c) and sera from 7 mice bearing BALB/c-ST cell-induced tumors (lane d-j) and analyzed by SDS-polyacrylamide gel electrophoresis (PAGE). The labeled proteins were detected by fluorography. Lane a contained ^{14}C-amino acid labeled molecular weight marker proteins.

probably located intracellularly. None of the p86, gp73, or p19 appeared to be related to the structural polypeptides of MuMTV. It should be mentioned that, in general, the transformation specific proteins associated with retroviruses, like the p86, are also not expressed on the cell surface (40).

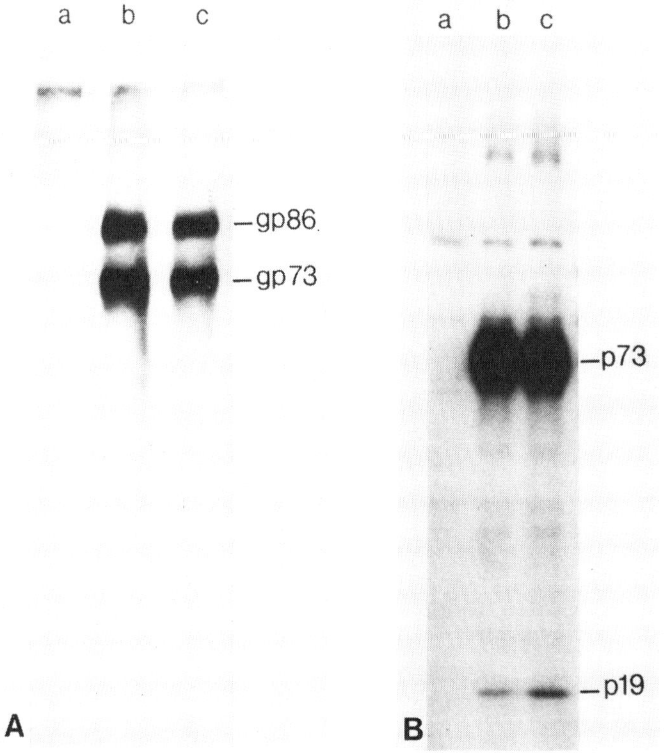

Fig. 5. (A) SDS-PAGE analysis of the immunoprecipitates obtained from extracts of ^{3}H-glycosamine labeled BALB/c-ST cells with a serum from a normal BALB/c mouse (lane a) and 2 sera from 2 BALB/c mice (lanes b and c) bearing BALB/c-ST cells induced tumors. (B) SDS-PAGE analysis of the immunoprecipitates obtained from extracts of ^{125}I labeled cells with a serum from a normal BALB/c mouse (lane a) and 2 sera from 2 tumor bearing mice (lanes b and c).

Because p86 is the predominant protein to be consistently immunoprecipitated from the extracts of BALB/c-ST cells by sera from mice bearing BALB/c-ST cell-induced tumors, we examined mammary tumor cells of BALB/cfC3H, GR and C3H mice, as well as normal mammary epithelial cells from BALB/c mice for the presence of p86, and sera from these strains of mice for antibody against p86. Antigen or antibody cross reactive to p86 was not detected in many of the samples examined. Thus, it appears that p86 may be a mammary tumor associated antigen unique to BALB/c mice. It should be pointed out that gp73 and p19 may also have important roles in the induction of mammary tumors in BALB/c mice.

As a part of our investigations into the host-virus interaction, we productively infected BALB/c-ST cells in vitro with C3H and RIII MuMTV and examined the effect of MuMTV infection on the tumorigenicity of the cells by transplanting them into syngeneic mice. It was found, in three separate experiments, that infection of the cells with milk borne MuMTV from RIII, but not from C3H mice resulted in a loss of tumorigenicity (Table 1) and an inhibition of the p86 production. These results provide additional evidence for the association between p86 and mammary tumorigenesis in BALB/c mice. The mechanism by which RIII-MuMTV infection abrogates the tumorigenicity of the BALB/c-ST cells must be complex and may involve different steps. However, it is possible that this phenotypic change may result from inactivation of the gene encoding p86 due to the insertion of RIII-MuMTV provirus in the vicinity of, or within, the p86 gene thus interupting transcription. There is a precedent for such a mechanism. The proviral DNA of the Moloney leukemia virus has been found to be integrated within the Rous sarcoma virus provirus containing the src gene and, as a result abrogates src gene expression (64,65).

It is known that retroviruses integrate at many sites in the host genome. Thus, if the mechanism responsible for the loss of tumorigenicity after RIII-MuMTV infection is a deletion or insertional mutation caused by the provirus, we have to assume (since we have not selected cells in which MuMTV proviruses are integrted specifically at the site of the p86 gene) that the RIII provirus must have an unique specificity of integration, in addition to random integration, in most of the BALB/c mammary tumor cells. Such a specificity of retroviral integration has yet to be observed and therefore our hypothesis is highly speculative.

Because of the possible role of p86 in the transformation of mammary cells in which no apparent MuMTV amplification is required as discussioned above, at least two important questions must be now be asked: 1) does endogenous MuMTV provirus(es) play a role in the expression of the p86 gene which is most likely a cellular gene, since p86 is not related to the structural polypeptides of MuMTV and since the MuMTV genome is not known to carry an oncogene? and 2) are antigens cross reactive to p86 present in human breast cancer?

The answer to the first question remains to be determined. However, it is conceivable that MuMTV may still be involved in the expression of p86. The recent analysis by Breznik and Cohen (66) of the DNA preparations from the D1 series of hyperplastic outgrowth cell lines from BALB/c mice indicated that one or all of the three endogenous proviral units were hypomethylated. Our preliminary results also suggest that the LTR region of the endogenous MuMTV proviruses in the BALB/c-ST tumor cells is hypomethylated. Thus, it is possible that hypomethylation of one of the proviral LTRs may have resulted in the derepression of the putative cellular oncogene that encodes p86. The enhanced expression of this gene, if it is linked in the orientation of transcription, may have been stimulated by downstream promotion from one of the viral promoters (especially the LTR), or if the gene is linked to the provirus in some other configuration, simply by the ability of the demethylated provirus to enhance transcription of the gene.

A promising answer to our second question has already been provided by Becker and her associates (67). These investigators transformed NIH-3T3 cells by transfecting them with DNA obtained from the human breast tumor cell line MCF-7. They found that the transformed cells produced tumors in mice and that sera obtained from tumor bearing mice reacted to three polypeptides (86, 70-75, and 19- kilodaltons) in the transformed cells; the 86-kilodalton protein (henceforth designated as 86K in order to differentiate this protein from the p86 that we found in the BALB/c-ST cells) being predominant. Becker et al. (67). found that the 86K protein was precipitated by sera from mice bearing tumors that were induced by NIH-3T3 cells after transfection with DNA from MuMTV-induced mammary tumors and by sera from mice bearing primary mammary tumors. These results suggest that 86K represents an antigen encoded by a transmissible transforming gene shared by both human and murine mammary carcinomas. At present we do not know if p86 and the 86K protein are structurally and/or functionally related. However, on the basis of their molecular weights and labeling characteristics, i.e. both are glycoproteins but cannot be detected by labeling the cell surface with ^{125}I (G. Cooper, personal communication) we speculate that the two 86-kilodaton proteins will be shown to be similar in regard to their biological functions.

CONCLUSION

Many factors are involved in the development of breast cancer. In the mouse, although we know that MuMTV plays the dominant role, the mechanism by which it induces mammary tumors is unknown at present. Recent findings that new proviruses are integrated into a specific site (31) or within a defined domain of the cell genome (30,68) in MuMTV-producing mammary tumors suggest that the integration of a new provirus may activate a neighbouring oncogene responsible for the transformation of normal mammary cells. In non-MuMTV producing mammary tumors,in which no new proviral integration occurs, mammary tumorigenesis may be linked to the expression of a putative oncogene as a consequence of hypomethylation of one of the existing endogenous proviruses. Thus, the current concept central to murine mammary tumorigenesis is the existence of a cellular oncogene(s) that is directly activated by MuMTV. However, it is possible that the interplay of several other factors such as genetic, hormonal, immunological, nutritional and physiological status of the host, and exposure to various chemical carcinogenes may in some instances result in the activation of this cellular gene.

It is not known whether or not a single oncogene or a small number of oncogenes are involved in the genesis of mammary tumors in various strains of mice. At present, the evidence suggests that more than one oncogene (30,68, N.H. Sarkar and P.R. Etkind, unpublished observations) is associated with mammary cancer. It is possible that one of the genes may be common to both MuMTV-producing and non-producing mammary tumors and even to some human breast tumor. Although such a gene has not been isolated, the product of such a gene may be the 86-kilodalton protein that I have discussed in this paper. The present realization that a gene may be specifically associated with human and mouse mammary carcinoma is a significant step towards an understanding of the etilogy not only of human breast cancer but a general mechanism of mammary tumorigenesis in mice and other mammals including humans.

REFERENCES

1. Sarkar, N. H. In: The Role of Viruses in Human Cancer, Vol. I, G. Giraldo and E. Beth (eds.), Elsevier North Holland, New York, 1980, p. 207.

2. Majors, J. and Varmus, H.E. : Nature: 289 : 253, 1981.

3 Buetti, E. and Diggelman, H. : Cell: 23: 335, 1981.

4 Majors, and Varmus, H.E: Symp. Mol. Cell. Biol. 18: 241, 1980.

5 Hynes, N.E., Kennedy, N., Rahmsdorf, U., and Groner, B.: Proc. Natl. Acad. Sci. USA. 78: 2038, 1981.

6. Donehower, L.A., Huang, A.L., and Hager, G.J.: J. Virol. 37: 226, 1981.

7. Shih, C., Shilo, B.Z., Goldfarb, M.P., Dannenberg, A., and Weinberg, R.A. Proc. Natl. Acad. Sci. USA 76: 5714, 1979.

8. Weinberg, R.A.: Adv. Cancer Res. 35: 149, 1982.

9. Cooper, G.M., Oberquist,S. and Silverman, L. : Nature(London) 284: 418, 1980.

10. Cooper, G.M. : Science 218: 801, 1982.

11. Day, N.K., Witkin, S.S., Sarkar, N.H., Kinne, D., Jussawalla, D.J., Levin,A., Hsia, C.E., Geller, N., and Good, R.A. : Proc. Natl. Acad. Sci. USA 78: 2483, 1981.

12. Marcus, S.L., Kopelman, R. and Sarkar, N.H. : J. Virol. 31: 341, 1979.

13. Levine, P. Mourali, N. , Tabbane, F., Costa, J., Mesa-Tejada, R., Spiegelman, S., Muenz,R. and Bekesi, J.G. : Proc. Am. Assoc. Cancer Res. 21 : 170, 1980.

14. Shimkin, M.B.: In: Recent Results in Cancer Research; Breast Cancer: A Challenging Problem, Vol. 42, M.L. Griem, E.V. Jensen, J.E. Ultman and R.W. Wissler (eds.), Springer-Verlag, New York, 1973, p.6.

15. Anderson, D.E.: Cancer 34:1090, 1974.

16. SEER Program: Cancer Incidence and Mortality in the United States 1973-1976, J.L. Young, A.J. Asire, and E.S. Pollack (eds), DHEW Pub. No. NIH 78-1837, Bethesda, Md., 1978.

17. Nandi, S., and McGrath, C.M.: Adv. Cancer Res. 17:353, 1973.

18. Mesa-Tejada, R., Keydar, I., Ramanarayanan, M., Ohno, T., Fenoglio, C., and Spiegelman, S.: Proc. Natl. Acad. Sci. USA 75:1529, 1978.

19. Lloyd, R.V., Rosen, P.P., Sarkar, N.H., Jiminez, D., Kinne, D.W., Menendez-Botet, C., and Schwartz, M.K.: Cancer 51:654, 1983.

20. Bishop, J.M., Quintrell, N., Medeiros, E. and Varmus, H.E.: Cancer 34:1421, 1974.

21. McGrath, C.M., Furmanski, P., Russo, J., McCormick, J.J. and Rich, M.M.: In: Tumor Virus Infections and Immunity, R.L. Crowell, H. Friedman and J.E. Prier (eds), University Park Press, Baltimore, Maryland, 1976, p.63.

22. Das, M.R., and Mink, M.M.: Cancer Res. 39: 5106, 1979.

23. Varmus, H.E., Quintrell, N., Medeiros, E., Bishop, J.M., Nowinski, R.C., and Sarkar, N.H.: J. Mol. Biol. 79:663, 1973.

24. Moore, D.H., Long, C.A., Vaidya, A.B., Sheffield, F.G., Dion, A.S., and Lasfargues, E.Y.: Adv. Cancer Res. 28:347, 1979.

25. Bentvelzen, P. and Hilgers, J.: In: Viral Oncology, G. Klein (ed), Raven Press, New York, 1980, p. 311.

26. Cohen, J.C., Shank, P.R., Morris, V.L., Cardiff, R. and Varmus, H.E.: Cell 19:333, 1979.

27. Fanning. T.G., Puma, J.P. and Cardiff, R.D.: J. Virol. 36:109, 1980.

28. Groner, B., Buetti, E., Diggelman, H., and Hynes, N.E.: J. Virol. 36:734, 1980.

29. Morris, V.L., Vlasschaert, J.E., Beard, C.L., Milazzo, M. and Bradbury, W.C.: Virology 100:101, 1980.

30. Nusse, R., and Varmus, H.E.: Cell 31:99, 1982.

31. Etkind, P.R. and Sarkar, N.H.: J. Virol. 45:114, 1983.

32. Callahan, R., Drohan, W., Tronick, S., and Schlom, J.: Proc. Natl. Acad. Sci, USA. 79:5503, 1982.

33. May, F.E.B., Westley, B.R., Rochefort, H., Buetti, E., and Diggelmann, H.: Nucl. Acids Res. 11:4127, 1983.

34. Muller, M.. Zotter, S., and Kemmer, C.: J. Natl. Cancer Inst. 56:295, 1976.

35. Hendrick, J.C., Francois, C., Calberg-Bacq, C.M., Colin, C., Franchimont, P., Gosselin, L., Kozma, S., and Osterrieth, P.M.: Cancer Res. 38:1826, 1978.

36. Witkin, S.S., Sarkar, N.H., Kinne, D.W., Breed, C., Good, R.A., and Day, N.K.: J. Clin. Invest. 67:216, 1980.

37. Redmond, S.M.S., and Dickson, C.: EMBO J. 2:125, 1983.

38. Majors, J.E., and Varmus, H.E.: J. Virol. 47:495, 1983.

39. Kurth, R., Fenyo, E.M., Klein, E., and Essex, M.: Nature 279:197, 1979.

40. Bishop, J.M., and Varmus, H.E.: In: RNA Tumor Viruses, R. Weiss, N. Teich, H. Varmus, and J. Coffin (eds), Cold Spring Harbor Laboratory, New York, 1982, p. 999.

41. Weiss, D.W.: In: Current Topics in Microbiology and Immunology, Springer-Vertag, New York; 1982 Vol. 89, p. 1.

42. Shiku, H., Takahashi, T., Herbert, F., Oettgen and Old, L.J.: J. Exp. Med. 144:873, 1976.

43. Garrett, T.J., Takahashi, T., Clarkson, B.D. and Old, L.J.: Proc. Natl. Acad. Sci. USA 74:4587, 1977.

44. Pfreundschuh, M., Shiku, H., Takahashi, T., Ueda, R., Ransohoff, J., Oettgen, H.F. and Old, L.J.: Proc. Natl. Acad. Sci. USA 75:5122, 1978.

45. Ueda, R., Shiku, H., Pfreundschuh, M., Takahashi, T., Li, L.T.C., Whitmore, W.F., Oettgen, H.F. and Old, L.J.: J. Exp. Med. 150:564, 1979.

46. Kohler, G., and Milstein, C.: Nature 256:495, 1975.

47. Lavrin, D.H., Blair, P.B. and Weiss, D.W.: Cancer Res. 26:929, 1966.

48. Vaage, J.: Cancer Res. 28:2477, 1968.

49. Morton, D.L.: J. Natl. Cancer Inst. 42:311, 1969.

50. Blair, P.B.: Isr. J. Med. Sci. 7:161, 1971.

51. Vaage, J. and Medina, D.: Cancer Res. 34:1319, 1974.

52. Stutman, O.: Cancer Res. 36:739, 1976.

53. Vaage, J.: Cancer Res. 38:231, 1968.

54. Sarkar, N.H., Pomenti, A.A., and Dion, A.S.: Virology 77:12, 1977.

55. Ringold, G., Lasfargues, E.Y., Bishop, J.M., and Varmus, H.E.: Virology 65: 135, 1975.

56. Fine, D.L., Plowman, J.K., Kelley, S.P., Arthur, L.O., and Hillman, H.E.: J. Natl. Cancer Inst. 52:1881, 1974.

57. Westerbrink, F. and Koornstra, W.: Eur. J. Cancer 16:763, 1980.

58. Ihle, J.N., Arthur, L.O. and Fine, D.L.: Cancer Res. 36:2840, 1976.

59. Arthur, L.O., Bauer, R.F., Orme, L.S. and Fine, D.L.: Virology 87:266, 1978.

60. Schochetman, G., Artur, L.O., Long, C.W. and Massey, R.: J. Virol. 32:131, 1979.

61. Moore, D.H., Sarkar, N.H., Holben, J.A., and Sheffield, J.B.: Int. J. Cancer 23:713, 1979.

62. Squartini, F.: In "Pathology of Tumors in Laboratory Animals," IARC Monograph Vol. 2:43, 1973.

63. Sarkar, N.H., Etkind, P.R., Lasfargues, E.Y., and Whittington, E., unpublished results.

64. Varmus, H.E., Quintrell, N., and Ortiz, S.: Cell 25:23, 1981.

65. Varmus, H.E.: Science 216:812, 1982.

66. Breznik, T., and Cohen, C.J.: Nature 295:255, 1982.

67. Becker, D., Lane, M., and Cooper, G.M.: Proc. Natl. Acad. Sci. USA 79: 3315, 1982.

68. Peters, G., Brookes, S., Smith, R., and Dickson, C.: Cell 33:369, 1983.

The Role of Viruses in Human Cancer. Volume II
G. Giraldo and E. Beth editors

INTERFERON AND HUMAN CANCER

GABRIEL EMOEDI
VIROGEN INC., BASLE, SWITZERLAND

Human interferons (IFs) are proteins with antiviral and antitumoral properties as well as a diversity of other biological activities. The IFs are generally classified in three groups: The alpha-type (leukocyte) and beta-type (fibroblast) which are both acid-stable, and the acid-unstable gamma-type (immune).

Alpha-IF is mainly secreted by leukocytes and also by several lymphoblastoid cell lines by viral induction. The beta-IF is released by fibroblast after induction of viruses, or using double-stranded RNAs such as Poly I:C and various synthetic compounds. The production of beta-IF is enhanced by a segmential treatment of the human fibroblasts with Poly I:C and metabolic inhibitors such as cyclohexamide and actinomycin D. The gamma-type is mainly secreted by lymphocytes stimulated with mitogens (1, 2).

Naturally there has been a great interest in the therapeutic potential of IF since it was described over 25 years ago. A major problem was that the supply of IF for clinical trials has been very limited until a few years ago. On the other hand, the short supply of IF over the past 25 years allowed more fundamental background knowledge to be acquired about IF than about any other therapeutic agent previously studied in humans. Many studies were carried out using in vitro and animal models of human disease. However, there were only few clinical studies and some of these were carried out on a small scale. Those early studies using crude or partially purified crude material showed the activity of IF in human in a limited number of viral infections and neoplastic diseases. The best effects were observed in animals when IF was administered as early as possible in the disease process, that is before any virus induced pathological development occurred or when the tumor mass or the number of tumor cells were still limited to a minimal level. In addition to the antiviral effects of IF it is evident that IF is of importance in host defenses against malignant diseases, possibly at a local level, which indicates therapeutic possibilities. On the other hand, IF may have a regulatory function in the cell division of normal cells in vivo; this could in fact cause a possible limitation of its clinical use (3).

With the development of new technologies, advances have occurred in large scale production of the naturally produced IFs. A new dramatical improvement for IF production resulted from the recombinant DNA-technology. This resulted in an expanded definition of IF as a family of proteins with differing functions and in its purification to homogenicity (4, 5). These methods bring about the commercial production of IFs at a reasonable cost and will make large clinical studies possible.

IFs differ both physicochemically and biologically from other antitumor and antiviral compounds.
It is being increasingly realized that IFs are involved in immunoresponses, not only in virus infections but also in response to other antigens, including neoplastic diseases. The antiviral effect is perhaps better understood than the antitumor effect. On the other hand, we are just beginning to understand the immunoregulatory effects of IF which may play a role in the clinical course of the disease. IF affects the cells in different ways: inhibiting growth, activating macrophages, altering cell membranes, activating natural killer cells and null cells, increasing cytotoxicity of lymphocytes, stimulating "antiviral stage", and in other ways which have not yet been defined. The effects of IF may, at least partially, be mediated through modulation of the functions of the intercellular network of the immunosystem which is operated and regulated by the cellular and humoral (6) immune system.

In addition to the part that IF plays in viral infections and neoplastic diseases, it has been proven to have an important function in other diseases such as various autoimmune diseases and also in multiple sclerosis.

IF enhances the release of histamine and stimulates the production of prostata glandine E. IF potentiates inflammatory reactions in asthma, arthritis, systemic lupus erythematosis and autoimmune disease in general. Analyses of the serum IF from patients with lupus erythematosis suggest that patients who have measurable IF positive serum levels are mostly those in an active phase of their disease stage. Serum samples from patients with rheumatoid arthritis are mostly negative. However, the synovial fluid of these patients contains measurable levels of IF, and also the patients whose joint fluid is positive are usually those with an acute exacerbation of the disease. The exact function of IF in these diseases ought to be worked out in additional studies.

It is of major interest to study the possible significant importance of IF in the acquired immune deficiency syndrom (AIDS). It has been demonstrated, as we learned at this meeting, that patients with AIDS have a high level of a deficient or acid-unstable alpha-IF (7). IF is not only a part of the host's defense mechanism, it may also be of importance in the pathogenesis.

Clinical Studies:

The early preliminary studies with naturally produced alpha-IF have been carried out in various virus infections. Pharmakokinetic dates of this type of IF became available and preliminary studies suggested beneficial antiviral effects in some viral infections (8, 9, 10, 11).

Local application:

Topical application of alpha- or beta-IF in patients with acute dendritic herpes keratitis (12) in combination with either minimal debridement or superficial thermocautery in double-blind, placebo-controlled trials show a clear clinical effect in terms that the healing time of the IF-treated patients was considerably shorter. IF in combination with trifluorothymidine has also proved to be more effective compared to the group treated only with trifluorthymidine (13, 14, 15).

In the treatment of adeno-virus keratoconjunctivitis with alpha- and beta-IF in a double-blind, controlled trial, a significant sensitivity to IF, a decrease of the duration of the keratoconjunctivitis and of a subsequent eruption of chronic severe keratitis have been reported (16, 17).

Merigan et al. reported the efficacy of large doese of alpha-IF administered before and after the challenge with rhinovirus with a decrease in acquisition of virus and symptoms related to the virus infection (18). A low dos of beta-IF did not produce any antiviral effect (19). However, using a mor purified material in a spray form, a significant inhibition of the rhinovirus replication and of clinical symptoms was achieved. High titer recombinant alpha-IF administered prior to rhinovirus challenge proved also effective in the prevention of rhinovirus infection. However, quite significant side effects were reported (20).

Systemic administration:

Already the very early observation using partially purified alpha-IF in patients with herpes zoster infection showed a clear sensitivity to IF (8). These observations were evaluated and established in a randomized, double-blind study (21). The sensitivity of herpes zoster viruses to IF action is of great interest in both immunosuppressed and normal patients.

We even saw that cytomegalovirus (CMV) has some sensitivity to IF in terms of the cessation of urinary virus shedding, however, no definite clinical effect on the course of the disease including tissue-destructive process has been demonstrated so far (9, 10, 11).

Bone marrow transplanted patients are at high risk for CMV infections due to a major case of mortality in these patients. IF treatment of bone marrow recipients after they developed a proved CMV pneumonia failed. It seems that in bone marrow transplanted patients with CMV infections, once the infection is established, therapeutic intervention with IF has no significant efficacy (22).

In two studies, the effect of alpha-IF on the natural history of CMV after kidney transplantation was evaluated. A delay in the onset of shedding and a decrease in the number of patients with viremia was observed (23).

In a double-blind controlled trial in children with varicella, with underlying leukemia, treated with IF as soon as possible after the onset of varicella a significant influence on the course of the disease was reported (24).

In patients operated on the trigeminal ganglia for relief of the symptoms of tic douloureux there is a high incidence of a re-activation of a latent herpes simplex infection. Administration of alpha-IF significantly reduced the incidence of reactivation and viral shedding in these controlled trials (25).

Summary:

In randomized double-blind trials using IF in the treatment of several human herpes infections including herpes simplex and varicella zoster a clear antiviral effect was demonstrated.

The development of technologies to measure viral markers such as hepatitis-B virus-specific DNA-polymerase has made possible a treatment study with IF. Merigan and his coworkers have demonstrated that using alpha-IF, alone and in combination with adenine-arabinoside, has a clinical effect on the course of the chronic hepatitis-B virus infections related viral markers including DNA-polymerase activity (26, 27).

There are other investigators administering beta-IF to patients with acute or chronic hepatitis-B infections. These are ongoing trials; however, the preliminary results are promising. There are also ongoing trials using alpha- or beta-IF with a surfactant substance for herpes genitalis infections.

IFs differ both physicochemically and biologically from other antitumor compounds in use over the last years. Since the beginning of the American Cancer Society trials, a marked expansion in our understanding of IF effects on neoplastic diseases has occurred. Based on results to date, IFs will prove effective adjuncts to existing approaches to cancer management (28,29,30).

Important questions still must be answered. Which of the three types will prove most effective in the cancer treatment? The ongoing studies are still focused on defining optimal doses and schedules for clinical use and on evaluating the dose-dependent side effects with partially purified IFs. Side effects related to naturally and recombinant DNA-technology produced IFs among patients of the same qualitative nature have been almost equivalent (29).
Fever between 38°C-4o°C are common at the start of the initial therapy, but abates with further administration.
With high doses, anorexia, fatigue and weight loss occur and are dose limiting.
Elevation of SGOT could occur, however, it has been reversible.
Leukocytopenia and thrombocytopenia seldom occurred in earlier treated patients and led to an interruption of the therapy.

The highest clinically tolerated single dose of alpha-IF (natural) was 3o x lo^6 units given by the i.m.route. An upper limit was not yet determined. In a schedule for 6o days we gave a daily dose of lo x lo^6 units by the i.m.route.

Dr. Borden reported a single dose of $alpha_A$-IF (recombinant DNA prod.) of up to 198 x lo^6 units which was clinically tolerated. However, on a daily schedule the highest tolerated dose of $alpha_A$-IF was 18 x lo^6, of $alpha_2$-IF 3o x lo^6 units.
In 37% of these patients anorexia and fatigue were dose-related limiting effects (31). Purifications have eliminated a number of side effects, except fever.

A cumulating date on antitumoral effects in phase II trials has resulted from studies using naturally produced alpha-IF. The dose of these trials was mostly 3 million units.

These results are representing a panorama of other centers' studies as well and support that naturally produced alpha-IF has an antitumor effect in patients with breast carcinomas, hypernephromas, multiple myelomas, lymphomas, leukemias, bladder carcinomas and prostatic carcinomas (31)..

Table.

Ongoing trials using beta- or gamma-IF and different types of recombinant IF are underway. The today available results strongly suggest that IFs will likely prove an important new modality in cancer treatment.

Table I

MAXIMALLY TOLERATED REPETITIVE CLINICAL DOSES OF INTERFERON

IFN	Route	Schedule	Dose
α natural (Cantell)	im	daily x 56d	>9 x 10^6*
natural (Cantell)	ia	daily (10 min) x 21d	20 x 10^6
α-2	iv	daily (10 min) x 5d	30 x 10^6
rA	im	daily x 56d	18 x 10^6
rD**	im	daily x 21d	>720 x 10^6*
β natural	iv	daily (combined 10 min and 3 hr infusion)	10 x 10^6

* upper limit not yet determined
** based on interferon units determined in bovine cells

Published by Dr. Borden

REFERENCES

1. Havell, E.A. and Vilcek, J.(1972) Production of high titered interferon in cultures of human diploid cells. Antimicrob. Agents Chemother. 2:476-484

2. Baron, S., Dianzani F.(1978) The interferon system, Tex. Rep. On. Biol. and Med. Vol.35

3. Merigan, T.C. (1981) Present appraisal of and future hopes for the clinical utilization of human interferons. In: (I.Gresser, Ed.), Interferons 1981, vol.3, p.135-154. Academic Press

4. Rubinstein,M.,Rubinstein,S.,Familetti,P.C.,Miller,R.S., Waldman,A.A. and Pestka,S.(1979) Human leukocyte interferon: production, purification to homogeneity and initial characterization. Proc.Nat.Acad.Sci.USA 76:64o-644

5. Derynck,R.,Content,J.,De Clercq,E.,Volckaert,G.,Tavernier,J., Devos,R. and Fiers,W. (198o) Isolation and structure of a human fibroblast interferon gene. Nature 285:542-547

6. Borden,E.C.,Ball L.A. Interferons:biochemical, cell growth inhibitory and immunological effects. In Brown EB (ed): "Progress in Hematology," Vol XII, New York: Grune and Stratton, Inc., pp 299-339

7. Buimovici-Klein,E. (1984). The role of viruses in human cancer. Elsevier, Holland. In press

8. Emodi,G., Rufli T. et al. (1975) Human interferon therapy for herpes zoster in adults. Scand.J.Inf.Dis. 7, 1-5

9. Emodi,G.,Just,M. et al.(1975) Circulating interferon in man after administration of exogenous human leukocyte interferon. J.Nat.Cancer Inst. 54, 1o45-1o49

1O. Emodi,G.,O'Reilly,R.,Muller,R.,Everson,L.K.,Binswanger,U., Just,M. (1976) Effect of exogenous leukocyte interferon in cytomegalovirus infections. J. Infect.Dis. 133 (Suppl.): 199-2o4

11. Arvin,A.M., Yeager,A.S.,Merigan,T.C. (1976) Effect of leukocyte interferon on urinary excretion of cytomegalovirus by infants. J.Infect.Dis. 133 (Suppl.):2o5-21o

12. Kobza,K., Emodi,G. et al. (1975) Treatment of herpes infection with human exogenous interferon. The Lancet, I, 1343-1344

13. Sundmacher,R., Neumann-Haefelin,D., Cantell,K. (1976) Successful treatment of dendritic keratitis with human leukocyte interferon. Albrecht von Graefes Archiv für klin. und experim. Ophtalmologie 2o: 39-45

14. Jones,B.R., Coster,D.J., Falcon,M.G.,Cantell,K. (1976) Topical therapy of ulcerative herpetic keratitis with interferon. Lancet 2:128

15. DeKoning,E.W.J., VanBijsterveld,O.P.,Cantell,K. (1982) Combination therapy for dendritic keratitis with human leukocyte interferon and trifluorothymidine. Brit.J. Ophtal. 66:5o9-512

16. Negoro,Y., Imanishi,J., Matsuo,A., Kishida,T. (198o) Treatment of epidemic keratoconjunctivitis by human leukocyte interferon. Jap.J.Ophthal. 24:125-127

17. Romano,A., Revel,M., Guarari-Rotman,D., Blumenthal,M., Stein,R. (198o) Use of fibroblast-derived (Beta) interferon in the treatment of epidemic keratoconjunctivitis. J. Interferon Res. 1:95-1oo

18. Merigan,T.C., Hall,T.S., Reed,S.E., Tyrrell,D.A.J. (1973) Inhibition of respiratory virus infection by locally applied interferon. Lancet 1:563-567

19. Scott,G.M., Reed,S., Cartwright,T., Tyrrell,D. (198o) Failure of human fibroblast interferon to protect against rhinovirus infection. Arch.Virol. 65:135-139

2o. Scott,G.M., Phillpotts,R.J., Wallace,J., Gauci,C.L., Greiner,J., Tyrrell,D.A.J. (1982) Prevention of rhinovirus colds by human interferon alpha-2 from Escherichia coli. Lancet 2:186-187

21. Merigan,T.C., Rand,R.H., Pollard,R.B.,Abdallah,P.S., Jordan, G.W., Fried, R.P. (1978) Human leukocyte interferon for the treatment of herpes zoster in patients with cancer. N.Engl.J.Med. 298:981-987

22. Meyers,J.D., McGuffin,R.Wl, Neimann,P.E., Singer,J.W., Thomas, E.D. (198o) Toxicity and efficacy of human leukocyte interferon for treatment of cytomegalovirus pneumonia after bone marrow transplantation. J.Infect.Dis. 141: 555-562

23. Cheeseman, S.H., Rubin,R.H., Stewart,J.A., Tolkoff-Rubin,N.E. Cosimi,A.B., Cantell,K., Gilbert,J., Winkle,S., Herrin,J.T., Black,P.H., Russell,P.S., Hirsch,M.S. (1979) Controlled clinical trial of prophylactic human leukocyte interferon in renal transplantation. N.Engl.J.Med. 3oo:1345-1349.

24. Arvin,A.M., Kusher,J.H., Feldman,S., Bachner,R.L., Hammond,D., Merigan,T.C. (1982) Human leukocyte interferon for the treatment of varicella in children with cancer. N.Engl.J.Med. 3o6: 761-765

25. Pazin,G.J., Armstrong,J.A., Lam,M.T., Tarr,G.C., Janetta,P.J. Ho,M. (1979) Prevention of reactivated herpes simplex infection by human leukocyte interferon after operation on the trigeminal root. N.Engl.J.Med. 3o1: 225-23o

26. Greenberg,H.B., Pollard,R.B., Lutwick,L.I., Gregory,P.B., Robinson,W.T., Merigan,T.C. (1976) Effect of human leukocyte interferon on hepatitis B virus infection in patients with chronic active hepatitis. N.Engl.J.Med. 295:517-522

27. Scullard,G.H., Andres,L.L., Greenberg,H.B., Smith,J.L., Sawhney,V.K., Mahal,A.S., Popper,H., Merigan,T.C., Robinson,W.S., Gregory,P.B. (1981) Antiviral treatment of chronic hepatitis B virus infection: Improvement in liver disease with interferon and adenine arabinoside. Hepatology 1: 228-232

28. Gutterman,J.U., Quesada,J. (1983) Clinical investigation of partially pure and recombinant DNA derived leukocyte interferon in cancer. Texas Reps. Biol.Med.41:in press

29. Hawkins,M.J., Borden,E.C., Fein,S., Simon,K.J. (1983) Evaluation of the clinical and biological effects of interferon arD. Proc.Amer.Soc.Clin.Oncol.

3o. Mellstedt,H., Ahre,A., Bjorkholm,M., Holm,G., Johansson,B., Strander,H. (1979) Interferon therapy in myelomatosis. Lancet L-245-247

31. Borden,E.C. (1983) Personal communication.

This paper was supported by the AMRASC Foundation, South Carolina, U.S.A.

© 1984 Elsevier Science Publishers B.V.
The Role of Viruses in Human Cancer. Volume II
G. Giraldo and E. Beth editors

ATAXIA-TELANGIECTASIA - A HUMAN AUTOSOMAL RECESSIVE DISORDER PREDISPOSING TO CANCER

YECHIEL BECKER, MEIRA SHAHAM, EYNAT TABOR AND YOSEF SHILOH
Departments of Molecular Virology and Human Genetics, Hebrew University - Hadassah Medical Center, 91010 Jerusalem, Israel

INTRODUCTION

In studying the role of viruses in human cancer, the genetic background of affected individuals must be taken into consideration. Transformation of human cells in vivo might be due to exogenous factors like radiation, chemical carcinogens or viruses, as well as mutations in genes governing the control of cellular proliferation. Hence, genetic disorders in man might result in a predisposition to cancer. This means that in vivo conditions (unknown as yet) are changed, and additional damage to some cells (due to radiation, carcinogens and viruses) might lead to cell transformation and to the development of a premalignant cell clone. McCaw et al. (1) reported that a clone of lymphocytes with a chromosomal translocation 14q appeared in an ataxia-telangiectasia (A-T) patient before leukemia was diagnosed. This clone appears to have given rise to chronic leukemic cells in the patient. This presentation covers studies on A-T in our laboratory which started about six years ago. We hoped to gain information on the importance of the genetic defect in the development of leukemia and the possible role of viruses in cancer in genetically affected individuals.

A-T is a human,multisystem, genetic disorder determined by a pleiotropic, autosomal, recessive gene that involves cerebellar degeneration, leading to severe progressive motor dysfunction, as well as oculocutaneous telangiectases, absent or degenerated thymus, and impaired function of both humoral and cell-mediated immune systems. The latter is accompanied in some of the patients by recurrent respiratory infections (2-5). The serum of A-T patients contains significantly higher than normal levels of two "oncofetal" proteins, α-fetoprotein and carcinoembryonic antigen (6, 7). These patients also develop lymphoreticular malignancies at a frequency approximately one hundredfold higher than that of the general population, and their close relatives, either obligatory or presumed A-T heterozygotes, are also reported to be predisposed to a variety of cancers (8-10).

The frequency of the A-T gene in the population is about one percent, and it has been estimated that between six and ten percent of patients with certain cancers are carriers of this mutation (9). The geographic distribution of A-T is not even: for example, a higher frequency of A-T cases was reported in Israel, where the A-T patients are mostly Jews of Moroccan origin (11), as well as Arabs (our unpublished data). One out of 45 individuals in the community of Moroccan Jews is estimated to be an A-T heterozygote (11).

Cultured lymphocytes and skin fibroblasts from A-T patients display chromosomal instability, expressed as a high rate of breakage, and aberrations in which there is preferential involvement of chromosome No. 14 (12, 13). For this reason, A-T is classified as one of the "chromosomal breakage syndromes", together with xeroderma pigmentosum, Bloom's syndrome and Fanconi's anemia. The phenomenon of chromosomal instability may be explained by the finding from our laboratory of a clastogenic factor present in the serum of A-T patients, in the amniotic fluid surrounding an A-T fetus, and in culture medium conditioned by A-T skin fibroblasts (14, 15). This factor, which has a molecular weight of 500-1000 and is sensitive to proteolytic digestion (16), induces chromosomal breaks in normal human lymphocytes or fibroblasts in culture. The clastogenic factor was purified by means of high performance liquid chromatography (HPLC) (Shaham et al., to be published).

A-T patients are extremely sensitive to the cytotoxic and oncogenic effect of ionizing radiation (17) and one of the most striking laboratory hallmarks of A-T cells is their in vitro sensitivity to the cytotoxic and clastogenic effects of this type of radiation (18, 19).

There have been conflicting reports on the sensitivity of A-T skin fibroblasts to various radiomimetic drugs and chemical carcinogens (18-22), but it appears from recent studies that they are hypersensitive - primarily, if not exclusively - to DNA breaking agents such as bleomycin (20, 21, 23), neocarzinostatin (NCS) (24, 25), adriamycin and hydrogen peroxide (Y. Shiloh et al., to be published). A recent finding from our laboratory is that A-T heterozygous fibroblasts consistently show a degree of sensitivity to NCS which is clearly intermediate between that of A-T homozygous fibroblasts and normal cells (25). This finding may serve as a method for the laboratory identification of heterozygous individuals who constitute a cancer-prone section of the population.

It has been suggested that a defect in the repair of certain radiogenic DNA damage may account for the radiation hypersensitivity in A-T cells. The reason is that certain A-T fibroblast strains show reduced DNA repair synthesis and a reduced ability to remove certain γ-ray induced DNA lesions, the chemical nature of which has not been identified (19). The nature of the critical DNA lesion in A-T cells is, however, still unclear, since A-T cells seem to efficiently repair DNA strand breaks induced by either X-rays (18, 19) or NCS (26). The only biochemical anomaly which is clearly correlated with cellular sensitivity in A-T is reduced inhibition of DNA synthesis. This was found following treatment with all the DNA breaking agents tested so far (25-30). The exact relationship between the anomaly in DNA replication and the presumed defect in DNA repair is still unclear. It has recently been shown that the defect in DNA synthesis can be "corrected" by fusion of A-T fibroblasts from different patients (31, 32), and this is taken as an indication for genetic heterogeneity in A-T.

Another feature of the in vitro phenotype in A-T recently found in our laboratory is accelerated cellular aging and an increased demand for growth factors, which is indicative of an abnormality in binding or processing of these factors by A-T cells (33).

A-T PATIENTS AND CELLS

Our laboratory has been in contact with physicians, hospitals and various institutions in the country regarding all the known A-T families in Israel. An A-T registry has been established and is continuously being updated. The collection of A-T cells in the Department of Human Genetics in our institution now includes fibroblast strains and lymphoblastoid lines from 24 A-T patients, 14 obligatory heterozygotes, and 14 healthy siblings of A-T patients ("possible heterozygotes") of the following ethnic groups: Moroccan Jews, Israeli Arabs, an Arab family from Saudi Arabia, Iranians and Turks. All these cell lines were characterized cytogenetically, and all the A-T homozygous fibroblast strains that were tested were found to produce A-T clastogenic factor.

GROWTH PROPERTIES AND OTHER FEATURES OF A-T FIBROBLASTS

We found that the doubling time of A-T fibroblast strains did not differ, on the average, from that of normal strains, although the interstrain variability was greater for A-T strains. Collagen production and secretion, as well as the ability to support the growth of herpes simplex virus type 1 were also found to be normal in our A-T strains (34, 35). In experiments performed at low passage levels and with very good batches of fetal calf serum, the colony-forming ability of our A-T strains was not substantially lower than that of normal strains. However, the life span in culture was found to be significantly shorter for A-T homozygous strains, as compared to normal cells; correspondingly, their colony-forming ability rapidly deteriorated at earlier passage levels than in normal cells (33). This may explain previous reports on poor plating efficiencies obtained with A-T strains. A-T heterozygous strains had an intermediate life span roughly between that of normal and A-T homozygous strains.

Using colony formation by single cells as a quantitative criterion for cellular growth capability, we found that A-T homozygous fibroblasts had an increased demand for certain growth factors that are present in varying amounts in different types or batches of serum. These factors could be replaced by fibroblast growth factor or epidermal growth factor, but not by insulin (33).

Following these studies, optimal conditions were selected for obtaining maximum plating efficiency in our survival studies. The cells were used at low passage levels in combination with serum batches that had excellent growth promoting ability. Under such conditions, we obtained plating efficiencies of 25%-45% with our A-T and normal strains without the use of feeder layers.

A-T homozygous fibroblasts have receptors for insulin. The observation that fibroblasts from A-T homozygotes do not respond to insulin as a growth factor in the colony formation assay required further clarification. It was of interest to know whether A-T fibroblasts have receptors to insulin or not, or, alternatively, if they have receptors, the organization of the cell membrane may be such that the binding of insulin does not induce further enzymatic processes in the cell as it does in normal cells. In collaboration with Dr. N. Keiser (Department of Hormone Research, our Medical Center), we carried out preliminary experiments which showed that A-T cells bind radiolabeled insulin to the same extent as normal cells.

SENSITIVITY OF A-T CELLS TO DNA DAMAGING AGENTS AND REPAIR OF DNA ALKYLATION ADDUCTS

There are conflicting reports on the sensitivity of A-T cells to various alkylating agents (18). With our A-T fibroblast strains, we found normal sensitivity to the following alkylating mutagens: methyl methanesulphonate (MMS), N-methyl-N'-nitro-N-nitrosoguanidine (MNNG), mitomycin C and chloroacetaldehyde (22, our unpublished results). The repair of purine methylation adducts induced by MNNG was found to be normal in A-T fibroblasts and lymphoblastoid cells (22, 36, 37). Our conclusion from these studies is that alkylation adducts are not the critical DNA lesions in A-T cells. However, all the A-T fibroblast strains we tested were hypersensitive to the cytotoxicity of X-rays and bleomycin (38), in accordance with previous reports based on the American and European A-T strains (18-21). Since these two agents induce primarily DNA strand breaks, we proceeded to test the response of A-T cells to DNA breaking agents and found marked and consistent hypersensitivity of A-T homozygous strains to neocarzinostatin (24, 25), streptonigrin, adriamycin and hydrogen peroxide (39). This hypersensitivity was accompanied by reduced inhibition of DNA synthesis compared to normal cells after treatment with these agents. A limited degree of cellular hypersensitivity that was not sufficient to allow for definition of a separate sensitivity range was shown by A-T heterozygous cells. On the other hand, the A-T cells showed a normal response to paraquat, saframycin A and ellipticine. All these results indicate that the critical DNA lesion in A-T cells is a strand break caused by deoxyribose destruction following the action of free radicals targeted into the DNA (39).

Ten A-T heterozygous strains displayed an intermediate degree of NCS hypersensitivity between that of eight A-T homozygous and six normal strains, while their X-ray sensitivity did not always differ from that of certain normal strains (23, 25). On the basis of these findings, NCS sensitivity of fibroblast strains from healthy siblings of A-T patients was tested. Out of five such strains tested, two had the typical intermediate NCS hypersensitivity of obligatory heterozygous strains, and three showed normal survival following NCS treatment. These findings indicate that we might have a laboratory method for the identification of A-T gene carriers.

NCS sensitivity of fibroblast strains from cancer patients. The first experiment in this series was done with a fibroblast strain from a breast carcinoma of a Jewish patient of Moroccan origin (obtained from Prof. Y. Keydar, Department of Microbiology, Tel Aviv University). The NCS sensitivity test (three repeats) revealed that the fibroblast strain from this patient resembled the heterozygous strains in its response to NCS.

Through our collaboration with Dr. L. Kopelovich (Memorial Sloan-Kettering Cancer Center, New York), we obtained fibroblasts from patients with adenomatosis of the colon and rectum (ACR). The sensitivity of three cell lines to NCS resembled normal cells.

DNA repair synthesis. To determine whether a defect in a major excision repair pathway exists in A-T cells, DNA repair synthesis was measured in our A-T and normal fibroblasts. The benzoylated-naphthoylated-DEAE-cellulose (BND-cellulose) method was used, following treatment with ionizing radiation and alkylating agents. Normal rates of DNA repair synthesis were induced in six A-T strains tested by γ-rays, NCS, MNNG and mitomycin C, although considerable interstrain variability was found. One normal strain exhibited a diminished extent of DNA repair synthesis after treatments with MMS and γ-rays (25, 40). Similar results were obtained with our strains in another laboratory, using different methodology (41). There was a correlation between inhibition of DNA synthesis and another parameter measured at high treatment doses, namely the inhibition of induced ornithine decarboxylase. The latter parameter was also abnormal in the cell strain with reduced DNA repair synthesis; however, the sensitivity of this strain to MMS and γ-rays, which is measured at considerably lower doses, was normal (41). It was concluded that variations in DNA repair synthesis do not necessarily correlate with cellular survival and that our A-T strains do not show any apparent deficiency in this parameter.

Inhibition of DNA synthesis by DNA-damaging agents. We found reduced inhibition of DNA synthesis following treatment with X-rays and bleomycin in all of our A-T fibroblast strains which were tested, as compared to normal strains (25), while the extent of inhibition following treatment with MMS and MNNG was normal. We extended these studies to include NCS and also found reduced inhibition of DNA synthesis with this agent (25). Using alkaline sucrose gradient analysis, we observed that both replicon initiation and strand elongation were inhibited by NCS, and both components were inhibited to a lesser extent in A-T cells (42). More recently, we obtained similar results with adriamycin (39). These results indicate reduced inhibition of DNA synthesis is a biochemical parameter, which is well correlated with reduced survival. With A-T heterozygous cells, we observed a small reduction in the inhibition of DNA synthesis induced by NCS, but this difference from normal strains was not significant enough to make it a diagnostic aid (25).

Strand breakage and repair following NCS treatment. Since both A-T homozygous and heterozygous cells are particularly sensitive to NCS, we studied the time course of strand breakage induction and repair in A-T skin fibroblast strains treated with NCS using the sensitive method of alkaline and neutral elution. We demonstrated the rapid action of NCS in human fibroblasts ($t_{1/2}$ for strand breakage induction = 1 min); NCS-induced inhibition of DNA synthesis followed the same kinetics. The ratio of double-stranded breaks to single-stranded breaks obtained with NCS was four times higher as compared to X-rays. We found that the kinetics for strand breakage repair were biphasic with a rapid phase ($t_{1/2}$ = 2 min for single-stranded breaks and 10 min for double-stranded breaks), followed by a slow phase extending to several hours. There was no difference between A-T and normal cells with regard to the rate of DNA rejoining or the relative proportion of breaks remaining open. We concluded that there is no defect in the rejoining of NCS-induced strand breaks in A-T cells (26). This is similar to results obtained in other laboratories following treatment with X-rays (17-19).

Potentially lethal damage repair to DNA. We investigated the extent and time course of potentially lethal damage repair and sublethal damage repair after NCS treatment of human skin fibroblasts (43). The kinetics of potentially lethal damage repair in normal cells followed a rapid time course with a half-life of 20-40 min. There was a lack of potentially lethal damage repair in quiescent A-T homozygous cells after NCS treatment which correlated with the results obtained with X-rays. On the other hand, we observed a limited degree of sublethal damage repair after treatment of these cells with NCS, while this process was absent in A-T homozygous cells treated with X-rays. A-T heterozygous cells which show an intermediate degree of NCS sensitivity could perform both damage repair processes, but with somewhat reduced efficiency as compared to normal cells (43).

THE A-T CLASTOGENIC FACTOR

The existence of the clastogenic factor in the plasma of A-T patients, in medium conditioned by A-T fibroblast strains, and in the amniotic fluid surrounding an A-T fetus, was reported by Shaham et al. (14, 15). The molecular weight of the factor is in the range of 500-1000 according to its ability to pass through Amicon Diaflo membrane filters of various pore sizes (16), and its activity is sensitive to proteolytic digestion but not to RNase (16). It is stable at 37°C and upon freezing and thawing. Following ultrafiltration through membrane filters, the medium conditioned by A-T fibroblasts was passed through a Lichrosorb RP8 column, the eluates were lyophilized, and the factor was redissolved. The A-T factor was found to adsorb to this column and was completely eluted with buffered 20% methanol, but not with buffer only. When the conditioned medium was fractionated, following ultrafiltration, using a high performance liquid chromatography (HPLC) system with a Dupont-C8 column, the clastogenic activity was found in specific fractions, using 20% methanol as the eluent. These preliminary results indicate that the clastogenic factor can be purified to homogeneity by the HPLC procedure, making possible the chemical analysis of the factor (M. Shaham, M. Chorev and Y. Becker, to be published).

Prenatal diagnosis of a fetus at risk for A-T showed that the clastogenic factor was present in the cell-free amniotic fluid of the fetus. A higher level of chromosome breaks was obtained in normal human peripheral lymphocytes cultured in amniotic fluid from the A-T fetus than in normal control amniotic fluids. There was also an increased rate of spontaneous chromosome damage in cultured amniotic cells of the A-T fetus. In addition, in one of the A-T cultures, a clone with a translocation in chromosome 14 (14/5) was observed in 15 cells of 100 analyzed (15).

PREDISPOSITION TO CANCER IN A-T

A-T homozygotes have an increased tendency to develop lymphoreticular malignancies. McCaw et al. (1) described an A-T patient with a clone of preleukemic cells with a translocation in chromosome 14q. It is of interest that Shaham et al. (15) noted a translocation in chromosome 14/5 in amniotic cells of a fetus suspected of being homozygous for A-T. It is possible that the clastogenic polypeptide produced by A-T skin fibroblasts, which is also present in the serum of A-T homozygote patients, is the cause of chromosomal aberrations that lead to the appearance of lymphocyte clones with translocations in chromosome 14. These subclones may be able to activate cellular oncogenes due to the translocation event (44), resulting in the appearance of malignantly transformed clones that lead to leukemia or lymphoma.

The increased sensitivity of A-T homozygotes to radiation and radiomimetic drugs might also be explained on the basis of the underlying effect of the clastogenic factor. If tissue cells have a low level of proliferation, treatment with radiation or carcinogens may induce changes in the cell DNA in such a way that cell proliferation is initiated without host control, leading to the development of cancer in the affected tissue.

The role of viruses and carcinogens in the induction of cancer in genetically affected individuals is still to be explored.

ACKNOWLEDGMENTS

The study on A-T was supported by a grant from the Leukemia Research Foundation, Inc., Chicago, U.S.A.; the Israel Cancer Association; the United States-Israel Binational Foundation; the Leonard Wolfson Foundation for Scientific Research; and a special research grant from the Authority for Research and Development, The Hebrew University of Jerusalem. The authors are indebted to Dr. Nurit Keiser, Department of Hormone Research, Hadassah University Hospital, for her collaboration.

REFERENCES

1. McCaw, B.K., Hecht, F., Harnden, D.G., and Teplitz, R.L.: *Proc. Natl. Acad. Sci. U.S.A.* 72:2071, 1975.
2. Miller, R.W.: In: *Ataxia-Telangiectasia: A Cellular and Molecular Link Between Cancer, Neuropathology and Immunodeficiency,* B.A. Bridges and D.G. Harnden (eds.), John Wiley & Sons, New York, 1982, p. 13.
3. Sedgwick, R.P.: *ibid,* p. 23.
4. Waldmann, T.A.: *ibid,* p. 37.
5. Sedgwick, R.P., and Boder, E.: In: *Handbook of Clinical Neurology,* P.J. Linken and G.W. Bruyn (eds.)Elsevier/North-Holland Biomedical Press, Amsterdam, 1972, Vol. 14, p. 267.
6. Waldmann, T.A., and McIntire, K.R.: *Lancet* 2:1112, 1972.
7. Sugimoto, T., Sawada, T., Tozawa, M., Kidowaki, T., Kusunoki, T., and Yamaguchi, N.: *J. Pediat.* 92:436, 1978.
8. Spector, B.D., Filipovich, A.H., Perry, G.S.III,and Kersey, J.H.: *op. cit.* (ref. 2) p. 103.
9. Swift, M.: *ibid,* p. 355.
10. Swift, M., Sholman, L., Perry, M., and Chase, C.: *Cancer Res.* 36:209,1976.
11. Levin, S., and Perlov, S.: *Israel J. Med. Sci.* 7:1535, 1971.
12. Oxford, J.M., Harnden, D.G., Parrington, J.M., and Delhanty, J.D.: *J. Med. Genet.* 12:251, 1975.
13. Cohen, M.M., Shaham, M., Dagan, J., Shmueli, E., and Kohn, G.: *Cytogenet. Cell Genet.* 15:338, 1975.
14. Shaham, M., Becker, Y., and Cohen, M.M.: *Cytogenet. Cell Genet.* 27:155, 1980.
15. Shaham, M., Voss, R., Becker, Y., Yarkoni, S., Ornoy, A., and Kohn, G.: *J. Pediatr.* 100:134, 1982.
16. Shaham, M., and Becker, Y.: *Hum. Genet.* 58:422, 1981.
17. Cunliffe, P.N., Mann, J.R., Cameron, A.H., Roberts, K.D., and Ward, H.W.C.: *Br. J. Radiol.* 48:374, 1975.
18. Huang, P.C., and Sheridan, R.B.III: *Hum. Genet.* 59:1, 1981.
19. Paterson, M.C., and Smith, P.J.: *Ann. Rev. Genet.* 13:291, 1979.
20. Lehmann, A.R.: *op. cit.* (ref. 2.) p.33.
21. Lehmann, A.R., James, M.R., and Stevens, S.: *ibid,* p. 347.
22. Shiloh, Y., Tabor, E., and Becker, Y.: *Mutat. Res.* 112:47, 1982.
23. Shaham, M., Voss, R., Lerer, I., and Becker, Y.: *Cancer Res.* 43:4244,1983.
24. Shiloh, Y., Tabor, E., and Becker, Y.: *Cancer Res.* 42:2247, 1982.
25. Shiloh, Y., Tabor, E., and Becker, Y.: *Carcinogenesis* 3:815, 1982.
26. Shiloh, Y., van der Schans, G.P., Lohman, P.H.M., and Becker, Y.: *Carcinogenesis* 4:917, 1983.
27. Lavin, M.F., Ford, M.D.,and Houldsworth, J.: *op. cit.* (ref. 2), p. 319.
28. Edwards, M.J., and Taylor, A.M.R.: *ibid,* p. 327.

29. Jaspers, N.G.J., De Wit, J., and Bootsma, D.: *ibid,* p. 339.

30. Painter, R.B., and Young, B.R.: *Proc. Natl. Acad. Sci. U.S.A.* 77:7315,1980.

31. Murnane, J.P., and Painter, R.B.: *Proc. Natl. Acad. Sci. U.S.A.* 79:1960, 1982.

32. Jaspers, N.G.J., and Bootsma, D.: *Proc. Natl. Acad. Sci. U.S.A.* 79:2641, 1982.

33. Shiloh, Y., Tabor, E., and Becker, Y.: *Exp. Cell Res.* 140:191, 1982.

34. Becker, Y., Stevely, W., Tabor, E., Asher, Y., Hamburger, Y., and Hadar, J.: *J. Gen. Virol.* 51:201, 1980.

35. Becker, Y., Stevely, W., Hamburger, Y., Tabor, E., Asher, Y., and Hadar, J.: *Conn. Tissue Res.* 8:77, 1981.

36. Shiloh, Y., and Becker, Y.: *op. cit.* (ref. 2) p. 177.

37. Shiloh, Y., and Becker, Y.: *Cancer Res.* 41:5114, 1981.

38. Becker, Y., Shaham, M., Shiloh, Y., and Voss, R.: In: *Biochemical and Biophysical Markers of Neoplastic Transformation*, P. Chandra (ed.), *Proceedings of a NATO Advanced Study Institute, Corfu Island, Greece, 1981,* Plenum Press, New York, 1983, p. 171.

39. Shiloh, Y., Tabor, E., and Becker, Y.: *Carcinogenesis* 4 (in press), 1983.

40. Shiloh, Y., Cohen, M.M., and Becker, Y.: In: *Chromosome Damage and Repair. NATO/EMBO Advanced Study Institute, Bergen, Norway, 1980,* Plenum Press, New York, 1981, p. 361.

41. Ben-Hur, E., Kol, R., Heimer, Y.M., Shiloh, Y., Tabor, E., and Becker, Y.: *Radiat. Environ. Biophys.* 20:21, 1981.

42. Shiloh, Y., and Becker, Y.: *Biochim. Biophys. Acta* 721:485, 1982.

43. Shiloh, Y., Tabor, E., and Becker, Y.: *Biochem. Biophys. Res. Commun.* 110: 483, 1983.

44. Klein, G.: *Cell* 32:311, 1983.

The Role of Viruses in Human Cancer. Volume II
G. Giraldo and E. Beth editors

SUMMING UP

BERNARD ROIZMAN The Marjorie B. Kovler, Viral Oncology Laboratories, The University of Chicago, 910 East 58th Street, Chicago Il 60637 USA

The research laboratory resembles in some respects a musical assembly. It starts out as a solo, a duo, or a trio, in which the head also plays. It blossoms into a symphony in which the head of the laboratory conducts. Everyone plays in tune with neither pause nor applause between movements. So ingrained is the drive to perform that more and more heads of large research laboratories crown their scientific careers as visiting conductors giving solo recitals... But if research laboratories resemble an instrumental assembly, scientific meetings resemble an opera. In this instance, the invited participants do the singing usaully in a language which is foreign to either the speaker or the audience, and there is applause after every aria. But there is a difference. In the conventional opera there is a plot, rehearsals are frequently public, and the heroine is expected to die. Scientific meetings are claimed to be unrehearsed, the heroine never dies (although an apocryphal story claims that on one occasion a sleeping member of the audience was seriously injured when he fell off his chair), and while the subject of the plot is mentioned in the title, the plot itself is at times difficult to fathom. Although something analogous to a libretto is ultimately published, scientific meetings frequently employ itinerant speakers to sum up the action. This has been my task.

Summations of scientific meetings range from homages to each speaker to a public venting of one's biases, but neither is appropriate on this occasion. For me this has been an exciting meeting. It was a very rare combination of excellent science, superb intellectual stimulation in a very colorful, historical ambiance of the Castel dell'Ovo. In the very best traditions of Napoletanian Opera and Science, there was a theme - the role of viruses in human cancer. To be more specific, several families of human viruses are objects of intensive investigation for their putative roles as causative agents of human cancer. The problem which we have recognized a long time ago is that the mere presence of a virus in a cancer cell is not *prima facie* evidence that the virus causes cancer. Even less compelling is the evidence that a virus can morphologically transform cells since viruses which appear to transform cells with high efficiency are not invariably proven to be oncogenic agents outside the laboratory. It has also become apparent that cancer is the cumulative expression of many rare events. While proof of association does not require the knowledge of the specific events triggered by the virus, understanding of the mechanisms by which viruses cause cancer is one of our most important objective.

The notion that herpes simplex viruses 1 and 2 (HSV-1 and HSV-2) might cause human cancer dates back to 1968 and has survived many vicissitudes. It has been kept alive primarily by the resolution and perseverence of a small group of investigators and especially by F. Rapp, L. Aurelian, D.A. Galloway, and J.K. McDougall against the prevailing opinion of an influential, if not a major fraction, of the scientific community. A pertinent example is Denis Burkitt's list of oncogenic agents; in his lecture he mentioned everything but HSV. The exciting findings reported at this meeting by L. Aurelian and J. K. McDougall centered on the role of HSV in cells transformation. It is apparent however that (i) while cell lines transformed by inactivated HSV-1 yielded stable transformed cell lines which retained viral DNA, the transformed phenotype induced by HSV-1 DNA fragments appears unstable, (ii) HSV-2 DNA may have two transforming regions neither of which is colinear with the one transforming region reported to be present in HSV-1 DNA, and (iii) neither of the HSV-2 transforming regions encodes a complete gene. The studies reported by J. K. McDougall that one of the transforming regions in HSV-2 DNA may reside in a sequence less than 200bp containing a large inverted repeat are a major achievement in molecular biology, but they raise questions which will fuel much of the future work in this area. Paramount is the question whether the second transforming region contains a similar sequence and whether DNA probes consisting of these sequences will more consistently detect viral DNA sequences in human neoplasias suspected to be caused by HSV. It is noteworthy that Human Cytomegalovirus (CMV), a rather distant cousin of HSV has a similar transforming sequence as revealed by B. Fleckenstein and his associates. The similarities between HSV and CMV were further stressed by E.S. Huang's review of the molecular biology of CMV.

We tend to forget that to HSV-1 and HSV-2 patients the virulence of the infection and the misery caused by recurrences of lesions are far more life-threatening than the potential of developing cancer. The discussions by Y.M. Centifanto-Fitzgerald and by Y. Becker centered on the genetic basis of virulence. It should be noted however that virulence is frequently related to the capacity to multiply and that mutations which affect the capacity to multiply also affect virulence. The operational definition of the genetic determinants of virulence should be acquisition of virulence by wild-type strains rather than the loss of virulence because of mutations whether or not it can be proved that the mutations affects ability to multiply in an organism. In a measured review on prevention and chemotherapy, H.E Kaufman dispelled any illusions that we might be close to either prevention or cure. The facts are that HSV maintained in a latent state is insusceptible to chemotherapy and shielded from the host immune response. The long range solution to the "herpetic plague" will come from a concerted attack on the biologic foundations of herpes latency rather than from novel immunogens or drugs inhibiting viral DNA metabolism.

The presentations by M.A. Epstein, E.Kieff, D.T. Purtilo and D.A. Thorley-Lawson underscored the extraordinary achievements in the molecular biology of Epstein-Barr virus (EBV) and the involvement of this virus and of human onco-genes in malignancies ascribed to it. The rapid progress in the identification of the major EBV glycoprotein and in the mapping of its gene should permit in the near future the testing of Epstein's cherished hypothesis that immunization against EBV should not only attenuate infectious mononucleosis but also prevent the malignancies associated with EBV. Less promising or convincing is the association of EBV with Acquired Immune Deficiency Syndrome (AIDS) espoused by D.T. Purtilo. Herpesviruses are opportunists par excellence and they are more likely to benefit from immune deficiency than to cause it. Equally improbable is the role of induced immune response to sperm; the epidemiology of AIDS in Zair and the apparent absence of the disease from women with infertility caused by induced immune response to sperm argue against this hypothesis. In a different category is the association of CMV with Kapossi's sarcoma which frequently accompanies AIDS. G. Giraldo, E. Beth, and G.G. Giordano, the original discoverers of CMV footprints in Kapossi's sarcoma, presented unambiguous evidence that only early viral gene products are present in cancer tissues. It could be expected that all viral gene products made in lytic infections would accumulate in cancer tissues if CMV were an adventitious passenger.

Lucid and timely accounts of the molecular biology, pathogenesis, immunology, and oncology of Hepatitis B viruses were given by J. Summers, P. Tiollais, A.M. Goudeau, and F. Barin. Hepatitis B is a clear example of a virus strongly associated with cancer of the liver in which the progression of the disease is evident and where the association is strengthened by the integration of viral DNA into human chromosomes. It can be expected that immunization against the virus will reduce the incidence of liver cancer further strengthening the association. Yet, Hepatitis B is also an example of a DNA virus which does not transform cells in culture and which does not require the expression of its genes for the maintenance of the oncogenic state. In contrast, we still do not know the role of papova and papilloma viruses in human malignancy notwithstanding their ability to transform cells and the apparent requirement for continued viral gene expression for the maintenance of the transformed state.

The presentations on the biology of retroviruses and, in particular, the beautiful studies on the role of retroviruses in human malignancy by Y. Hinuma and by N.H. Sarkar were the highlights of the conference. If Microbe Hunters were written today, it would surely include a detailed account of the travails and vicissitudes of the search for the human retroviruses.

The three day Congress was all too short. The rapid succession of results and ideas presented at the Congress will serve as intellectual sustenance for years to come. As the participants left Castel dell'Ovo for the last time, it must have occurred to many that in its long history the castle has been a silent witness of war, oppression, pestilence and misery, but also of some of the finest creative accomplishments of the Western civilization. There is no doubt that the Congress on the Role of Viruses in Human Cancer dwelled on and nurtured human accomplishments and aspirations as lofty as any encountered in the history of Castel dell'Ovo.

AUTHOR INDEX

SUBJECT INDEX